THERMODYNAMICS OF MOLECULAR SPECIES

Ernest Grunwald

Brandeis University, Waltham, Massachusetts

A WILEY-INTERSCIENCE PUBLICATION

JOHN WILEY & SONS, INC.

New York / Chichester / Brisbane / Toronto / Singapore / Weinheim

This text is printed on acid-free paper.

Library of Congress Cataloging in Publication Data:

Grunwald, Ernest, 1923–
Thermodynamics of molecular species / Ernest Grunwald.
p. cm.
"A Wiley-Interscience publication."
Includes bibliographical references and index.
ISBN 0-471-01254-8 (cloth; alk. paper)
1. Thermodynamics. I. Title.
QD504.G78 1996
541.3′69–dc20 96-14103

Printed in the United States of America

10 9 8 7 6 5 4 3 2 1

Courtesy of Judith Grunwald

ENTROPY

To Esther
With lasting love and affection

CONTENTS

PREFACE

When I retired from teaching I decided to write one more book and to call it *Thermodynamics of Molecular Species.* Molecular species are at the very core of mechanistic chemistry: They are the macroscopic or near-macroscopic ensembles of molecules that represent, in the real world, the molecules symbolized in chemical equations.

Chemistry has outgrown the operational practices current at the time of J. Willard Gibbs and G. N. Lewis, when the traditional concepts of chemical thermodynamics were formulated. We now accept physical and spectroscopic evidence on a par with chemical evidence, we no longer require that physically well-characterized new compounds be isolated as pure components, and we can control observational time scales over an enormous range. The time seems ripe for an updating of the concepts, and in this book I try to do it.

The issue is urgent. Although thermodynamics is a very general science, chemical–thermodynamic phenomena keep cropping up that traditional concepts seem unable to decipher. These chronic "puzzles" include the magnitudes of partial entropies and heat capacities in hydrogen-bonding solvents; the occasional (but not infrequent) enthalpy–entropy compensation ascertained for reactions in liquid solutions; and, in a different vein, the broad question of why linear free-energy relationships of the type discovered by L. P. Hammett work so well. The seeming intractability of such puzzles is giving chemical thermodynamics a black eye and is compromising its status as a tool for solving chemical problems.

The key concept in my scheme is the flexible composition model, which specifies composition on at least two levels. The first level lists the formal components of the system. The second level lists the molecular species that stem from the components. If necessary, there is a third level that lists any subspecies derived from the molecular species. The composition model is flexible because it lists only those molecular species that are chemically or stoichiometrically significant, and this listing can be changed as our understanding of the chemistry in the system improves.

The scheme involving composition models lets me derive specific relationships and general theorems. The basic strategy is to insist that the thermodynamic properties of a system are independent of whether composition is specified in terms of formal components or molecular species. A second strategy is to insist that the laws of thermodynamics are independent of the observational time scale. By judicious application of these strategies I have derived perhaps 20 general theorems concerning the thermodynamic properties of molecular species and their relationships to those of the formal components. The most important of these I have named, for the sake of clarity. The Correspondence Theorems and the Stability Theorem are well known, although not by name. One of my favorites is the Tolerance Theorem, a generalization based on thermodynamic error tolerance. It lets me understand the remarkable success of L. P. Hammett's linear free-energy relationships, and it provides a qualitative entry into thermodynamic enthalpy–entropy compensation. Corollaries of the Tolerance Theorem then provide the formal entry.

It turns out that the traditional chemical thermodynamics, which deals with processes among equilibrium states, is incomplete because it takes the existence and maintenance of equilibrium for granted and fails to specify it explicitly. This is a surprising omission in a science that prides itself on the precision of its specifications. By contrast, the thermodynamics of molecular species includes departures from equilibrium as a matter of course, and the maintenance of equilibrium is specified explicitly. The specifying mathematics then adds specific terms—I call them *molar-shift terms*—that are missing from the traditional formulations.

The saving grace for traditional chemical thermodynamics is that the molar-shift terms are mostly small or zero. For equilibrium processes at constant T and P, the molar-shift terms vanish identically for partial free energies, by the Second Law. They do not vanish for other partial properties (such as partial enthalpies and entropies) but are often negligible. Yet I will cite situations in which the molar-shift terms are not negligible, and they correspond to the chronic "puzzles" of traditional thermodynamics.

I believe that sooner or later, any chemist planning to use thermodynamics will have to know the limitations of traditional chemical thermodynamics and, when appropriate, switch to the thermodynamics of molecular species. This book attempts to make that as painless as possible. It uses the familiar nomenclature of Lewis and Randall [1923] and standard symbols recommended by the IUPAC's Committee on Physical Chemistry. The few technical terms it introduces for the sake of clarity are phrased in descriptive English. The mathematics is consistent with that used in junior- and senior-level college courses in physical chemistry and thermodynamics.

Naturally I have quite a bit to say about molecular species and about the distinguishability that is their essence. I have a long-standing interest in solvation in liquid solutions and in the related topic of environmental isomers—isomers whose distinguishability results from differences in the liquid environment, which can be predicted by Schrödinger wave mechanics. This interest includes borderline cases of distinguishability—cases in which the thermodynamics loses some of its bite because the molecular species are not sharply discrete. Examples include the domains of encounter pairs, linkage pairs, and molecular complexes, as well as the domains of dissociated ions and ion pairs. It is not possible to include these topics without risking controversy, but my models serve at least as plausible vehicles for thermodynamic explorations.

I am pleased to acknowledge the help of colleagues and friends, especially those whose papers I cite. Their writings have taught me a lot. I also thank those colleagues, at Brandeis University and elsewhere, who listened to my ideas and tried seriously to answer my questions. Those were acts of real friendship. Finally, I thank my late teachers: Saul Winstein, James B. Ramsey, and Louis P. Hammett. I also thank my intimate collaborators: John E. Leffler, Saul Meiboom, and Colin Steel. Their influence is patent throughout this book.

ERNEST GRUNWALD

Brandeis University

GLOSSARY OF SYMBOLS

$a, a_i, a_{\mathrm{J}}, a_{\pm}$ activity
a distance of closest approach
A Helmholtz free energy
A prefactor in Arrhenius equation
b cage radius; $\mathsf{a}(1 + \beta)$
$c, c_i, c_{\mathrm{J}}, c_{\mathrm{A\backslash x}}$ concentration
$c_{\rightarrow}$ c of forward-moving transition state
c molar concentration
c (subscript) stands for "complex formation"
C_P heat capacity at constant P
$C_{P2}, C_{P,\mathrm{J}}$ partial heat capacity
D, D_i, D_{J} coefficient of linear diffusion
$\mathcal{D}$ electric displacement
e unit (proton) charge
e (subscript) environmental
eq (subscript) at equilibrium
E cell emf
$\mathcal{E}$ electric field
E potential energy
$E^{(2)}$ per mole pairwise molecular interaction energy
$E^{(\mathrm{C})}$ per mole interaction energy of a molecule with its cage
E_{act} per mole Arrhenius activation energy
(ECP) electrochemical potential
f, f_i formal concentration, per kilogram of solvent
f packing fraction
$\mathcal{F}$ Faraday constant, per equivalent
$g(r)$ radial distribution function
(g) gas phase
G Gibbs free energy
$G_i, G_{\mathrm{J}}, G_{\mathrm{A\backslash x}}$ (Gibbs) partial free energy
$(G_i)_{\alpha}$ isomolar partial free energy
h Planck's constant
H enthalpy

$H_i, H_J, H_{A\backslash x}$	partial enthalpy
$(H_i)_\alpha$	isomolar partial enthalpy
$\mathcal{H}_i, \mathcal{H}_J, \mathcal{H}_i(M)$	Henry's law constant
i (subscript)	any ith component
I	ionic strength [Eq. (12.11)]
J (subscript)	denotes any molecular species J
$J(r)$	Fuoss ion-pair distribution function
k	rate constant
k	Boltzmann constant
$k_a \cdots k_m$	force constant
L	Avogadro's number
$\mathcal{L}$	molecular weight
m, m_i, m_J	molality (Table 3.1)
(M)	solvent medium
n (subscript)	nominal
n_1, n_i	formula-weight number
$n_A, n_J, n_{X\backslash a}$	mole number
n	refractive index
$\langle n \rangle$	mean number of collisions (per event)
n_a, n_b	number of a, b molecules [Eq. (9.2)]
N_A, N_J	mole fraction
$N_{1,x}$	colligative fraction of solvent (Table 3.1)
0 (superscript)	standard value of the given property
O	order of magnitude
p	probability
P	pressure
P_i, P_J	partial pressure
P	electrical polarization
q	displacement from equilibrium position
q_{RT}	value of q when $E = RT$
r	any distance or displacement
r (subscript)	stands for "reaction"
r_B	Bjerrum distance, $z_+z_-e^2/2 \in \mathsf{k}T$
(rad) (ch. 12)	ionic crystal radius, ionic hard-sphere radius
R	gas constant per mole
s, s_x	mean neighbor number
S	entropy
$S_i, S_J, S_{A\backslash a}$	partial entropy
$(S_2)_\alpha$	isomolar partial entropy
S_γ	Debye–Hückel limiting slope for γ (Chapter 12)
t (subscript)	stands for "transfer"
T	temperature
U	energy
v	velocity
V	volume

$V_i, V_J, V_{X\backslash a}$	partial volume
W	any kind of work
W(strain)	work to produce a molar strain
$W_{i,el}, W_{i,chem}$	electrical, chemical work per ion
x, y	stoichiometric coefficients [Eq. (12.18)]
X, Y	extensive properties
y, y_x	molar stress $(\partial G/\partial \alpha, \partial G/\partial s_x)$
z, z_+, z_-	charge number
z	molecular partition function
z	pairwise hard-sphere collision rate
Z	total hard-sphere collision rate
Z_1, Z_2	formula-weight fractions of the solvent components in a binary solvent
α	progress variable
α_a, α_j	fractional population
α_J	stoichiometric fraction of species J
β	fractional expansion
β	"isokinetic" temperature [Eq. (6.8)]
$\gamma, \gamma_i, \gamma_J, \gamma_\pm$	activity coefficient
$\gamma_{q\backslash h}$	quasi-harmonic force constant [Eq. (9.16)]
δ, Δ	change operator
$\delta_R, \delta_M, \delta_i$	perturbation operator
δ	transition-state distance (Fig. 5.6)
δ	screening distance
ϵ	energy of a quantum state
ε	molar extinction coefficient
ε	dielectric constant
η	coefficient of viscosity
η_0	permittivity of space (Table 12.2)
κ	radius of ion atmosphere [Eq. (12.15)]
Λ	equivalent conductance
μ	electric dipole moment
μ_i, μ_j	chemical potential (same as G_i, G_J)
ν	frequency
ξ	distinguishability index [Eq. (10.7)]
ρ	density
ρ_q	charge density
σ	standard deviation
$\sigma_{(C)}$	σ of Brownian noise at cage center
$\sigma_{a,b}$	encounter distance
σ_{hs}	hard-sphere diameter
σ_{rms}	rms relative strain [Eq. (4.11)]
τ	mean lifetime, mean time of exchange
τ_{obs}	observational time scale
τ_{corr}	correlation time for Brownian motion

ϕ	space part of wave function
ϕ	electric potential
ϕ	any function of T, n_2, α [Eq. (7.6)]
ψ	Schrödinger wave function
{ }	indicates a "set"
$\backslash, \backslash a, \backslash x$	indicates the "environment of a molecule"

1

THERMODYNAMIC COMPONENTS AND MOLECULAR SPECIES

Chemical thermodynamics is a hybrid science that combines the inductive, macroscopic approach of classical thermodynamics with molecular theory. Its major role is to serve as a framework for organizing, correlating, and understanding facts concerning chemical reactivity. However, the addition of molecular theory spoils the inductive purity of classical thermodynamics and introduces concepts whose logical status must be defined. In this chapter we will examine the key concepts of thermodynamic components and molecular species.

FORMAL COMPONENTS

In traditional thermodynamics, a component is a tangible substance whose preparation is described and whose properties are characterized so that the substance can be reproduced at will. When several components are used, the components are said to be *independent* if none of them can be generated from others by simple mixing. The set of mixtures formed from two components is called a two-component system. Similarly, the set of mixtures formed from n independent components is called an n-component system.

The preferred components in chemical thermodynamics are pure substances with known chemical formulas—pure substances because chemists know how to purify substances reproducibly, and with known chemical formulas, so that chemists may devise models of the molecular species within the constraints of a known stoichiometry. A *chemical* formula, in this context, is not synonymous with a *molecular* formula. A chemical formula is the symbol under which the substance is listed in a chemical-formula index (e.g., $C_2H_4O_2$ for acetic acid), whereas a molecular formula characterizes an actual molecular species (e.g.,

$(CH_3CO_2H)_2$ for acetic acid dimer in benzene solution). For most substances under most conditions, the chemical formula represents the molecular composition of the chief molecular species and of its isomers. However, for electrolytes the chemical formula is usually the empirical formula, and for regular polymers it might be the formula of the repeat unit.

A thermodynamic component that is a pure substance with known chemical formula is called a *formal component*. The adjective *formal* in this context means "with known chemical formula." Because the chemical formula also defines the stoichiometric formula weight, alternative meanings for *formal* are "by formula weight" or "per formula weight."

Formal components, being a class of thermodynamic components, must of course be macroscopic in all aspects of their definition. The limitation to pure substances creates no difficulty, because pure substances can be defined operationally as tangible substances whose properties, after repeated purification, have reached a constant limit. The requirement of a known chemical formula can take us into the realm of atomic theory if we so desire, but there is no logical necessity for that because the determination of the chemical formula is based on the macroscopic operations of elemental analysis and gas density or osmometry. Atomic theory enters only to the extent that it suggests the algorism by which the data are processed. The computed result—the chemical formula—is a function of macroscopic variables and thus macroscopic.

The chemical formula in turn defines the *formula weight*, which therefore is a macroscopic unit. Thermodynamics is permissive about units. If chemists wish to express the amounts of pure substances in formula-weight units, thermodynamics has no objection.

Perhaps the most remarkable feature of the traditional formal component is that it must be a pure substance. The pioneers who founded chemistry insisted on that, because chemistry had been an uncertain science until chemists began to resolve the materials of nature into pure substances. By insisting on chemical purity, they fostered a sense of discipline and personal responsibility: Chemists were expected to check and personally guarantee the purity of their materials. (The properties of a pure substance do not change upon repurification.) Psychologically, there is a reassuring concreteness about having a pure substance in hand, with reproducible attributes that can be described and demonstrated to colleagues. It is understandable, therefore, that the traditional identity of a formal component as a pure substance lingered on even after the limitations of this definition became apparent. These limitations will emerge as we discuss molecular species. Eventually, at the end of this chapter, we will give a more general definition of a chemical component.

MOLECULAR SPECIES

The molecular formulas in the chemical equations for macroscopic reactions symbolize molecular species. In general, a molecular species is a macroscopic or near-macroscopic ensemble of molecules that are characterized by a definite molecular formula, a definite and distinctive equilibrium geometry, and a distinctive set of molecular modes of motion and spectral properties. For example, the actors in chemical reaction mechanisms—the reactants, products, substrates, catalysts, reactive intermediates, and even the mechanically unstable transition states—all represent separate molecular species, as do any sets of molecules that become distinguishable in physical interaction mechanisms.

It is often convenient to divide a molecular species into distinguishable subsets or subspecies. Such division may be as fine-grained as desired, provided that all subspecies are distinguishable and that their state variables (P, V, T) can be defined.

Because chemical reactivity depends on molecular structure, each chemical species has a definite and distinctive equilibrium geometry. The "equilibrium geometry" described here denotes a state of *mechanical* equilibrium, characterized by a stable or metastable minimum or an unstable maximum on the potential-energy surface for the interacting atoms defined by the molecular formula. (Unless stated otherwise, the potential-energy surface will be that of the electronic ground state.) In principle, each minimum and each maximum represents a separate molecular species.

For example, Fig. 1.1 shows the reaction coordinate for the isomerization of acetylene to vinylidene on the potential-energy surface for C_2H_2 in the electronic ground state [Gallo et al., 1990].

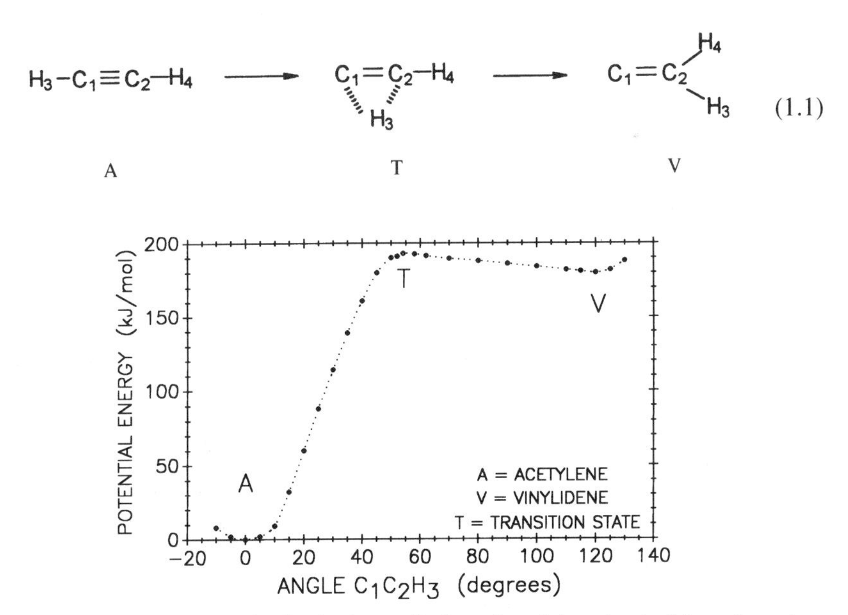

Figure 1.1. Potential-energy plot for the isomerization of acetylene to vinylidene, based on quantum-mechanical calculations [Gallo et al., 1990]. Molecular species, reaction-coordinate angle, potential energy: A (stable), 0°, 0 kJ; T (unstable), 54°, 193 kJ; V (metastable), 120°, 180 kJ.

The figure shows three molecular species: acetylene (A, stable minimum), vinylidene (V, metastable minimum), and the transition state (T, unstable maximum). In addition to defining equilibrium geometries, the figure also shows the truly enormous range of relative stabilities and half-lives that can be found among molecular species defined by the same potential surface. We estimate, from the energy values at the potential minima, that the mole ratio of [V]/[A] at equilibrium at 300 K is about 10^{-31}, or about 1 molecule of vinylidene in a billion liters(!) of acetylene at 1 atm. By pumping in the energy under nonequilibrium conditions, the amount of vinylidene can be made significant—enough so that normal vibrations of vinylidene can be measured [Ervin et al., 1989]. The energy barrier for conversion of vinylidene to acetylene is low. The half-life of vinylidene in this experiment is less than 1 ps(!), and it would be of the order of 1 ps at 300 K. The half-life of acetylene in the reverse process (A $\rightarrow$ V), on the other hand, is astronomical, exceeding the age of the earth. At 300 K it is of the order of tens of billions of years.

Associated with each of the equilibrium geometries is a distinctive set of molecular modes of motion so that, according to principles of statistical thermodynamics, each molecular species has a characteristic set of thermodynamic properties [Lewis et al., 1961]. Equally important, each set of molecular modes establishes a distinctive set of spectral properties so that the presence, and even the amount, of the molecular species can in principle be detected. In practice, many molecular species of theoretical interest are too short-lived and thinly populated to be spectroscopically detectable, but that situation is improving. In particular, the techniques of femtosecond spectroscopy have opened the door to very short-lived species, such as transition states in chemical reactions [Dantus et al., 1987; Levine and Bernstein, 1987; Kliner and Zare, 1990].

Distinguishability

Molecular species and subspecies must have distinct properties. But since their thermodynamics makes use of molecular theory, we must expect two kinds of distinguishability: real and ideal. Real distinguishability is based on observation; at least in principle it can be verified by experiment. Ideal distinguishability is based on theory and may not be verifiable by direct observation; but the underlying theory must be major, such as quantum mechanics. To illustrate the difference: The vibrational quantum states of polyatomic molecules in the gas phase become dense and exchange-broadened at moderate levels of excitation [Gilbert and Smith, 1990]. The energy levels then overlap strongly and form a quasi-continuum, so that distinct quantum properties are immeasurable for most levels. The distinguishability of most levels is thus ideal, but that is sufficient for *theories* requiring distinguishability, such as the statistical mechanics of molecular ensembles in the gas phase. When the quantum states in the quasi-continuum are reckoned as distinguishable, statistically predicted macroscopic properties agree with observation.

On the other hand, the distinguishability of molecular species is real, because chemistry is basically an operational science, and molecular species are basic to chemistry. All aspects of their characterization should therefore be real. But when we divide a molecular species into subspecies, the distinguishability of the subspecies may be real or ideal, depending on the fineness and aim of the subdivision. If the aim is to build a theoretical bridge to statistical mechanics, the subspecies become the populations of the individual quantum states of the molecular species at the given T and P, and the distinguishability of most of them then is ideal.

Returning to more ordinary subspecies, the distinctive property might be molecular isotopic composition (e.g., $CDCl_3$ versus $CHCl_3$), quantum number in a spin multiplet or in a specific mode of motion of a polyatomic molecule, or nearest-neighbor environment in a rigid or dense fluid medium. In state-to-state chemistry (where one determines molecular reactivity in specific quantum states) a subspecies might be the population of an individual vibrational–rotational energy level, but only if the translational temperature and partial pressure can be defined thermodynamically [Reisler and Wittig, 1986; Hoffman et al., 1989].

Nomenclature: Molar Versus Formal

The adjectives *molar* and *formal* are not synonymous. *Molar*, as in "molar property," means "relating to molecular species." *Molar* is also used as a unit of measure, to mean "per mole of molecules" or "per gram molecular weight," as in "molar volume." By contrast, *formal* means "relating to formal components" or "per gram formula weight." The difference between "molar" and "formal" parallels that between molecular species and formal components.

ISOMERIC SPECIES

Isomeric species, or *isomers*, are distinct molecular species with the same molecular formula. They represent different equilibrium configurations on the same potential-energy surface. Most molecular potential-energy surfaces are multidimensional; for an n-atomic molecule, the potential energy is a function of $3n - 6$ molecular configuration coordinates. Many of us find it difficult to visualize entire surfaces in multidimensional space. We deduce their features, imperfectly, from two- or three-dimensional projections or from physical arguments. Two features are especially worth noting.

First, perhaps the most obvious fact about isomers is that there are so many of them. The well-known varieties include structural isomers, geometric isomers, conformational isomers, chiral isomers including enantiomers and diastereomers, isomers based on proton tautomerism (such as acetone and its enol), isomers based on intramolecular proton transfer (such as p-aminobenzoic acid and its zwitterion), and isomers based on intramolecular hydrogen-bonding (such as *syn* and *anti* 2-chlorophenol). Altogether, the facts send a clear signal that molecular potential-energy surfaces are rich in equilibrium configurations.*

Second, the regions around minima on a potential surface are often called potential "wells," but the shapes of these potential "troughs" are often more complex. This is because the vertical distances between maxima and minima on a potential surface are highly variable, and short-lived isomers can be "wrapped up" in long-lived isomers. To illustrate this feature, we shall consider a classic example, the isomers of dichloroethane.

It has long been known that the molecular formula $C_2H_4Cl_2$ represents two stable structural isomers, 1,1-dichloro- and 1,2-dichloroethane (1,1-DCE and 1,2-DCE), whose topological formulas are given in Eq. (1.2). These isomers are distinct substances that can be separated, purified at leisure, and stored for years without decomposition. We may infer that their potential "troughs" are separated by a high barrier.

```
Cl         H              H         Cl
  \       /                \       /
H—C1 —— C2—H             H—C1 —— C2—H        (1.2)
  /       \                /       \
Cl         H              Cl        H

   1,1-DCE                   1,2-DCE
```

During the 1930s it was discovered that the structural isomers shown in (1.2) can be further subdivided into conformational isomers due to restricted rotation about the C—C bond. The discovery, mostly due to the work of Mizushima and coworkers [Mizushima, 1954; Mizushima and Morino, 1938; Mizushima et al., 1934, 1936, 1975] represents one of the early uses of infrared and Raman spectroscopy and of molecular symmetry to prove the existence of new molecular species. In the case of 1,2-DCE, the spectrum of the pure solid agreed with prediction for the *trans* geometry indicated in Eq. (1.2), but the gas and liquid showed a composite spectrum whose peculiar infrared and Raman features indicate the presence of two 1,2-DCE isomers differing in molecular symmetry about the C–C bond. The conformational structures of these isomers, and some of the key properties associated with rotation of the C–C bond, are shown in Fig. 1.2 [Mizushima et al., 1975].

The conformational isomers in Fig. 1.2 correspond to the staggered conformations: One isomer is the centrosymmetric *trans*, and the other is the mirror-image pair *gauche*(+) and *gauche*(−). These conformations are separated by transition states with eclipsed conformations. The differences in potential energy involving these species are quite small: 5 kJ/mol for *trans* versus *gauche*, and 14 kJ/mol for the barrier between them. (The *gauche*(+)/*gauche*(−)

*Excited electronic states have discrete potential-energy surfaces and thus are not isomers of the electronic ground state. Each excited electronic state defines its own family of isomeric species.

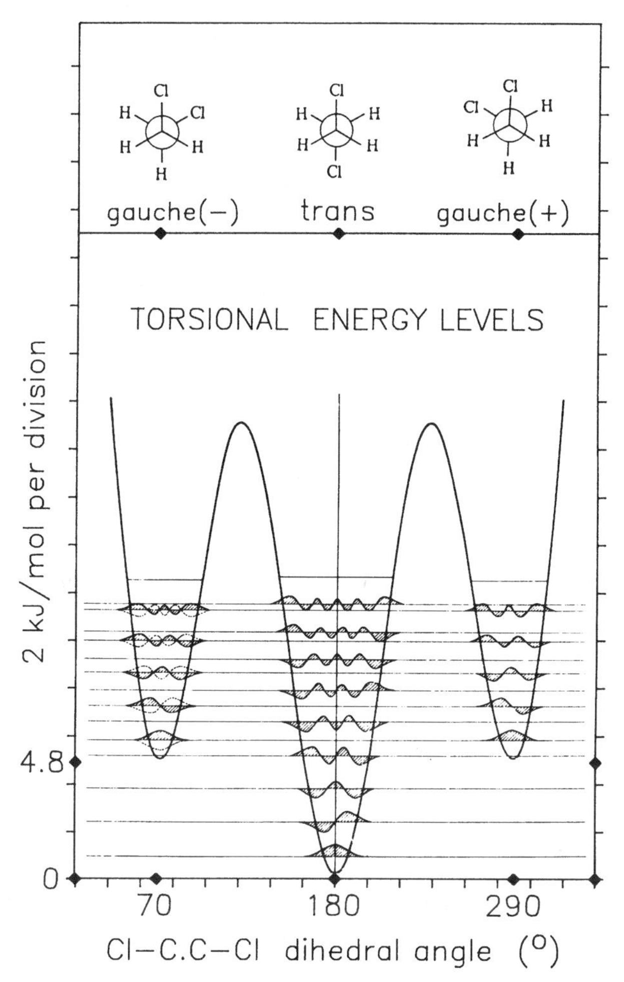

Figure 1.2. Potential energy for restricted rotation about the C–C bond in 1,2-dichloroethane. The figure also shows the torsional energy levels and torsional wave functions [Mizushima et al., 1975].

barrier could not be deduced from the spectroscopic data, but was believed to be less than 40 kJ/mol [Mizushima, 1934].)

For 1,1-DCE, the analogous staggered conformations are equivalent, owing to the threefold symmetry of the CH_3 group. The potential barrier for rotation about the C–C bond is 15 kJ/mol, practically equal to the 14-kJ/mol barier in 1,2-DCE [Daasch et al, 1954; Li and Pitzer, 1956; Wulff, 1963]. The values of these barriers, and more recent estimates of interconversion rates based on infrared spectral line shapes [Cohen and Weiss, 1983], indicate that the mean residence time of a DCE molecule in a staggered conformation is less than 100 ps.

It seems clear that for 1,1- and 1,2-DCE, the depressions in the potential surface occupied by the overall structural isomers are not shaped like simple wells but more nearly resemble elongated structures such as a gorge or a ravine. A diagram of the region for the 18-dimensional configuration space of 1,1-DCE is shown in Fig. 1.3. Imagine that a superbeing with multidimensional vision looks at the scenery from a low point on such a potential surface. In 17 dimensions it will see potential walls rising to great heights. But in the eighteenth dimension, which describes rotation about the C–C bond, it will see a gorge whose floor shows gentle sinusoidal undulations. The potential minima in the gorge represent the stable staggered conformations, whereas the maxima represent the unstable eclipsed conformations. The entire population of the gorge represents the formal component, 1,1-DCE.

Formal components that are equilibrium mixtures of rapidly interconverting molecular species are quite common. There are many modes of interconversion. They include:

- Conformational changes—by rotation about a bond [Lowe, 1968], by inversion of pucker in a covalent structure, or as steps in the pseudorotation of medium-sized rings [Hendrickson, 1967; Anet and Yavari, 1978]
- Racemizations—by inversion at the nitrogen atom in amines and ethyleneimines [Lowe, 1968]
- Formation and dissociation of weakly bound molecular complexes [Eigen, 1964]
- Hydrogen-bonding [Grunwald and Ralph, 1971; De Maeyer, 1971]

In addition, modes of interconversion are generated when molecular species are subdivided into subspecies. Examples include populations of different energy levels whose state variables (P, V, T) can be defined and also include—in a con-

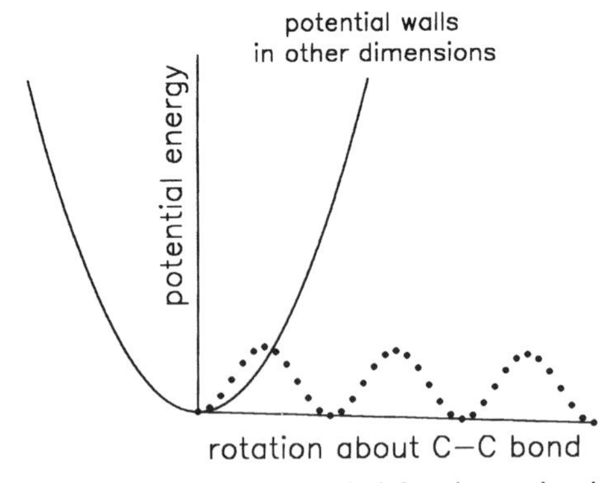

Figure 1.3. Portrait of potential gorge occupied by the molecular population of 1,1-dichloroethane.

densed phase—subspecies that are distinguishable by virtue of different local environments.

COMPOSITION MODELS

Neither chemistry nor thermodynamics encourages unnecessary complexity. A *composition model* expresses composition in terms of molecular species and subspecies, but as simply as possible. It omits all species that are insignificant both stoichiometrically and chemically and includes subspecies only if they elucidate a particular facet of reactivity.

A molecular species is significant *stoichiometrically* if its concentration is above tolerance limits for analytical error. A species is significant *chemically* if it is a reactant, product, transition state, intermediate, oxidation state, catalyst, inhibitor, chain carrier, or other actor in the mechanism of a pertinent reaction, or if the species interferes in the physical measurement because it is surface-active, light-absorbing, paramagnetic, or compromises accuracy in any way.

The number of molecular species included in a composition model varies with the nature of the problem. Species whose presence might compromise accuracy need be specified only in descriptions of actual experiments. Transition states and other stoichiometrically negligible actors in reaction mechanisms need be specified only in mechanistic or kinetic analyses. Thus, formal components may often be modeled as single species, although that representation is completely accurate only for monatomic components.* For all higher-atomic components the single-species model neglects at least one species and ignores all subspecies.

A brief reference to diatomic hydrogen as a formal component will indicate what might be neglected. The potential curve for dihydrogen, H_2, is shown in Fig. 1.4 [Herzberg, 1939]. The potential energy, which is a function of the internuclear distance, shows a single minimum that defines the equilibrium geometry of stable H_2. At long distances the potential curve approaches an asymptotic maximum that represents a pair of dissociated hydrogen atoms. However, the dissociation energy is so high that at equilibrium under ordinary conditions, the hydrogen atoms are stoichiometrically insignificant and the component "hydrogen" may be modeled as a single species, H_2. This approximation breaks down for reactions in which the hydrogen atoms play an essential role as chain carriers—for instance, in the chain-propagating and chain-inhibiting steps of the thermal reaction of hydrogen with bromine, Eqs. (1.3) [Bodenstein and Lind, 1906]. Although no longer at equilibrium, the hydrogen atoms remain stoichiometrically negligible, but their chemical presence can no longer be ignored.

*For monatomic species the potential-energy function degenerates to a single point—or to a set of points if there are subspecies based on isotopic mass, nuclear spin, or electron spin.

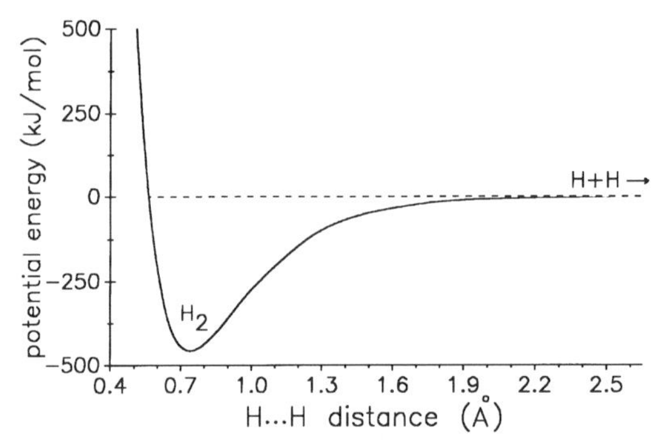

Figure 1.4. Potential energy versus H ··· H distance for the diatomic hydrogen molecule. The stable potential minimum is at 0.741 Å and −458.0 kJ/mol [Herzberg, 1939]. The metastable asymptote at 0 kJ/mol represents dissociated H + H.

$$
\begin{aligned}
Br_2 &\rightarrow 2Br \\
Br + H_2 &\rightarrow HBr + H \\
H + Br_2 &\rightarrow HBr + Br \\
H + HBr &\rightarrow H_2 + Br
\end{aligned}
\qquad (1.3)
$$

Some molecular species are negligible at chemical equilibrium, but they become *chemically* significant under kinetic steady-state conditions and equilibrate slowly enough to be *stoichiometrically* significant under other nonequilibrium conditions. For example, common acetone, C_3H_6O, is nominally an equilibrium mixture consisting of the dominant ketone and a trace of the enol [Eq. (1.4)].

$$
\underset{\text{Ketone}}{H_3C\text{—}\overset{\overset{\displaystyle O}{\|}}{C}\text{—}CH_3} \rightleftharpoons \underset{\text{Enol}}{(HO)(H_3C)C{=}CH_2} \qquad (1.4)
$$

At equilibrium the [enol]/[ketone] ratio is very small. For example, in water at 25°C it is a minuscule 4.7×10^{-9}, and the component "acetone" is essentially 100% ketone [Chiang et al., 1984, 1987; Keeffe et al., 1988]. On the other hand, the enol is chemically significant as a reaction intermediate in the halogenation of acetone [Lapworth, 1904], and it becomes significant as a stoichiometric component when generated photochemically at an appropriate pH [Chiang et al., 1987].

Errors of Omission

Composition models are useful because they delete irrelevant detail and leave the essentials. If a nonessential species is nonetheless included, some clarity is lost but no harm is done, because the equations of thermodynamics are general and apply to any number of molecular species.

If a species is innocently omitted, no harm is done if that species is negligible and would have been omitted deliberately had its existence been known. On the other hand, if the omitted species is real and significant, the omission can be detected because the composition model becomes inadequate. For example, the *P–V–T* relationship of nitrogen dioxide deviates from the ideal-gas law, even at low pressures until one assumes that nitrogen dioxide is really an equilibrium mixture of NO_2 and N_2O_4 [Bodenstein and Boes, 1922; Giauque and Kemp, 1938]. For another example, the heat capacity of gaseous 1,3-butadiene seems anomalous until one assumes the existence of a temperature-dependent equilibrium between *s-trans* and *s-cis* conformational species [Eq. (1.5)] [Aston et al., 1946; Bock et al., 1979; Mui and Grunwald, 1982].

s-trans   *s-cis*   (1.5)

The discovery of molecular species from seeming anomalies of thermodynamic data is a proven supplement to more direct observation by spectroscopy. The thermodynamic method, which is based on changing the composition model until it fits the facts, is broader in scope and becomes singularly useful when spectroscopic methods fail.

COMPOSITION TREES

A *composition tree* is the diagram of a composition model that traces in descending order the relationships from formal components to molecular species (including associated and dissociated products) and to subspecies. It includes chemical names, abbreviations, and acronyms. It indicates existing chemical equilibria. It assigns number labels to the components and assigns capital-letter labels to the molecular species and subspecies. It defines concentration variables, progress variables for chemical reactions, specific parameters, and stoichiometric relationships. In short, it outlines the essence of the chosen model.

As an example, we shall consider the relatively simple composition tree (T-1.1).

Composition Tree (T-1.1)

Carbon tetrachloride (1, n_1)	*N*-Methylacetamide (2, n_2)	
CCl_4 (A, n_A)	*cis*-NMA (C, n_C)	*trans*-NMA (T, n_T)
$n_A = n_1$	$n_C + n_T = n_2$	

The first line shows the components: carbon tetrachloride (indicated as component 1), and *N*-methylacetamide (indicated as component 2). The respective formula-weight numbers are n_1 and n_2.

The second line shows the descendant molecular species. Carbon tetrachloride is modeled as the single species CCl_4 and is assigned the label A. *N*-Methylacetamide is given the acronym NMA and is modeled as an equilibrium mixture, $C \rightleftarrows T$, of *cis* and *trans* geometrical isomers about the C–N bond [Eq. (1.6)] (*cis* and *trans* describe the geometry of the methyl groups). The isomers, C and T, are placed under a common vertical line to denote that they exist in equilibrium. The mole numbers of the molecular species are n_A, n_C, and n_T, respectively. In aqueous solution, the *cis*/*trans* equilibrium ratio is *ca.* 0.03, and at 60°C the relaxation time for *cis*/*trans* interconversion is ~9 ms [Baker and Boudreaux, 1967; Drakenberg and Forsen, 1971].

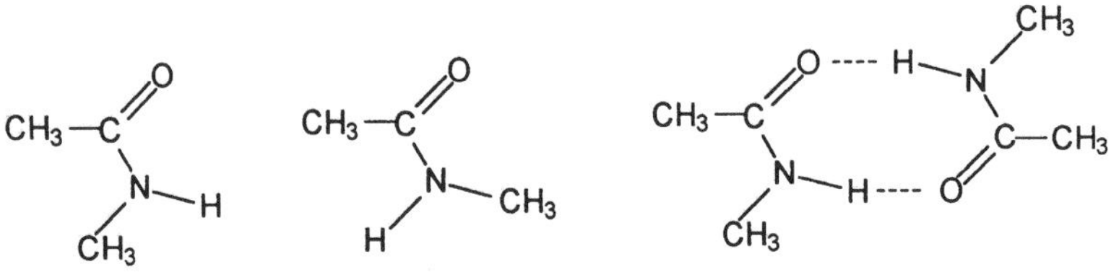

(1.6)

cis *trans* H-bonded dimer

Actually, the composition tree (T-1.1) gives a good representation only when the NMA component is a dilute solute. At higher concentrations, the *cis* isomer must be represented as an equilibrium mixture of an authentic monomer and a hydrogen-bonded dimer, $C \cdot C$, also depicted in Eq. (1.6). This more refined composition model is represented by composition tree (T-1.2).

Composition Tree (T-1.2)

Carbon tetrachloride $(1, n_1)$ — *N*-Methylacetamide $(2, n_2)$

CCl_4 (A, n_A) — total *cis* ("C", $n_{\text{“C”}}$) — *trans* (T, n_T)

$C\,(n_c)$ — $C\cdot C\,(n_{C\cdot C})$

$$n_1 = n_A, \qquad n_2 = n_{\text{“C”}} + n_T, \qquad n_{\text{“C”}} = n_C + 2n_{C\cdot C}$$

The format of (T-1.2) is typical of all but the simplest composition trees. The first line shows the components, the second line shows the "total," or all-inclusive, molecular species, and the third line subdivides any molecular species whose association, dissociation, or dissection into subspecies is chemically or stoichiometrically significant. "Total" or all-inclusive molecular species (such as "C") that need to be subdivided further are shown in quotation marks. Species that exist in chemical equilibrium, such as $C \rightleftarrows C \cdot C$, are connected by placement under a horizontal line.

When the chemistry is so rich that the composition model becomes quite complicated, the representation in terms of a composition tree may be replaced by a matrix representation, such as that described by Alberty [1992] for biochemical reactions.

For the sake of clarity we shall make frequent use of composition trees. This is because composition models are variable, even for systems of nominally constant composition. They depend on the values of the external variables (T, P, V), on the context in which the model is used, on the subtlety with which the chemical mechanisms are described, and—as we shall see presently—on the time scale of the experimental observations.

OBSERVATIONAL TIME SCALES AND COMPONENTS

The time scale on which observations are made is important because it limits the substances whose concentrations can be varied independently. We shall now consider this issue and, as a result, broaden the definition of a component.

In its classical definition, a thermodynamic component is a well-characterized, tangible, stable substance that can be reproduced anywhere on earth. This criterion is ideally satisfied by the formal components of chemical systems, which are stable pure substances with known chemical formulas. On the molecular level, formal components are often mixtures of molecular species whose presence is revealed spectroscopically. Because operationally the formal component is a pure substance, we may infer that the interconversion of the descendant species proceeds rapidly, so that the species remain in equilibrium and are not separated by the purification procedures. As a corollary, the same species remain in equilibrium during all laboratory procedures that are conducted on a comparable time scale, such as pouring, mixing, manual operations in general, and, indeed, nearly all operations that are controlled by humans with their unaided senses.

Given that the response time of human senses is of the order of 100 ms, we may infer that when the mean time of approach to equilibrium in a reaction is shorter than 10 ms, the reaction appears to be instantaneous. Actual rate measurements then require fast-acting robots and measuring instruments. Effective methods were developed mostly after the Second World War, as suitable technology became available. Most of the measurements are based on one of the following two approaches. In the first approach one measures the width and shape of spectral lines. The results are then analyzed by what is tantamount to Fourier transformation from the spectral frequency domain to the time domain, to yield kinetic information. This approach will be further discussed in Chapter 9. In the second approach, which interests us at this point, one introduces a sudden, controlled displacement of the reaction from equilibrium and then measures the rate of return to equilibrium in real time. The initial displacement from equilibrium, and the subsequent kinetic measurements, are on time scales that are short compared to those of unaided manual operations. Because the initial displacement from equilibrium can be controlled, the concentrations of the affected molecular species become independent variables. Thus, on the given time scale, these species behave just like independent components.

Well-tested techniques for producing sudden, controlled displacements from equilibrium are as follows: mixing of fluids, sudden change of external conditions, and flash-photochemical generation of substrates in situ [Caldin, 1964; Hague, 1971; Bernasconi, 1976]. The rapid mixing of fluids to give a single homogeneous phase can be accomplished by proper hydrodynamic design. Reactions can be displaced from equilibrium in as little as a few milliseconds [Roughton and Chance, 1961].

Sudden, controlled changes in temperature can be produced by the sudden introduction of electrical energy from a condenser, or of infrared or microwave radiant energy from a laser pulse. Sudden pressure jumps can be produced by mechanical puncture of a pressure-bearing thin wall. Reactions can be displaced from equilibrium in as little as a few microseconds [Eigen and De Maeyer, 1961].

The fastest technique for producing displacements from equilibrium employs powerful radiant laser pulses of an appropriate wavelength to photochemically generate new species or subspecies, or to change the amount of existing species. The laser pulses can be made as short as 10^{-13} s [Porter, 1961; Weller, 1968; Rentzepis, 1970; Dantus et al., 1987; Trautman et al., 1990].

In theory, the scope of the fast kinetic methods is quite general, being limited only at imponderably high interconversion rates when the time-dependent molecular wave functions lose their separate identities.* In practice, however, the fast kinetic methods are not child's play, and their scope, though expanding, is limited by the current state of technology. When the methods *do* apply, the results can be quite as accurate as those obtained on more leisurely time scales and quite as convincing. For example, in Fig. 1.5 a transient anion, an electronic triplet generated by a 10-ns laser flash, is characterized by its visible absorption spectrum. The reaction of this species with an electron donor to produce a short-lived radical dianion is then followed on a time scale of 0.1 to 10 μs, by recording the time evolution of the absorption spectrum [Hurley et al., 1988; Loeff et al., 1991].

The Timewise Component

As shown in the preceding discussion, independent components can be controllably generated in solution and characterized on time scales that are too short to permit direct observations with the unaided human senses. These components can then be used in a wide variety of chemical and physical applications

*At sufficiently high interconversion rate, the time-dependent wave functions coalesce by exchange-averaging or overlap by short-lifetime uncertainty broadening. See Chapter 9.

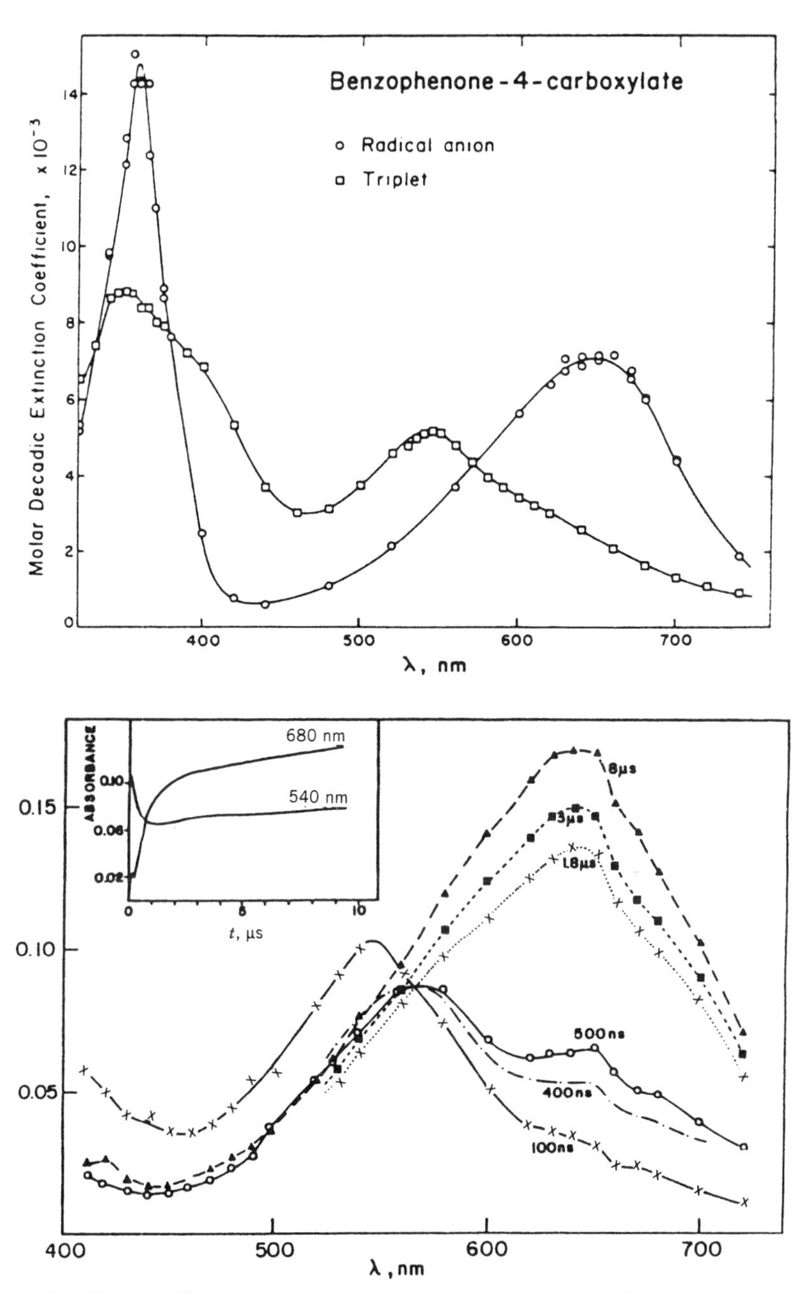

Figure 1.5. Absolute (*top*) and time-dependent absorption spectrum for the ultraviolet laser pulse-generated benzophenone-4-carboxylate triplet anion, $^3C_6H_5{\cdot}CO{\cdot}C_6H_4CO_2^-$(p), and for the radical dianion $(BC^{2-})^{\cdot}$ which is produced from it by electron transfer involving formate ion in water at pH 11.2. Optical absorption by formate ion and formyl radical is negligible under these conditions. (Reprinted from (*top*) Hurley et al. [1988] and (*bottom*) Loeff et al. [1991], with permission of the American Chemical Society.)

on similarly short time scales, much in the manner of traditional formal components on ordinary time scales. It is reasonable, therefore, to redefine the formal components of chemical systems to include such substances. The following definition includes traditional formal components and adds controllable short-lived substances in such a way that the operational thermodynamic character of the definition is preserved.

> A component in a chemical system on a given time scale is either a pure substance with a known chemical formula or a characterized, distinctive substance whose concentration can be varied independently by production in situ, displacement of a reaction from equilibrium, or other controllable means.

For example, the formal component *n*-propyl nitrite, C_3H_7ONO (b.p. 57°C), has been shown by its proton magnetic resonance spectrum in the liquid state to be an equilibrium mixture of *cis* and *trans* conformational isomers about the O–N bond [Eq. (1.7)].

$$\underset{cis}{CH_3CH_2CH_2\text{–}O\text{–}N{=}O} \qquad \underset{trans}{CH_3CH_2CH_2\text{–}O\text{–}N{=}O} \tag{1.7}$$

At 28°C, the *cis*/*trans* equilibrium ratio is 1.8, and the mean time for relaxation to equilibrium is 10 μs [Phillips et al., 1957]. Accordingly the *cis*/*trans* mixture, if displaced from equilibrium, consists of two quasi-stable components on time scales of less than 100 ns (less than about $\frac{1}{100}$ of 10 μs) but exists as, and remains, an equilibrium mixture of two molecular species on time scales above 100 μs (more than about 10×10 μs), the component then being *n*-propyl nitrite. In general, on time scales of less than 0.01 times the mean time for relaxation to equilibrium in a reaction, the reactants and products are essentially nonreactive components; on time scales of more than 10 mean times for relaxation to equilibrium, they emerge from thermodynamic processes essentially at equilibrium. At intermediate time scales, rates of approach to equilibrium become significant.

2

GIBBS FREE ENERGY, CHEMICAL EQUILIBRIUM, AND CORRESPONDENCE THEOREMS

In Chapter 1 we described composition both in terms of thermodynamic components and of models involving molecular species. In this chapter we will introduce the thermodynamics for this dual representation. We will assume a background in undergraduate physical chemistry or introductory thermodynamics, but we will refresh that background by treating some topics in detail.

When composition is a variable, thermodynamic descriptions are shaped by the underlying mathematics. We shall find that chemical nonequilibrium enters unavoidably. At equilibrium, the Correspondence Theorems relate the partial free energies of formal components to those of their descendant molecular species.

MATHEMATICAL ANALOGY

The work of proving thermodynamic theorems is simplified by the existence of mathematical analogies. The essence of a mathematical equation is in the operations it prescribes; the symbols used for the variables are arbitrary. For instance, if a dependent variable is defined as the square of an independent variable, we might use y for the former and x for the latter and write $y = x^2$. Or we might use TREE for the former and BIRD for the latter and write TREE = BIRD^2, to the consternation of naturalists. Mathematically, the two equations express exactly the same relationship.

Physical equations differ from this because the symbols have physical significance. For instance, by international recommendation the symbol U is reserved for internal energy, H for enthalpy, A for Helmholtz energy, S for entropy, and G for Gibbs energy or free energy [International Union, 1988]. Mathematical analogy exists when physically different equations are mathematically identical. For example, all equations for damped harmonic motion follow the same mathematics, even though they describe phenomena ranging from the vibration of mechanical structures to the amplitude of sound waves.

Legendre Analogy for Energy Functions

Mathematical analogy applies to equations involving the energy functions U, H, A, and G, which are related by Legendre transformations (named in honor of the eighteenth-century French mathematician A. M. Legendre). To explain the analogies we begin by defining some terms. The independent variables in a physical equation are those one must control so that the other variables become dependent—that is, determined by natural law. A *closed* system is a thermodynamic system containing a fixed amount of matter. A *chemical* system is a system whose exchange of work with the surroundings is limited to expansion work, $\int P dV$; other kinds of work such as electrical, surface, or photophysical work are negligible. The *state of a system* is defined when enough variables (such as T, P, V) have been specified so that all other variables become dependent.

The basic thermodynamic equation for changes of state in a closed chemical system is

$$dU = T\,dS - P\,dV \tag{2.1}$$

Mathematically, this equation defines two independent variables, S and V, and three dependent variables, U, T, and P. U is the primary dependent variable, while T and P are dependent derivatives: $T = (\partial U/\partial S)_V$ and $P = -(\partial U/\partial V)_S$. T is paired with S, and P is paired with V.

A Legendre transformation on U switches the dependency status of T and S, or of P and V, or of both. To switch the status of P and V, replace the energy U by the enthalpy $H = U + PV$. This is a purely bookkeeping operation that does not change the state of the system. Given $H = U+PV$, $dH = dU+P\,dV+V\,dP$ and using Eq. (2.1) to eliminate dU, we obtain Eq. (2.2):

$$dH = T\,dS + V\,dP \tag{2.2}$$

If we think of $\{H, T, S, P, V\}$ as mathematical symbols rather than physical properties, Eq. (2.2) becomes equivalent to Eq. (2.1). To see this, write U in place of H, V in place of P, and $-P$ in place of V, and behold! You obtain Eq. (2.1). Physically, there is a difference, of course, because the independent variables that must be controlled are S and P in Eq. (2.2) and are S and V in Eq. (2.1). Moreover, the energy function H is numerically different from U.

As a consequence of this mathematical analogy, if there exists a physically valid relationship $f(U, S, T, V, P) = 0$, the consistent substitution of H for U,

P for V, and $-V$ for P will generate an equally valid but physically different relationship $g(H, S, T, P, V) = 0$. S and T remain untransformed.

For example, the following is the famous "energy equation" of thermodynamics that was used in the nineteenth century to show that the energy of an ideal gas does not change upon expansion at constant temperature:

$$\left(\frac{\partial U}{\partial V}\right)_T = T\left(\frac{\partial P}{\partial T}\right)_V - P$$

By following the given recipe, mathematical analogy yields an equally valid equation for the enthalpy:

$$\left(\frac{\partial H}{\partial P}\right)_T = -T\left(\frac{\partial V}{\partial T}\right)_P + V$$

Note that *all* variables must be treated by the recipe, including those that are indicated as remaining constant in the partial derivatives. Thus $(\partial P/\partial T)_V$ becomes $-(\partial V/\partial T)_P$, the transformation being $P \rightarrow -V$ and $V \rightarrow P$.

To switch the dependency status of T and S, replace the energy U by the Helmholtz function $A = U - TS$, as in Eq. (2.3). To switch the dependency status of T and S, as well as P and V, replace the energy U by the Gibbs free energy G, as in Eq. (2.4).

$$A = U - TS, \qquad dA = -S\,dT - P\,dV \tag{2.3}$$

$$G = U + PV - TS, \qquad dG = -S\,dT + V\,dP \tag{2.4}$$

Then convince yourself by appropriate replacement of symbols that the resulting equations for dA and dG are mathematically equivalent to Eq. (2.1) for dU. Mathematical analogy therefore embraces the functions of U, H, A, and G. The recipes for generating the analogous equations are summarized in Table 2.1.

After application of a recipe, the resulting equation may benefit from minor neatening-up. For example, transformation of the energy equation from the original variables $\{U, S, T, P, V\}$ to new variables $\{G, S, T, P, V\}$ gives Eq. (2.5). The left-hand side of that equation can be rewritten in standard form by noting that $(\partial G/\partial P)_{-S} = (\partial G/\partial P)_S$, because S is constant when $-S$ is constant.

$$\left(\frac{\partial G}{\partial P}\right)_{-S} = -S\left(\frac{\partial(-V)}{\partial(-S)}\right)_P + V, \qquad \left(\frac{\partial G}{\partial P}\right)_{-S} = \left(\frac{\partial G}{\partial P}\right)_S = -S\left(\frac{\partial V}{\partial S}\right)_P + V \tag{2.5}$$

Table 2.1. Recipes for creating Legendre analogies in closed systems

New Variables	Recipe
	Original Variables $\{U, S, T, P, V\}$
$\{H, S, T, P, V\}$	$U \to H$; $S \to S$; $T \to T$; $P \to -V$; $V \to P$
$\{A, S, T, P, V\}$	$U \to A$; $T \to -S$; $S \to T$; $P \to P$; $V \to V$
$\{G, S, T, P, V\}$	$U \to G$; $T \to -S$; $S \to T$; $P \to -V$; $V \to P$
	Original Variables $\{H, S, T, P, V\}$
$\{U, S, T, P, V\}$	$H \to U$; $S \to S$; $T \to T$; $P \to V$; $V \to -P$
$\{A, S, T, P, V\}$	$H \to A$; $T \to -S$; $S \to T$; $P \to V$; $V \to -P$
$\{G, S, T, P, V\}$	$H \to G$; $T \to -S$; $S \to T$; $P \to P$; $V \to V$
	Original Variables $\{A, S, T, P, V\}$
$\{U, S, T, P, V\}$	$A \to U$; $S \to -T$; $T \to S$; $P \to P$; $V \to V$
$\{H, S, T, P, V\}$	$A \to H$; $S \to -T$; $T \to T$; $P \to -V$; $V \to P$
$\{G, S, T, P, V\}$	$A \to G$; $S \to S$; $T \to T$; $P \to -V$; $V \to P$
	Original Variables $\{G, S, T, P, V\}$
$\{U, S, T, P, V\}$	$G \to U$; $S \to -T$; $T \to S$; $V \to -P$; $P \to V$
$\{H, S, T, P, V\}$	$G \to H$; $S \to -T$; $T \to S$; $V \to V$; $P \to P$
$\{A, S, T, P, V\}$	$G \to A$; $S \to S$; $T \to T$; $V \to -P$; $P \to V$

In experiments on liquid solutions, the primary energy function is usually the Gibbs free energy G, because the independent variables T and P for dG in Eq. (2.4) are relatively easy to control. Most of the thermodynamics in this book therefore treats G as the primary function. But in view of the recipes in Table 2.1, this is scarcely a limitation, because each equation in G generates related equations in U, H, and A by simple mathematical analogy.

EULER EQUATIONS AND SOLUTION PROPERTIES

The variables used in the thermodynamics of physically homogeneous phases are either extensive or intensive. Extensive variables are proportional to the amount of matter in the phase; examples are the volume and heat capacity of the phase and the formula-weight numbers of the components. Intensive variables are independent of the amount of matter. Some are intrinsically independent of the system's size, such as T, P, and the refractive index. Others are intensive because they are *ratios* of extensive variables obtained for the same phase, such as volume per unit mass or formula-weight fraction of a component. The thermodynamic state of a system is defined intensively. That is, the independent variables whose number is given by Gibbs' phase rule are intensive variables.

An open system (in contrast to a closed system) exchanges matter with its surroundings. The formula-weight numbers of the components are variables, and extensive features are therefore important. In this section we shall consider the mathematics of extensive properties.

Definition

Let n_1, $n_2, \ldots, n_k$ denote the formula-weight numbers in a reference system, and let the size of the system vary by a finite scale factor λ without change of the thermodynamic state. A dependent variable or property Y then is extensive if

$$Y(\lambda n_1, \lambda n_2, \ldots, \lambda n_k) = \lambda \cdot Y(n_1, n_2, \ldots, n_k) \tag{2.6}$$

Euler Equations

Let n_1, $n_2, \ldots, n_k$ be extensive composition variables for a k-component phase at constant T and P, and let $\{n \neq n_i\}$ denote the set of all n's excluding n_i. An extensive property Y then is related to n_1, $n_2, \ldots, n_k$ by Eq. (2.7):

$$Y = n_1 \left(\frac{\partial Y}{\partial n_1} \right)_{\{n \neq n_1\}} + n_2 \left(\frac{\partial Y}{\partial n_2} \right)_{\{n \neq n_2\}} + \cdots + n_k \left(\frac{\partial Y}{\partial n_k} \right)_{\{n \neq n_k\}} \tag{2.7}$$

Equation (2.7) is known as the *Euler equation* in chemical thermodynamics, in honor of the eighteenth-century Swiss mathematician Leonhard Euler.

PROOF. It will be sufficient to give a proof for the volume V of a two-component system. We begin with the definition of an extensive property:

$$\lambda \cdot V(n_1, n_2) = V(\lambda n_1, \lambda n_2)$$

On the left, the independent variable is λ; n_1 and n_2 are properties of the reference system. On the right, the independent variables are λn_1, λn_2. Let $x_1 = \lambda n_1$, and let $x_2 = \lambda n_2$. Differentiate with respect to λ:

$$V(n_1, n_2) = (\partial V(x_1, x_2)/\partial x_1)_{x_2} (\partial x_1/\partial \lambda)_{n_1} + (\partial V(x_1, x_2)/\partial x_2)_{x_1} (\partial x_2/\partial \lambda)_{n_2}$$
$$V(n_1, n_2) = n_1 (\partial V(x_1, x_2)/\partial x_1)_{x_2} + n_2 (\partial V(x_1, x_2)/\partial x_2)_{x_1}$$

These equations are valid for any value of λ. If we let $\lambda = 1$, then $x_1 \to n_1$ and $x_2 \to n_2$, where n_1 and n_2 are now variables. Thus,

$$V(n_1, n_2) = n_1 (\partial V(n_1, n_2)/\partial n_1)_{n_2} + n_2 (\partial V(n_1, n_2)/\partial n_2)_{n_1}$$

as required by Eq. (2.7). Q.E.D.

Euler Analogy

Partial derivatives such as $(\partial Y/\partial n_i)_{\{n \neq n_i\}}$ and $(\partial V/\partial n_2)_{n_1}$ are *mathematically* (not physically) analogous to formal properties of pure components. The analogy is almost obvious, so we shall illustrate it only for the two-component volume.

For *pure* component 1 the volume is given by $V = n_1 \mathsf{V}_1$, where V_1 is the volume per formula weight (the formal volume) of the pure component. For two separate pure components the volume is a simple sum: $V = n_1 \mathsf{V}_1 + n_2 \mathsf{V}_2$. Hence $(\partial V/\partial n_1)_{n_2} = \mathsf{V}_1$ and $(\partial V/\partial n_2)_{n_1} = \mathsf{V}_2$, so that we may write $V = n_1(\partial V/\partial n_1)_{n_2} + n_2(\partial V/\partial n_2)_{n_1}$—the same expression as for a physically homogeneous solution. To symbolize the analogy, we shall denote $(\partial V/\partial n_1)_{n_2}$ for the *solution* by V_1 and denote $(\partial V/\partial n_2)_{n_1}$ for the solution by V_2, so that the solution volume becomes $V = n_1 V_1 + n_2 V_2$. Note that V_1 and V_2 have the dimensions of V/n and thus are intensive properties per formula weight.

The analogy of V_1 to V_1, and of V_2 to V_2, is especially useful for the calculation of volume changes. For example, the volume change $\Delta_{\text{mix}} V$ on forming a homogenous solution from the separate components is calculated as follows:

$$
\begin{aligned}
\Delta_{\text{mix}} V &= V(\text{solution}) - V(\text{separate components}) \\
&= (n_1 V_1 + n_2 V_2) - (n_1 \mathsf{V}_1 + n_2 \mathsf{V}_2) \\
&= n_1(V_1 - \mathsf{V}_1) + n_2(V_2 - \mathsf{V}_2)
\end{aligned}
$$

In the final equation, the mathematics operates as if on forming the solution the formal volume of component 1 changes from V_1 to V_1, and that of component 2 changes from V_2 to V_2. That is, V_1 and V_2 are surrogates of formal volumes for the respective components in the homogeneous solution.

Analogous mathematical operations, in which partial derivatives in Euler equations serve as surrogates for the respective formal component properties in solution, give correct answers whenever a component is transferred into or out of a homogeneous solution. In other words, for purposes of thermodynamic calculations, we may think of $\partial V/\partial n_i$, $\partial G/\partial n_i$, $\partial H/\partial n_i$, or in general $\partial Y/\partial n_i$, as surrogates for the volume per formula weight, free energy per formula weight, enthalpy per formula weight, and the generalized property Y per formula weight, of the ith component in the given solution.

Nomenclature and Physical Significance

Because of their obvious importance in solution thermodynamics, the partial derivatives in Euler's equations need a name. In the text by Lewis and Randall [1923] they are called *partial molar properties*. For instance, $\partial V/\partial n_i$—that is, V_i—is the *partial molar volume*. The adjective *partial* indicates that V_i is a partial derivative, and *molar* indicates that V_i is an intensive property per mole or per formula weight. In this book the adjective *molar* is redundant and will be omitted, because we hardly ever use partial properties that are *not* intensive per mole or per formula weight. Thus, when we use the adjective *partial* (as in partial volume or partial free energy), we mean

- A partial derivative that appears in Euler's equation for the given solution property
- That is an intensive property per mole or per formula weight
- and That, in the given solution, serves as a surrogate for a real property of the dissolved component

Accordingly, we will extend the notation in which $\partial V/\partial n_i$ is denoted by V_i to other partial derivatives as well. For example, $G_i = \partial G/\partial n_i$, the partial free energy; and $S_i = \partial S/\partial n_i$, the partial entropy.

Physically, a partial property is a property of the *entire solution—not* an intrinsic property for a single solution component. For example, in case of the partial volume $V_2 = \partial V/\partial n_2$, the differential volume change ∂V includes not only the space taken up by the ∂n_2 added molecules themselves, but also any volume change in the original liquid due to changes in packing density induced by the addition. Complexities such as this, in the physical significance of partial solution properties, are the grist of the thermodynamics of molecular species. A good part of this book will be devoted to their study.

The Gibbs–Duhem Equation

Euler's equations have a corollary that is sometimes useful in chemical thermodynamics. Using our simplified notation, Eq. (2.7) is cast in the form (2.8a), where $Y_i = (\partial Y/\partial n_i)_{T,P,\{n \neq n_i\}}$:

$$Y = n_1 Y_1 + n_2 Y_2 + \cdots + n_k Y_k \tag{2.8a}$$

The general form of the differential is therefore (2.8b):

$$dY = Y_1\, dn_1 + n_1\, dY_1 + Y_2\, dn_2 + n_2\, dY_2 + \cdots + Y_k\, dn_k + n_k\, dY_k \tag{2.8b}$$

But also, Y is a function of $(n_1, n_2, \ldots, n_k)$—hence (2.8c):

$$dY = Y_1\, dn_1 + Y_2\, dn_2 + \cdots + Y_k\, dn_k \tag{2.8c}$$

Subtraction of Eq. (2.8c) from Eq. (2.8b) yields Eq. (2.8d), which is a generalized form of the Gibbs–Duhem equation:

$$n_1\, dY_1 + n_2\, dY_2 + \cdots + n_k\, dY_k = 0 \tag{2.8d}$$

Equation (2.8d) is useful because it shows that in a change of state, only $k-1$ of the k changes $(dY_1, dY_2, \ldots, dY_k)$ are independent, for any extensive property Y. In particular, for a two-component system, $dY_1 = -(n_2/n_1)dY_2$.

THE FREE ENERGY

For a closed homogeneous phase, $dG = -S\ dT + V\ dP$ [Eq. (2.4)], where T and P are the independent variables, and $G = G(T, P)$. Combining this with Euler's Eq. (2.7) (by letting $Y = G$), we find that for an open system with k components, $G = G(T, P, n_1, n_2, \ldots, n_k)$. Physically, the independent variables are the variables one controls. Since it is convenient, when working with liquid solutions, to control $(T, P, n_1, n_2, \ldots, n_k)$ as independent variables, the free energy G is a convenient energy function for liquid solutions.

Beginning with G, other thermodynamic properties are obtained by partial differentiation. Thus from Eq. (2.4), $S = -(\partial G/\partial T)_{P,\{n\}}$ and $V = (\partial G/\partial P)_{T,\{n\}}$, where $\{n\}$ denotes the set of formula-weight numbers in the phase. Other properties are given by higher derivatives of the free energy. For instance, $C_p = T(\partial S/\partial T)_{P,\{n\}} = -T(\partial^2 G/\partial T^2)_{P,\{n\}}$. The partial free energy of the ith component is given by $G_i = (\partial G/\partial n_i)_{T,P,\{n \neq n_i\}}$.

When composition is specified in terms of the number Q of molecular species rather than the number k of components, G becomes a function $G(T, P, n_{\mathrm{A}}, \ldots, n_{\mathrm{J}}, \ldots, n_Q)$. As before, we are using capital-letter subscripts for molecular species. Mole numbers are extensive variables (see below); hence Euler's equation (2.9a) applies. The partial free energies $G_{\mathrm{A}}, \ldots, G_Q$ are thus defined in Eq. (2.9b):

T and *P* = constant in Eqs. (2.9a) and (2.9b)

$$G = G_{\mathrm{A}} n_A + \cdots + G_{\mathrm{J}} n_{\mathrm{J}} + \cdots + G_Q n_Q \tag{2.9a}$$

$$G_{\mathrm{A}} = (\partial G/\partial n_{\mathrm{A}})_{\{n \neq n_{\mathrm{A}}\}}, \qquad G_{\mathrm{J}} = (\partial G/\partial n_{\mathrm{J}})_{\{n \neq n_{\mathrm{J}}\}}, \qquad G_Q = (\partial G/\partial n_Q)_{\{n \neq n_Q\}} \tag{2.9b}$$

Note the active and inactive (i.e., constant) mole numbers in (2.9b). In the case of G_{J}, for instance, the mathematics requires that all mole numbers except n_{J} are constant. Note also that T and P are constant.

In some composition trees, each component contains a single molecular species, so that $k = Q$. But, in general, k will be smaller than Q, with the difference $Q - k$ being due to molecular species produced in chemical equilibria. There are then $Q - k$ equations of constraint on the mole numbers, either in the form of equilibrium constants or through stoichiometry. These constraints are *intensive*, so that at constant formal composition the mole *ratios* are independent of the size of the phase. The mole *numbers* accordingly are extensive. Thus Eq. (2.9) expresses the free energy correctly, regardless of the presence of chemical equilibria.

Partial Free Energies and Chemical Potentials

Mathematically, the Gibbs chemical potential μ is identical to the partial free energy. That is, for the ith component, $\mu_i = (\partial G/\partial n_i)$ at constant $T, P, \{n \neq n_i\}$. However, while the partial free energy G_i is a property of the entire homogeneous phase, some practitioners view the chemical potential μ_i as a separate physical entity, with a physical life of its own. An example will show why this distinction is important.

Consider a three-component liquid solution. Component 1 is the solvent, and components 2 and 3 are solutes. Suppose that our practitioner is measuring $\partial G/\partial n_2$ as a function of the molality m_3 of the other solute, and that he wishes to interpret these results in terms of an internal pressure concept. If he regards $\partial G/\partial n_2$ as a property of the entire solution, his theory will begin with a model for the internal pressure and its effect on G for the entire solution, and then it will deduce G_2 (or G_1 or G_3) by partial differentiation of the model G with respect to n_2 (or n_1 or n_3).

By contrast, if he regards $\partial G/\partial n_2$ as a property μ_2 in its own right, his theory will model μ_2 directly as a function of the internal pressure and will concern itself with the change in internal pressure as a function of m_3. Both approaches have been tried, and the results are substantially different [McDevit and Long, 1952; Grunwald and Butler, 1960]. But the approach beginning with G is the correct one.

There are basically two arguments why the chemical potential μ *might* have a physical life of its own [Gibbs, 1948]. First, in heterogeneous systems as well as in single inhomogeneous phases, the partial derivative $\partial G/\partial n_i$ acts like a potential: When $\partial G/\partial n_i$ has a gradient, the ith component flows so as to reduce that gradient, just as electric charge flows so as to reduce a gradient of electric potential. Either motion stops when the gradient goes to zero. However, this argument does not *prohibit* μ_i from being subordinate to G.

Second, μ_i is not tied specifically to the free energy G, but can be derived equally from G, A, H, or U, as shown in Eq. (2.10).

Alternative expressions for μ_i

$$\mu_i = \left(\frac{\partial G}{\partial n_i}\right)_{T,P} = \left(\frac{\partial A}{\partial n_i}\right)_{T,V} = \left(\frac{\partial H}{\partial n_i}\right)_{S,P} = \left(\frac{\partial U}{\partial n_i}\right)_{S,V}$$

$$\{n \neq n_i\} = \text{constant} \tag{2.10}$$

However, these equalities are consequences of the Legendre mathematics of the energy functions and do not create physical content.

PROOF. To illustrate the Legendre mathematics, we shall show that $(\partial A/\partial n_i)_{T,V} = (\partial U/\partial n_i)_{S,V}$. We begin with the definition $A = U - TS$.

$A = U - TS$; $\{n \neq n_i\}$ = constant

$$\left(\frac{\partial A}{\partial n_i}\right)_{T,V} = \left(\frac{\partial U}{\partial n_i}\right)_{T,V} - T\left(\frac{\partial S}{\partial n_i}\right)_{T,V}$$

Change variables from $U(n_i, T, V)$ to $U(n_i, S, V)$ as explained in the next section.

$$\left(\frac{\partial A}{\partial n_i}\right)_{T,V} = \left(\frac{\partial U}{\partial n_i}\right)_{S,V} + \left(\frac{\partial U}{\partial S}\right)_{n_i,V}\left(\frac{\partial S}{\partial n_i}\right)_{T,V} - T\left(\frac{\partial S}{\partial n_i}\right)_{T,V}$$

When both n_i and $\{n \neq n_i\}$ are constant, the system is a closed chemical system. Hence $dU = T\,dS - P\,dV$, $(\partial U/\partial S)_{n_i,V} = T$, and the last two terms cancel:

$$\left(\frac{\partial A}{\partial n_i}\right)_{T,V} = \left(\frac{\partial U}{\partial n_i}\right)_{S,V} \qquad \text{Q.E.D.}$$

The author believes firmly that chemical potentials, in common with partial free energies, are properties of an entire physical phase. The two terms will be used synonymously, and model expressions for partial free energies will be derived from model expressions for G.

BEING COMFORTABLE WITH PARTIAL DERIVATIVES

It is clear that partial derivatives play a prominent role in the mathematics of thermodynamics. But why do we use such a cumbersome notation? For example, in the case of $(\partial U/\partial n_i)_{S,V,\{n \neq n_i\}}$, we must certainly show the dependent variable U and the active independent variable n_i. But must we also show S, V, and $\{n \neq n_i\}$, the independent variables that remain inactive in the partial differentiation?

The answer is "yes," because the value of a partial derivative depends on *all* independent variables, not just the active one. For example, suppose a physical property $f = t^2x$, where t and x are the independent variables. Then, for the physical state in which $t = 1$ and $x = 1$, we find that $f = 1$ and $(\partial f/\partial t)_x = 2tx = 2$. Now change the independent variables to (t, y), where $y = xt^3$. The physical property f becomes $f = y/t$, and $(\partial f/\partial t)_y = -y/t^2$. In the same physical state defined by $t = 1, f = 1$ in which $(\partial f/\partial t)_x = 2$, we now find that $y = 1$ and $(\partial f/\partial t)_y = -1$. Thus, by merely changing the *inactive* variable from x to y, we change in this case both the magnitude and the sign of the partial derivative, even though the active variable t and the physical state to which the partial derivative applies remain the same.

In chemical thermodynamics we are often obliged to change one or more of the independent variables. To define a partial derivative, there is really no way but to indicate all the independent variables. We shall try to make the notation less onerous by gathering up all variables that remain inactive throughout a development and listing them at the head of the block of equations.

Partial Derivatives as Infinitesimal Processes

Mathematically, partial derivatives are slopes in higher-than-two-dimensional space. Physically, they may be identified with infinitesimal processes, which is sometimes more revealing. For example, the partial derivative

$$G_2 = \left(\frac{\partial G}{\partial n_2} \right)_{T,P,\{n \neq n_2\}}$$

portrays an infinitesimal process in which ∂n_2 formula weights of component 2 is added to the mixture while $(T, P, n_1, n_3, \ldots, n_k)$ remains constant, thereby causing a change ∂G in the free energy. The constancy of $(T, P, n_1, n_3, \ldots, n_k)$ must of course be mathematical. Mere experimental constancy is not good enough, because on the infinitesimal scale of ∂n_2 any infinitesimal change in $T, P, n_1, n_3, \ldots$ or n_k, even though practically unobservable, is significant. Experimental constancy does become good enough when the added amount is scaled up from infinitesimal to small-but-finite—that is, from ∂n_2 to δn_2. Such scaling-up will make it more obvious whether the given partial derivative portrays an infinitesimal process at equilibrium, and hence an equilibrium property. This technique will be explored in the later section on the equilibrium status of thermodynamic partial derivatives.

Transformation of Variables

It will be sufficient to consider three kinds of transformation: (i) replacement of one or more of the independent variables by new ones, (ii) exchange of the dependent and the *active* independent variable, and (iii) exchange of the dependent and an *inactive* independent variable.

(i) To illustrate the replacement of a present independent variable we shall return to the proof after Eq. (2.10) and transform $U(n_i, T, V)$ to $U(n_i, S, V)$. The required result is

$$\left(\frac{\partial U}{\partial n_i} \right)_{T,V} = \left(\frac{\partial U}{\partial n_i} \right)_{S,V} + \left(\frac{\partial U}{\partial S} \right)_{n_i,V} \left(\frac{\partial S}{\partial n_i} \right)_{T,V}$$

To derive this result we will first use a formal mathematical procedure and then a shortcut recipe.

Formal Procedure. We are given $(\partial U/\partial n_i)_{T,V}$ and seek to relate it to $(\partial U/\partial n_i)_{S,V}$, replacing T by S. V is inactive before and after. In general, any variable that is inactive before and after is also inactive in the transformation and need not be shown. We shall therefore write V = constant and stop carrying V explicitly through the mathematics. The steps are as follows.

- We begin with $U(n_i, S)$: $dU = (\partial U/\partial n_i)_s\, dn_i + (\partial U/\partial S)_{n_i}\, dS$.
- We express S as a function $S(n_i, T)$: $dS = (\partial S/\partial n_i)_T\, dn_i + (\partial S/\partial T)_{n_i}\, dT$.
- We substitute dS in dU and collect terms:

$$dU = \left[\left(\frac{\partial U}{\partial n_i}\right)_S + \left(\frac{\partial U}{\partial S}\right)_{n_i}\left(\frac{\partial S}{\partial n_i}\right)_T\right] dn_i + \left[\left(\frac{\partial U}{\partial S}\right)_{n_i}\left(\frac{\partial S}{\partial T}\right)_{n_i}\right] dT$$

The coefficient of dn_i equals $(\partial U/\partial n_i)_T$; that of dT equals $(\partial U/\partial T)_{n_i}$.

$$\therefore \quad \left(\frac{\partial U}{\partial n_i}\right)_T = \left(\frac{\partial U}{\partial n_i}\right)_S + \left(\frac{\partial U}{\partial S}\right)_{n_i}\left(\frac{\partial S}{\partial n_i}\right)_T, \qquad V = \text{constant} \qquad \text{Q.E.D.}$$

Shortcut Recipe. V is inactive before and after. We shall therefore write V = constant and stop carrying V explicitly through the mathematics.

We begin with $U(n_i, S)$: $dU = (\partial U/\partial n_i)_S\, dn_i + (\partial U/\partial S)_{n_i}\, dS$.

We rewrite the differentials (dU, dn_i, dS) in the above as partial differentials $(\partial U, \partial n_i, \partial S)$ and divide each by $(\partial n_i)_T$ (noting that $\partial n_i/\partial n_i = 1$), to obtain the desired result:

$$\left(\frac{\partial U}{\partial n_i}\right)_T = \left(\frac{\partial U}{\partial n_i}\right)_S + \left(\frac{\partial U}{\partial S}\right)_{n_i}\left(\frac{\partial S}{\partial n_i}\right)_T, \qquad V = \text{constant.} \qquad \text{Q.E.D.}$$

The shortcut recipe is a time- and error-saver because it cuts the amount of intricate, error-prone mathematical writing in half.

(ii) The exchange of a dependent and an active independent variable will be illustrated for $(\partial T/\partial P)_V$. Here T is the dependent variable and P is the active independent variable. We seek to exchange their status; that is, we require $(\partial P/\partial T)_V$. Since V is inactive before and after, it is inactive in the transformation. This creates a mathematical analogy to the inversion of an ordinary derivative, and we may write

$$\left(\frac{\partial P}{\partial T}\right)_V = \frac{1}{(\partial T/\partial P)_V}; \quad \text{or, in general,} \quad \left(\frac{\partial y}{\partial x}\right)_z = \frac{1}{(\partial x/\partial y)_z} \tag{2.11}$$

(iii) The transformation of an inactive independent variable into a dependent variable is especially important in thermodynamics and will be illustrated for $(\partial T/\partial P)_V$. We shall seek an expression in which V is the dependent variable. We begin with the implicit function of state, $f(P, V, T) = 0$, which is neutral as to the dependency status of (P, V, T). Hence $df = (\partial f/\partial P)_{V,T}\, dP + (\partial f/\partial V)_{T,P}\, dV + (\partial f/\partial T)_{P,V} dT = 0$.

(a) Set $dP = 0$; then $(\partial V/\partial T)_P = -(\partial f/\partial T)_{P,V}/(\partial f/\partial V)_{P,T}$
(b) Set $dV = 0$; then $(\partial T/\partial P)_V = -(\partial f/\partial P)_{T,V}/(\partial f/\partial T)_{P,V}$
(c) Set $dT = 0$; then $(\partial V/\partial P)_T = -(\partial f/\partial P)_{T,V}/(\partial f/\partial V)_{P,T}$

It follows that $(a) \cdot (b)/(c) = -1$; or, explicitly [by applying Eq. (2.11)], that $(\partial V/\partial T)_P \cdot (\partial T/\partial P)_V \cdot (\partial P/\partial V)_T = -1$; note the cyclical symmetry! Finally

$$\left(\frac{\partial T}{\partial P}\right)_V = -\left(\frac{(\partial V/\partial P)_T}{(\partial V/\partial T)_P}\right); \quad \text{or, in general,} \quad \left(\frac{\partial x}{\partial y}\right)_z = -\left(\frac{(\partial z/\partial y)_x}{(\partial z/\partial x)_y}\right) \tag{2.12}$$

On the left-hand side, V (or z) is an *inactive* independent variable. In both partial derivatives on the right, V (or z) is the dependent variable [Margenau and Murphy, 1943].

CHEMICAL EQUILIBRIUM

When a phase is prepared from the components, the properties need time to settle down, but sooner or later they reach a constant limit. If this limit is independent of the order in which the components were added, if it subsequently shows no hysteresis after T or P are cycled, and if—in the case of chemical equilibria—the same limit is reached whether the approach is from reactants or products, we conclude that the phase exists at operational equilibrium.

Not mentioned so far is the time scale on which the equilibrium is reached, although it is important. Tables of standard free energies of formation show that most of the substances we call "stable," and most of their descendant molecular species, are in fact metastable. For example, liquid *n*-butyl alcohol, a chemical staple with a shelf-life of many months, is metastable with respect to its structural isomers and, in the presence of air, with respect to a long list of oxidation products. The rates at which chemical reactions go to equilibrium vary so widely that many metastable substances have half-lives that are long on the time scale of human activities and indeed of human lives, so that no matter how patient we are, the potential conversion to more stable substances waits longer.

On the opposite side, molecular species that reach equilibrium with half-lives of the order of hours or minutes become unreactive, with potentially controllable concentrations, on time scales of the order of milliseconds.

Traditional equilibrium thermodynamics was based on observations on time scales suitable for the unaided human senses. But as the technically attainable time scales expand, the Laws of Thermodynamics remain unfazed; one might say that they do not watch the clock. The necessary clock-watching must be done by us—by being time-conscious when we construct composition models and formulate composition trees.

Free Energy and Reaction Progress

As a chemical reaction in a homogeneous phase at constant T and P goes to equilibrium, the free energy G of the phase goes to a minimum. This theorem has far-reaching consequences and is verified every time an equilibrium constant is measured reproducibly. The theorem is ultimately based on the Second Law—that heat flows spontaneously from the higher to the lower temperature, but the proof is long and involved. On the other hand, if we begin with the Second Law's chemical version, the proof is easier. Here is that version: A chemical reaction on its way to equilibrium can be harnessed to produce useful work in the surroundings, but there is no way to get useful work after chemical equilibrium has been reached.

Useful work is the work done, not counting expansion work; that is, $W_{\text{useful}} = W_{\text{total}} - \int P dV$. Harnessing a reaction on its way to equilibrium requires a system of ingenuity and complexity. Familiar systems are electrochemical and mechanochemical. Purely chemical systems, which are of special interest to us, are too simple. By definition, chemical systems do expansion work only and thus cannot produce useful work. But never mind! For a given change in the progress of a chemical reaction, the associated change in free energy is the same, regardless of the nature (chemical, electrochemical, or whatever) of the system.

It can be shown by rigorous thermodynamics [Lewis et al., 1961] that at constant T and P, the maximum amount of useful work (when the harness is reversible) is equal to the decrease in the free energy G of the system. The free energy therefore decreases, at constant T and P, as long as the reaction continues on its way to chemical equilibrium. But after the reaction reaches equilibrium, no further useful work can be done, and the free energy is at a minimum.

Let α denote the progress variable for a reaction in a closed system at constant T and P, on a scale from 0 for the reactants to 1 for the products. The free energy $G(\alpha)$ then is a continuous function of α, with a minimum at $\alpha = \alpha_{\text{eq}}$, the value at chemical equilibrium. We now wish to prove that the derivative $\partial G/\partial\alpha$ is also continuous at α_{eq}, and that $\partial G/\partial\alpha$ at equilibrium must therefore be equal to zero.

Figure 2.1 shows the two modes in which a continuous function $G(\alpha)$ can

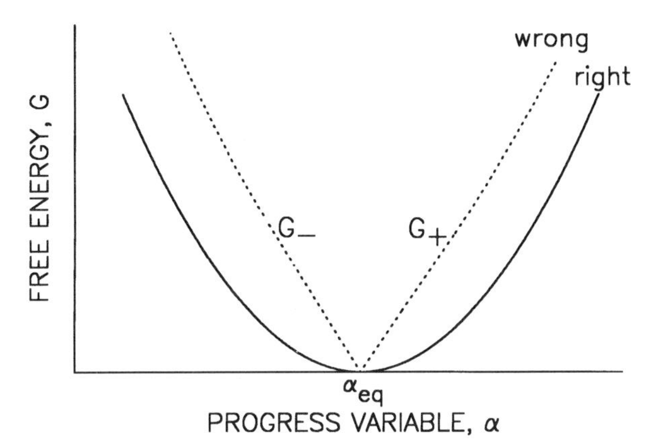

Figure 2.1. Two modes in which the free energy of a closed system at constant T and P can reach a minimum at chemical equilibrium.

have a minimum. In the solid curve $\partial G/\partial\alpha$ is continuous at the minimum, and $\partial G/\partial\alpha = 0$. In the broken curve $\partial G/\partial\alpha$ is discontinuous at the minimum, the slope being negative to the left and positive to the right.

The proof that the solid curve is physically valid requires only that G and its first derivatives, such as $(\partial G/\partial P)_T$, are state functions and therefore continuous throughout the range of α.

PROOF BY REDUCTIO AD ABSURDUM. For definiteness, let T and $\{n\}$ be constant and consider the volume $V = (\partial G/\partial P)_\alpha$, assuming (incorrectly) that the broken curve in Fig. 2.1 is right. Then G consists of two branches, G_- and G_+, which intersect at the equilibrium point.

Taylor's series expansions

$$\text{For } \alpha \le \alpha_{\text{eq}}, G_- = G_{\text{eq}} + g'_-(\alpha - \alpha_{\text{eq}}) + \cdots; \quad g'_- \le 0$$
$$\text{For } \alpha \ge \alpha_{\text{eq}}, G_+ = G_{\text{eq}} + g'_+(\alpha - \alpha_{\text{eq}}) + \cdots; \quad g'_+ \ge 0$$

The volume V, obtained from $(\partial G/\partial P)_\alpha$, therefore consists of two branches, V_- and V_+.

$$\text{For } \alpha \le \alpha_{\text{eq}}, V_- = V_{\text{eq}} + g'_-(\partial[\alpha - \alpha_{\text{eq}}]/\partial P)_\alpha + (\partial g'_-/\partial P)(\alpha - \alpha_{\text{eq}} + \cdots$$
$$\text{For } \alpha \ge \alpha_{\text{eq}}, V_+ = V_{\text{eq}} + g'_+(\partial[\alpha - \alpha_{\text{eq}}]/\partial P)_\alpha + (\partial g'_+/\partial P)(\alpha - \alpha_{\text{eq}}) + \cdots$$

Because V is a function of state, V_- and V_+ must become equal as α approaches α_{eq}. The last term on the right vanishes at α_{eq}, but the terms involving g'_- and g'_+ need not vanish because we have assumed that the broken curve in Fig. 2.1 is right. Yet to achieve continuity in V, g'_- and g'_+ must be equal. The broken curve in Fig. 2.1 is therefore untenable.

On the other hand, the solid curve, where the slopes g'_- and g'_+ at the minimum are both zero and the slope of the combined curve is continuous, remains acceptable. Hence $(\partial G/\partial \alpha)_{P,T,\{n\},\alpha=\alpha_{eq}} = 0$. Q.E.D.

Partial Free Energies at the Equilibrium Point

We are now ready to begin the dual representation of the free energy G of a homogeneous phase at constant T and P in terms both of components and molecular species. The mathematics will be based on Euler's equation (2.7), with $Y \equiv G$, and the corollary Gibbs–Duhem equation (2.8d). For definiteness, we shall use the composition tree (T-2.1), which is a slight expansion of (T-1.1) and just elaborate enough to illustrate the issues.

Composition Tree (T-2.1)

Carbon tetrachloride (1, n_1)	*N*-Methylacetamide (2, NMA, n_2)	
CCl_4 (A, n_A)	*cis* (C, n_C)	*trans* (T, n_T)
$n_A = n_1$	$n_C = (1 - \alpha)n_2$	$n_T = \alpha n_2$

Liquid carbon tetrachloride (1) is modeled as a single species, and *N*-methylacetamide (2, NMA) is modeled as a mixture of two isomers, *cis* (C) and *trans* (T). At ordinary temperatures, C and T interconvert on a time scale of a few milliseconds [Drakenberg and Forsen, 1971], so that on the time scale of manual laboratory operations, C and T are at chemical equilibrium. The progress variable α is defined as the fraction of *trans* in the NMA component; that is, $\alpha = n_T/n_2$, an intensive variable. The formula-weight numbers (n_1, n_2) and the mole numbers (n_A, n_C, n_T) are extensive variables.

When composition is expressed in terms of the two components, Euler's equation takes the form (2.13a), with the partial free energies (G_1, G_2) defined in (2.13b). When composition is expressed in terms of the three molecular species, the free-energy equations are (2.14a) and (2.14b).

***T* and *P* = constant in Eqs. (2.13) and (2.14)**

$$G = n_1 G_1 + n_2 G_2 \tag{2.13a}$$

$$G_1 = (\partial G/\partial n_1)_{n_2}, \qquad G_2 = (\partial G/\partial n_2)_{n_1} \tag{2.13b}$$

$$G = n_A G_A + n_C G_C + n_T G_T \tag{2.14a}$$

$$G_A = (\partial G/\partial n_A)_{n_C, n_T}, \qquad G_C = (\partial G/\partial n_C)_{n_A, n_T}, \qquad G_T = (\partial G/\partial n_T)_{n_A, n_C} \quad (2.14b)$$

Mathematically, Eqs. (2.13) and (2.14) are *not* in harmony, because the former shows two composition variables, while the latter shows three. To achieve consistency, a constraint must be imposed on the mole numbers in (2.14).*

The constraint will be that C and T exist at chemical equilibrium. But before we impose it, let us restate the conventions used in writing a composition tree for thermodynamic purposes.

(i) The first line lists the components, with the tacit understanding that on the time scale of the thermodynamic operations, the molecular species (and subspecies) in each component are at chemical equilibrium.

(ii) The second line lists the molecular species descended from the components, with clear indication of their component source. (Subspecies are similarly listed under the molecular species.) More than one molecular species is listed under a component only if such modeling adds significantly to our ability to rationalize and predict the chemical and thermodynamic properties.

(iii) By listing more molecular species than there are components and by showing the descendant relationships, the composition tree indicates the significant chemical reactions. Each independent reaction then requires an independent progress variable [such as $\alpha = n_T/n_2$ in (T-2.1)]. This is required even when the reaction is always observed at equilibrium and even if there is no practical, controllable way by which the reaction might be displaced from equilibrium. The progress variable is required because the thermodynamic constraint for chemical equilibrium—that is, that $\alpha = \alpha_{eq}$—cannot be applied unless the variable α *exists.*

Returning to Eqs. (2.14a) and (2.14b) in order to specify that C and T are at chemical equilibrium, it is helpful to transform the composition variables from (n_A, n_C, n_T) to (n_1, n_2, α), as follows: $n_A = n_1$, $n_C = n_2(1 - \alpha)$, and $n_T = n_2\alpha$. The expression for G then becomes Eq. (2.14c):

$$G = n_1 G_A + n_2(1 - \alpha)G_C + n_2\alpha G_T \qquad (2.14c)$$

Equation (2.14c) is the same Euler equation as Eq. (2.14a), so (G_A, G_C, G_T) continue to be defined by Eq. (2.14b), in terms of (n_A, n_C, n_T). But we now have the option to keep (n_1, n_2) constant and thus let α vary in a closed chemical system at constant T and P. Now, as the *cis/trans* reaction goes to equilibrium, $G(\alpha)$ goes to a minimum, and $(\partial G/\partial \alpha)_{T,P,\{n\}} = 0$. Imposition of this constraint on Eq. (2.14c) gives Eq. (2.15a); substitution of the result in Eq. (2.14c) gives Eq. (2.15b).

*See also the following section on the equilibrium status of thermodynamic partial derivatives.

T, P, and $\{n\}$ = constant in Eqs. (2.15a) and (2.15b)

$$G_T - G_C = 0, \text{at equilibrium} \tag{2.15a}$$

$$G = n_1 G_A + n_2 G_T \tag{2.15b}$$

The formula-weight numbers n_1 and n_2 in Eq. (2.15b) are of course the same as those in Eq. (2.13a), so the coefficients must be equal. Thus, we obtain the following:

When equilibrium exists at constant (T, P), for any composition (n_1, n_2)

$$G_1 = G_A, \qquad G_2 = G_T, \qquad G_C = G_T = G_2 \tag{2.16}$$

Equations (2.16) show that there are correspondences between the partial free energies of the formal components and those of their isomerically descended molecular species. These correspondences will be amplified and generalized in the next section. At this point we shall state, however, that the correspondences are unique to partial free energies, because their derivation requires a minimum in the thermodynamic property at the precise point where $\alpha = \alpha_{eq}$. For a closed chemical system at constant T and P, such a minimum is mandated by the Second Law only for the free energy G. Most other properties are monotonic functions of α, as illustrated for the enthalpy H in Fig. 2.2. Those properties that *do* show a minimum (or maximum) do not show it exactly at α_{eq}.

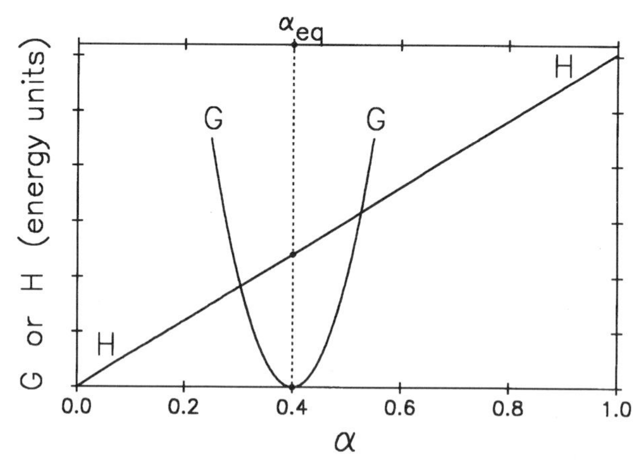

Figure 2.2. Schematic plots of the free energy $G(\alpha)$ and the enthalpy $H(\alpha)$ for a closed chemical system at constant T and P. Only $G(\alpha)$ has a minimum exactly at α_{eq}.

THE EQUILIBRIUM STATUS OF THERMODYNAMIC PARTIAL DERIVATIVES

We saw that to prove chemical equilibrium we must allow progress variables such as α in Fig. 2.2 to be displaced from equilibrium. We also saw that partial derivatives such as G_2 in Eq. (2.13b) and G_T in Eq. (2.14b) may be identified with very small physical processes. If both the initial and final state in the very small process are equilibrium states, then the partial derivative is an equilibrium property. If not—if either or both states are off equilibrium, on the very small scale of the process—then the partial derivative is a nonequilibrium property. We shall find that if the composition tree includes even one chemical equilibrium, the thermodynamic partial derivatives of the molecular species are nonequilibrium properties. This is true not only for the reactants and products at equilibrium, but (at least in principle) also for molecular species that remain inert. On the other hand, the thermodynamic partial derivatives of the formal components are equilibrium properties.

In the case of molecular species, consider the partial free energy G_T of *trans*-NMA (T) in (T-2.1):

$$G_T = \left(\frac{\partial G}{\partial n_T} \right)_{T, P, n_A, n_C}$$

In the very small physical process representing G_T, δn_T moles of *trans*-NMA is added while T, P, n_A, and n_C remains strictly constant. Now suppose that the initial state is an equilibrium state. Then the initial ratio, n_T/n_C, is equal to the equilibrium constant, while the final ratio, $(n_T + \delta n_T)/n_C$, departs from the equilibrium constant by $\delta n_T/n_C$, which is significant on the very small scale of the process. The final state is therefore off equilibrium. Thus, even though the partial derivative is measured at an equilibrium point, it is a nonequilibrium property.

Next, consider a molecular species that is modeled as unreactive on the composition tree, such as carbon tetrachloride (A) in (T-2.1):

$$G_A = \left(\frac{\partial G}{\partial n_A} \right)_{T, P, n_C, n_T}$$

Although carbon tetrachloride does not enter the *cis*/*trans* equilibrium of NMA directly, it might affect it indirectly by exerting a neutral medium effect on the equilibrium constant, thereby moving the n_T/n_C ratio off equilibrium. In most real cases the neutral medium effect is small enough to be neglected and the thermodynamic partial derivative is tantamount to an equilibrium property, but in principle the effect is always present. The thermodynamic partial derivatives of molecular species thus are genuine equilibrium properties only if the accepted composition tree shows no chemical equilibria whatsoever.

The preceding examples have been limited to the partial free energies of molecular species in composition tree (T-2.1), but reflection will show that analogous proofs can be given for *any* thermodynamic partial derivative and for *any* composition tree. For example, the partial entropy S_T of *trans*-NMA is defined by

$$S_T = \left(\frac{\partial S}{\partial n_T} \right)_{T,P,n_A,n_C}$$

Except for the substitution of S for G, S_T is mathematically analogous to G_T and may be identified with the same very small process.

By contrast, the partial thermodynamic derivatives of *formal components* are equilibrium properties *by definition*. Accordingly, one must be careful to avoid inconsistencies with the nonequilibrium properties of molecular species.

Actually, inconsistencies can arise only when the composition tree includes chemical equilibria, because in the absence of chemical equilibria the thermodynamic partial derivatives of molecular species are equilibrium properties. However, from a modern point of view, the composition trees of traditional thermodynamics are unrealistically sparse in chemical equilibria, because the intrusion of molecular theory was deliberately kept to a minimum.

In the absence of chemical equilibria, the set of independent variables consisting of T, P, and the formula-weight numbers $\{n\}$ is sufficient. Thus, in the absence of chemical equilibria, the partial free energy of component 1 is defined as follows:

$$G_1 = \left(\frac{\partial G}{\partial n_1} \right)_{T,P,n_2,n_3\cdots}$$

Physically, G_1 represents a very small process in which δn_1 formula weights of component 1 is added, while $T, P, n_2, n_3, \cdots$ remain mathematically constant. In the absence of chemical reactions, operational equilibrium exists in both the initial and final states, and G_1 is an equilibrium property.

On the other hand, when there is no bias against the intrusion of molecular theory, composition trees with significant chemical equilibria become common, but then the set of independent variables $(T, P, \{n\})$ is inadequate because each independent reaction now requires one independent progress variable to specify its equilibrium. For example, if there is just one reaction with progress variable α, we must specify that the free energy $G(\alpha)$ is locked into a minimum, at constant T and P. That is, $\partial G/\partial\alpha$ must remain equal to zero, even though the value of α at the minimum may change during the very small process—from α_{eq} in the initial state to $\alpha_{eq} + \delta\alpha_{eq}$ in the final state. A constraint on α must be specified because without it the value of α is ambiguous and the system is not defined. The full expression for the partial free energy of component 1 thus becomes

$$G_1 = \left(\frac{\partial G}{\partial n_1} \right)_{T, P, \{n \neq n_1\}, \partial G/\partial \alpha = 0}$$

Physically, G_1 now represents a very small process in which δn_1 formula weights of component 1 is added, while T, P, n_2, $n_3, \ldots$ remain constant, and $\partial G/\partial \alpha$ remains constant at zero so that the phase remains at chemical equilibrium. Since the small incremental addition of component 1 alters the composition, the equilibrium value of α is liable to change. If the incremental addition causes only a medium effect, the change of α_{eq} is probably small; but when it produces an unequivocal mass-action effect, the change of α_{eq} is significant.

For example, the composition tree (T-2.2) for sodium dihydrogenphosphate in water shows four molecular species: H_2O, Na^+, $H_2PO_4^-$, and the dimeric anion $(H_2PO_4)_2^{2-}$. The presence of the latter is based on Raman and other evidence; a suggested structure, in which the $H_2PO_4^-$ ligands are joined by three hydrogen bonds, is shown in Eq. (2.17) [Preston and Adams, 1979].

```
       O · · · H—O          2–
      /           \
O–P–O–H· · ·O–P–O–H
      \           /
       O—H · · · O
```
(2.17)

The progress variable α_{eq} for the dimerization increases significantly with the molality m_2 up to the solubility limit, through the mass-action expression, $\alpha_{eq}/(1 - \alpha_{eq})^2 = 2Km_2$.

Composition Tree (T-2.2)

Water $(1, n_1, m_1)$	Sodium dihydrogenphosphate $(2, n_2, m_2)$		
H_2O $(W, m_W = m_1)$	Na^+ (m_2)	$H_2PO_4^-$ $([1 - \alpha]m_2)$	$(H_2PO_4)_2^{2-}$ $(\alpha m_2/2)$

$K = (\alpha_{eq})/(2[1 - \alpha_{eq}]^2 m_2) = 0.6 \pm 0.3$ at $25°C$

Concerning the Very Small Processes

It is reasonable to ask how one would actually conduct the very small processes that represent thermodynamic partial derivatives. One conceivable method for G_T of *trans*-NMA (see above) might begin with the initial equilibrium state, add δn_T formula weights of an *equilibrated cis/trans mixture*, and then plan a T-jump so as to restore the initial value of n_C and produce the final nonequilibrium state. Whether this process is feasible in fact is beside the point, because thermodynamics will be satisfied as long as the process is feasible *in principle*. Most students are aware that thermodynamics is comfortable with physically possible but technically unrealized procedures or devices such as membranes that are truly semipermeable, pistons that are truly frictionless, and changes of state that are truly reversible; indeed, thermodynamics gains much of its generality in this way. In the present case, displacements from equilibrium are always possible in principle. This follows from the basic definition that the reactants and products must be distinguishable, since they represent different molecular species, and, furthermore, that the consequent real difference in properties can in principle be made the basis of nondestructive physical methods—perhaps thermal, electrical, magnetic, mechanical, or photophysical—to produce displacements from equilibrium.

THE CORRESPONDENCE THEOREMS

Direct measurement yields the properties of *components*. The properties of *molecular species* must be deduced from those of the components. For a closed system at chemical equilibrium at constant T and P, the relationships are predictably simple only for the partial free energies. Here they can be stated in the form of two Correspondence Theorems, which require only a knowledge of the stoichiometries involved in the composition tree. The relationships for other partial properties are more complicated and, as a rule, need to be specifically derived.

The Correspondence Theorems for partial free energies may be stated as follows:

First Correspondence Theorem. The partial free energy of a formal component at constant T and P is equal to that of any of its descendant isomeric molecular species and subspecies with which it exists in chemical equilibrium. Thus, if "i" is any component with chemical formula $\mathcal{L}_i$ and J is any molecular species with molecular formula $\mathcal{L}_\mathrm{J}$ with which "i" is in equilibrium; then if $\mathcal{L}_\mathrm{J} = \mathcal{L}_i$, $G_\mathrm{J} = G_i$.

Second Correspondence Theorem. The change in partial free energy, $\Delta_r G$, for the conversion of reactants to products in a chemical reaction vanishes when equilibrium is reached.

These Correspondence Theorems apply regardless of the relative concentrations of the components and the molecular species. For example, it is reasonable to assume that the *cis/trans* isomers of *N*-methylacetamide in (T-2.1) actually exist in equilibrium with small but chemically significant amounts of the proton tautomers shown in (2.18).

$$CH_3-C(-O-H)(=\ddot{N}-CH_3) \qquad CH_3-C(-O-H)(=N:-CH_3) \tag{2.18}$$

proton tautomer of *trans*-NMA (*tauto*-T)

proton tautomer of *cis*-NMA (*tauto*-C)

Although the amounts of these species are too small to be stoichiometrically significant, when chemical equilibrium exists at constant T and P, the first Correspondence Theorem states that the following partial free energies are all equal:

$$\begin{aligned} G_2(\text{for the formal component, NMA}) &= G_T(\text{for } trans - \text{NMA}) \\ &= G_C(\text{for } cis - \text{NMA}) \\ &= G_{\text{tauto-T}}(\text{for } tauto\text{-T}) \\ &= G_{\text{tauto-C}}(\text{for } tauto - \text{C}) \end{aligned}$$

Note that the partial free energies are equal even though the concentrations at equilibrium are very different. We shall find, however (Chapter 3), that the differences in stability show up as differences in the *standard* partial free energies.

PROOF OF THE FIRST THEOREM. To make the proof sufficiently general, we shall use the symbolic composition tree (T-2.3), in which component 2 consists of an arbitrary number, $k+1$, of isomeric molecular species, $U_1, U_2, \ldots, U_k, V$. Component 1 is modeled simply as a single molecular species A.

Composition Tree (T-2.3)

Component 1 (n_1)	Component 2 (n_2)						
$A; n_A = n_1$	$U_1\,(\alpha_1)$	$U_2\,(\alpha_2)$	$\cdots$	$U_i\,(\alpha_i)$	$\cdots$	$U_k\,(\alpha_k)$	$V\,(\alpha_V)$

$$n_{U_i} = \alpha_i n_2; \quad \sum_{i=1}^{k} \alpha_i = 1 - \alpha_V; \text{ V is the common reactant}$$

It is convenient to let V be the common reactant, so that the k isomerization reactions become $V \to U_1$; $V \to U_2; \ldots; V \to U_k$.

It is further convenient to let the progress variables be the mole fractions *in component 2* of the designated products U_1, U_2, ..., U_k; that is, $\alpha_1 = n_{U_1}/n_2$, $\alpha_2 = n_{U_2}/n_2, \ldots, \alpha_k = n_{U_k}/n_2$. These α's are independent variables, so that the Gibbs free energy is a function $G(T, P, n_1, n_2, \{\alpha\})$, where $\{\alpha\}$ denotes the set $\{\alpha_1, \alpha_2, \ldots, \alpha_k\}$.

To constrain component 2 to exist at equilibrium at constant (T, P, n_1, n_2), we must minimize the Gibbs free energy with respect to variations in any of the α's by solving k equations of the form $(\partial G/\partial \alpha_i)_{\{\alpha \neq \alpha_i\}} = 0$. The procedure is similar to that for the specific case described in Eqs. (2.13) and (2.14). Equation (2.19) expresses G in terms of the formal components, while Eq. (2.20a) expresses it in terms of the molecular species.

$(T, P, n_1, n_2) =$ constant in Eqs. (2.19)–(2.20d)

$$G = n_1 G_1 + n_2 G_2 \tag{2.19}$$

$$G = n_1 G_A + n_2 \alpha_v G_V + n_2 \sum \alpha_i G_{U_i}, \qquad i = 1, 2, \ldots, k \tag{2.20a}$$

Equation (2.20b) introduces the constraint that $\alpha_v = 1 - \sum \alpha_i$, and Eqs. (2.20c) are the k solutions of $(\partial G/\partial \alpha_i)_{\{\alpha \neq \alpha_i\}} = 0$. Finally, Eq. (2.20d) results from substitution of Eqs. (2.20c) in Eq. (2.20a). Comparison of Eqs. (2.20c) and (2.20d) with Eq. (2.19) then yields the correspondence (2.21). Q.E.D.

$$G = n_1 G_A + n_2 G_V + n_2 \sum \alpha_i (G_{U_i} - G_V), \qquad i = 1, 2, \ldots, k \tag{2.20b}$$

$$G_{U_1} = G_V, \quad G_{U_2} = G_V, \quad \ldots, \quad G_{U_k} = G_V; \qquad \{\alpha\} = \{\alpha_{eq}\} \tag{2.20c}$$

$$G = n_1 G_A + n_2 G_V \qquad \text{at chemical equilibrium} \tag{2.20d}$$

$$\therefore \quad G_1 = G_A, \quad G_2 = G_V = G_{U_i} \qquad \text{for any } i \text{ at chemical equilibrium} \tag{2.21}$$

PROOF OF THE SECOND THEOREM. To make the proof sufficiently general, we shall use the symbolic reaction:

$$a\mathrm{A} + b\mathrm{B} \rightarrow p\mathrm{P} + q\mathrm{Q}$$

whose stoichiometry is general enough to indicate the behavior of any single-step reaction. The free-energy change $\Delta_r G$ for this reaction is

$$\Delta_r G = pG_P + qG_Q - aG_A - bG_B$$

We will name the components "A", "B", and "Q", specify that B is in stoichiometric excess over A, and define the progress variable α to express the fraction of "A" that is converted to product. On this basis, the composition tree becomes (T-2.4).

Composition Tree (T-2.4)

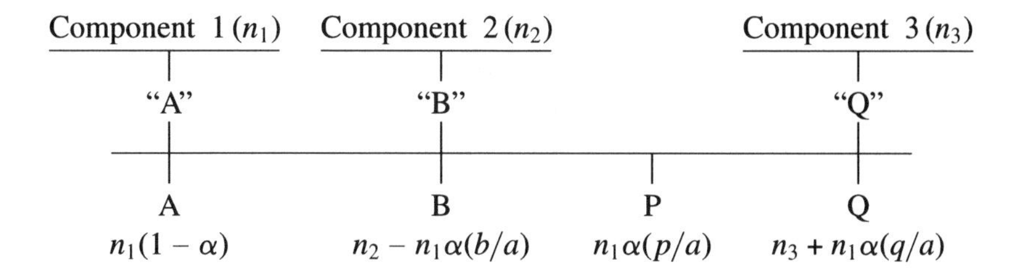

The free energy in terms of molecular species is expressed by Eq. (2.22a). The partial derivative $\partial G/\partial\alpha$, which vanishes at equilibrium, is given by Eq. (2.22b).

***T*, *P*, n_1, n_2, n_3 = constant in Eqs. (2.22a) and (2.22b)**

$$G = n_1(1-\alpha)G_A + (n_2 - n_1\alpha b/a)G_B + (n_1\alpha p/a)G_P + (n_3 + n_1\alpha q/a)G_Q \tag{2.22a}$$

$$(\partial G/\partial\alpha)_{T,P,n_1,n_2,n_3} = (n_1/a)(qG_Q + pG_P - bG_B - aG_A) = (n_1/a)\Delta_r G = 0 \quad \text{at equilibrium} \tag{2.22b}$$

Since $n_1/a \neq 0$, Eq. (2.22b) states that $\Delta_r G = 0$ at equilibrium. Q.E.D.

Application of the Second Theorem

The result that $\Delta_r G = 0$ for any reaction at equilibrium in a closed system at constant T and P allows us to obtain correspondence for molecular species resulting from association, dissociation, or chemical interaction of the components. For (T-2.4), for example, we know from the first Correspondence Theorem that $G_A = G_1$, $G_B = G_2$, and $G_Q = G_3$. The theorem that $\Delta_r G = 0$ establishes that $G_P = (bG_2 + aG_1 - qG_3)/p$—a *composite* of the partial free energies of the formal components.

For an example of the dissociation of a formal component, consider an ionic solution of sodium chloride (component 2) in water (component 1) in which the process $NaCl \rightarrow Na^+ + Cl^-$ is significant. At equilibrium, $G_2 = G_{NaCl}$ according to the first Correspondence Theorem, and $G_{NaCl} = G_{Na^+} + G_{Cl^-}$, because $\Delta_r G = 0$. Thus G_2 of "sodium chloride" is equal to the sum, $G_{Na^+} + G_{Cl^-}$, of the partial free energies of the ions. This is true even when sodium chloride behaves as a strong electrolyte and the concentration of NaCl molecules (ion pairs) is very small. In that case the sodium ions and chloride ions react as nearly independent reagents, but their partial free energies remain coupled in the expression for G_2.

The Correspondence Theorems and the Operational Time Scale

In designing a composition model, we consider whether the time scale is long enough to allow a given reaction to proceed to equilibrium, or whether it is so short that progress towards equilibrium is negligible. In the latter case, we say that the reaction is "frozen" in its nonequilibrium state.

Molecular species that are frozen in a nonequilibrium state are operationally indistinguishable from components. For an example we shall return to composition tree (T-2.3). Since (T-2.3) shows k independent reactions, the kinetics of relaxation to equilibrium involves k independent relaxation times, which are typically spread over a range. Now suppose that the operational time scale is so short that only one process, say $U_1 \rightarrow V$, is fast enough to go to equilibrium. The other processes ($U_2 \rightarrow V, \ldots, U_k \rightarrow V$) are "frozen," so that the concentrations of $U_2, \ldots, U_k$ remain constant during the observations and $U_2, \ldots, U_k$ are tantamount to independent components. The original component 2 is therefore cut down to the species U_1 and V. The resulting composition tree is (T-2.5).

Composition Tree (T-2.5)

[Same as (T-2.3), but on a time scale where $U_2, \ldots,$ U_k are "frozen"]

Component $1(n_1)$	Component 2		"U_2"	"U_3"	$\cdots$	"U_k"
$A; n_A = n_1$	$U_1\,(1-\alpha)$	$V\,(\alpha)$	U_2	U_3		U_k

$[V]/[U_1] = \alpha/(1-\alpha) = K_1$

There is now just one progress variable α, and operational equilibrium exists when $(\partial G/\partial\alpha) = 0$, at constant T and P. The first Correspondence Theorem applies only to component 2 and its descendant molecular species U_1 and V; that is, $G_2 = G_{U_1} = G_V$. $U_2, U_3, \ldots, U_k$ are no longer descendant species in component 2, and their partial free energies depend on the respective "frozen" concentrations. The argument that, on a longer time scale, component 2 will expand and encompass $U_2, U_3, \ldots, U_k$ is correct but irrelevant, since we must work with the composition tree as it appears on the operational time scale and *not* as it might appear under other conditions.

So far we have assumed that the rates of approach to chemical equilibrium are fixed and definite. In fact, however, reaction rates can be accelerated by specific catalysts and inhibited by specific inhibitors. The addition of an appropriate catalyst or inhibitor thus gives us another tool for changing the place of a species in the composition tree.

3

DILUTE SOLUTIONS, STANDARD PARTIAL FREE ENERGIES, AND THE STABILITY THEOREM

The partial free energy of a component in a homogeneous phase is a function of the composition of the phase. In this chapter we explore this relationship, especially for dilute solutions where it is simple. One feature of the relationship is that it introduces a parameter called the *standard partial free energy*, which effectively normalizes the partial free energy to unit concentration. When consistent concentration units are used, the standard partial free energy provides a thermodynamic index of relative stability and reactivity at constant T and P.

A gas phase consisting of unreactive components is said to be dilute when the P–V–T relationship differs only negligibly from that for an ideal gas. A liquid solution is dilute when solute–solute interactions are negligible—that is, when virtually all solute molecules interact only with solvent molecules.

PARTIAL FREE ENERGIES IN IDEAL-GAS MIXTURES

The ideal-gas equation of state, $PV = nRT$, expresses Avogadro's hypothesis (together with Boyle's and Charles' law) and thus operates on the level of *molecular species*. For gaseous solutions, the mole number n is summed over all molecular species: $n = \sum n_J$. Hence the mole fraction N_J equals n_J/n, and the partial pressure P_J is defined by $P_J = P \cdot N_J$. Multiplying both sides of the ideal-gas equation by N_J then yields $P_J V = n_J RT$, for any molecular species in the quasi-ideal gaseous solution.

The free energy G of n_J moles of a quasi-ideal *pure gas* at T and P is given by Eq. (3.1):

$$G = n_J G_J^0 + n_J RT \ln P \tag{3.1}$$

G_J^0 is the *standard free energy* of the pure gas at the given T. If unit pressure ($P = 1$) is in the quasi-ideal range, G_J^0 equals the molar free energy, G/n_J, when $P = 1$. But even when unit pressure is not in the quasi-ideal range, the standard free energy G_J^0 in principle is a function of T and represents the pressure-independent limit reached by $(G/n_J - RT \ln P)$ as the pressure drops into the quasi-ideal range.

The *partial* free energy G_J of a molecular species J dissolved in a quasi-ideal gas is given by Eq. (3.2), where P_J is the partial pressure. G_J^0 is the *standard partial free energy*. We shall show presently that G_J^0 is identical to the standard free energy of the pure gas.

$$G_J = G_J^0 + RT \ln P_J \tag{3.2}$$

PROOF OF EQ. (3.1). At constant T, Eq. (2.4) gives $dG = VdP = (n_J RT/P)\, dP$. Hence $G = n_J RT \ln P$ + a constant of integration. Since T is an inactive independent variable, this "constant" is a function of T; and since G is extensive, it is proportional to n_J. On writing the constant of integration in the form "constant" $= n_J G_J^0$ (where G_J^0 is the standard free energy, an intensive function of T), we obtain Eq. (3.1). Q.E.D.

PROOF OF EQ. (3.2). It will be sufficient to let the gas mixture consist of n_J moles of J and n_K moles of K. The free energy at pressure P then is the sum of (i) the separate free energies of J and K [Eq. (3.1)], each at pressure P, and (ii) the free energy of mixing J and K at pressure P:

$$\begin{aligned} G = {} & \underbrace{n_J G_J^0 + n_J RT \ln P}_{\text{for pure J}} + \underbrace{n_K G_K^0 + n_K RT \ln P}_{\text{for pure K}} \\ & + \underbrace{n_J RT \ln N_J + n_K RT \ln N_K}_{\text{free energy of mixing}} \end{aligned}$$

$$\begin{aligned} (\partial G/\partial n_J)_{n_K} &= G_J^0 + RT \ln P + RT \ln N_J + n_J/N_J(\partial N_J/\partial n_J) \\ &\quad + n_K/N_K(\partial N_K/\partial n_J) \\ &= G_J^0 + RT \ln [P \cdot N_J] + n(\partial [N_J + N_K]/\partial n_J); \qquad [N_J + N_K] = 1 \\ \therefore \quad (\partial G/\partial n_J)_{n_K} &= G_J = G_J^0 + RT \ln P_J \qquad \text{Q.E.D.} \end{aligned}$$

The preceding development is based on the ideal-gas equation, which in turn is based on Avogadro's hypothesis, which is the granddaddy of molecular theory. Equations (3.1) and (3.2) therefore apply to molecular species. In order to obtain the partial free energies of gaseous *components* under ideal-gas conditions, we simply apply the Correspondence Theorems in reverse. Thus, for any molecular species J at equilibrium with its isomeric parent component "i", $G_J = G_i$. For any molecular species M at equilibrium with its isomeric species

N, $G_M = G_N$. And for any compound XY at equilibrium with its reactants X and Y, $G_{XY} = G_X + G_Y$. In particular, if "i" is the isomeric parent component for X and "k" is the isomeric parent component for Y, then

$$G_i = G_X, \quad G_k = G_Y, \quad \text{and} \quad G_{XY} = G_i + G_k$$

DILUTE LIQUID SOLUTIONS: HENRY'S LAW

A component "i" is said to be *self-contained* if its molecular species do not react with molecular species descended from other components. The component is *isomerically self-contained* if it is self-contained and if its descendant molecular species are all isomers whose molecular formula matches the chemical formula of the component. There are then two equations of conservation: (i) $n_i = \sum(n_{J,i})$; that is, the number of formula weights of the component equals the number of moles of its molecular species; and (ii) when the component is in a quasi-ideal gas phase, $P_i = \sum(P_{J,i})$; that is, the partial pressure of the component equals the sum of the partial pressures of its molecular species.

Experimental laws for isomerically self-contained components are stoichiometrically simple. For example, the distribution of such a component between a low-pressure gas phase and a dilute solution at equilibrium follows a simple 1 : 1 ratio, according to Henry's law:

$$c_i/P_i = \mathcal{H}_i(\text{M}) \tag{3.3}$$

Here, M denotes the liquid solvent, c_i is the formal concentration of the component at gas–liquid equilibrium, and $\mathcal{H}_i(\text{M})$ is the Henry's law constant. Equation (3.3) and its corollaries were tested so thoroughly during the nineteenth century that modern tests are relatively rare. Figure 3.1 shows one such test, namely,

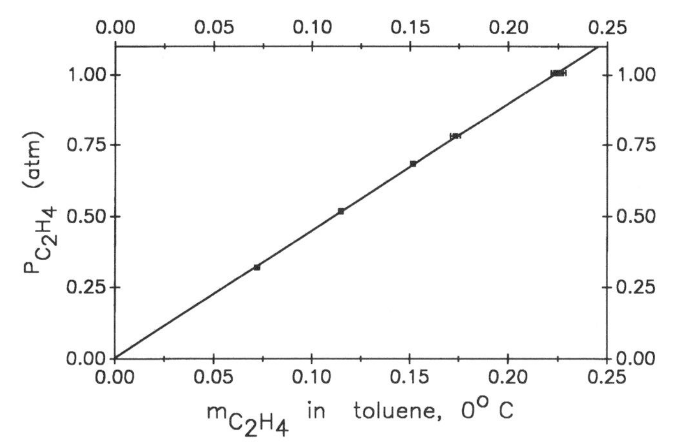

Figure 3.1. Henry's law plot for dilute solutions of ethylene in toluene at 0°C. The error bars correspond to ±1%. (Data from Waters et al. [1970].)

the variation of P_i with c_i for the distribution of ethylene, C_2H_4, between the gas phase and liquid toluene [Waters et al., 1970]. The error bars correspond to ±1%.

At first sight, Eq. (3.3) looks undramatic, but its impact on the molecular theory of liquids is remarkable. When a molecule transfers from a low-density gas to a condensed phase, the environment changes from one where the molecules can move freely between collisions, to one where they huddle together so that intermolecular interactions are continually important. Although the molecules in a condensed phase thus do not move freely, they would still be separate entities if their motions were *uncorrelated* (rather than coupled). Lack of motional correlation may not be taken for granted: For crystalline solids, the existence of phonon waves and crystal vibrations suggests that the molecular motions are coupled. But for liquids, Henry's law suggests that the solute molecules retain a degree of dynamic independence that permits 1 : 1 equilibrium relationships with the dynamically independent molecules of the same species in the gas phase.

In the following we shall regard liquids as ensembles of independently moving molecules. In Chapter 9 we shall examine the cage model of liquids to rationalize this approach.

The Partial Free Energy of a Dilute Liquid Solute

Because phase equilibrium exists, the partial free energies G_i in the gas phase and $G_i(\mathrm{M})$ in the liquid phase are equal. It then follows from Eqs. (3.2) and (3.3) that $G_i(\mathrm{M})$ is a logarithmic function of c_i:

$$G_i(\mathrm{M}) = G_i^0(\mathrm{M}) + RT \ln c_i \tag{3.4a}$$

$$G_i^0(\mathrm{M}) = G_i^0(g) - RT \ln \mathcal{H}_i(\mathrm{M}) \tag{3.4b}$$

PROOF. Write $G_i(\mathrm{M}) = G_i(g) = G_i^0(g) + RT \ln P_i$. Then substitute $\ln P_i = \ln c_i - \ln \mathcal{H}_i(\mathrm{M})$. Q.E.D.

As in Eq. (3.2), the temperature-dependent parameter $G_i^0(\mathrm{M})$ is called the *standard partial free energy* of the dilute solute "i" in the given solvent M. However, Eq. (3.4b) shows that $G_i^0(\mathrm{M})$ is a sum consisting of an intrinsic term $[G_i^0(g)]$ and a solvation term $[-RT \ln \mathcal{H}_i(\mathrm{M})]$. If the unit of concentration is chosen so that a solute at $c_i = 1$ is dilute, then $G_i^0(\mathrm{M})$ represents the partial free energy $G_i(\mathrm{M})$ at unit concentration. In general, $G_i^0(\mathrm{M})$ is the constant value reached by $(G_i(\mathrm{M}) - RT \ln c_i)$ at constant T as c_i enters the dilute range; thus $G_i^0(\mathrm{M})$ may be said to represent $G_i(\mathrm{M})$ normalized to unit concentration. An actual plot of $(G_i(\mathrm{M}) - RT \ln c_i)$ for ethylene in toluene at 0°C is shown in Fig. 3.2.

Of course, there is no reason why the scope of Henry's law should be limited to solutes as volatile as ethylene in toluene. Indeed, by using a highly sensitive

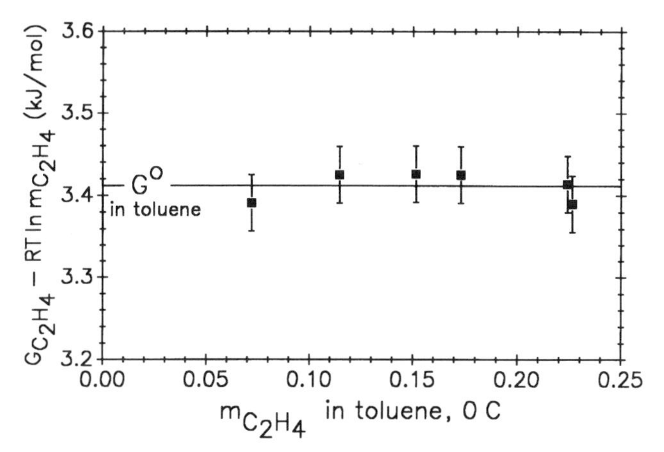

Figure 3.2. Plot of (G_i(M) − RT ln c_i) for ethylene in toluene at 0°C. Same data as in Fig. 3.1. The free energy scale is referenced to $G_i^0(g) = 0$ for ethylene in the gas phase.

^{14}C tracer technique, the Henry's law constant was measured for *N*-methylacetamide in aqueous solution, even though the vapor pressure under the given conditions was only 1.00 × 10^{-8} atm (7.6 μtorr). The data obtained in this measurement are reproduced in Fig. 3.3 [Wolfenden, 1978; Wolfenden et al., 1979]. The broad scope of Henry's law has also been tested indirectly, because the logarithmic concentration dependence predicted by Eq. (3.4) for the partial free energy has been confirmed for many solutes in a wide range of solvents. In such indirect tests, volatility is not a factor. It *is* necessary, however, that the solute component be dilute and isomerically self-contained; that is, all solute molecules must have the same molecular formula.

When these conditions are met, the logarithmic concentration dependence fits nearly all nonionic, nonmacromolecular solutes, in nearly all liquid solvents, up to total solute concentrations of the order of 0.1 M. The error of prediction increases with solute concentration and typically reaches 10–100 J/mol at 0.1 M. When the solute includes nonionic macromolecules, the dilute-solution model fits up to about 1 wt% of total solute.

Equation (3.4) is considerably less appropriate for dilute ionic solutes. For univalent ions in water, deviations from the logarithmic concentration dependence reach 50 J/mol near 0.0003 M ionic strength. The same error is reached at even smaller ionic strengths if the ions are multivalent, or if the dielectric constant is lower than that of water.

Extension to Molecular Species

The derivation of Eq. (3.4) has so far been limited to dilute solute *components*, but the Principle of Detailed Balance allows us to extend it directly to molecular species. According to Detailed Balance, when a chemical process exists at

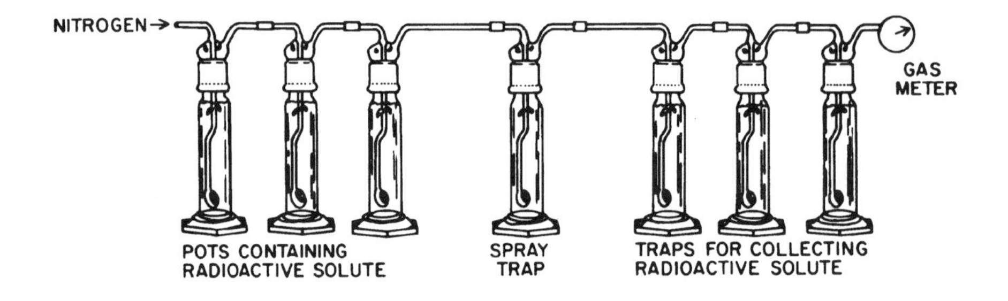

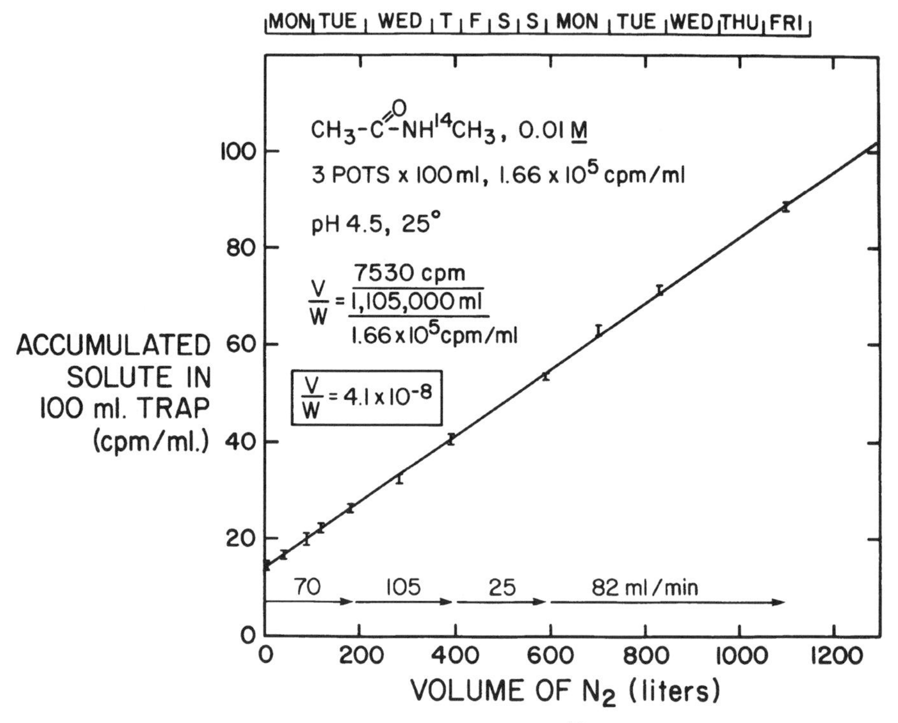

Figure 3.3. Dynamic vapor pressure measurement for ^{14}C-labeled 0.01 M aqueous *N*-methylacetamide (NMA) at 25°C. The equilibrium concentration of NMA in the gas phase is 4.1×10^{-10} M. (a) Apparatus. (Reproduced from Wolfenden [1978], with permission of the American Chemical Society.) (b) Graph of original data. (Courtesy of Professor R. Wolfenden.)

overall dynamic equilibrium, equilibrium exists simultaneously along all possible macroscopic and microscopic paths [Tolman, 1938]. In the case of Henry's law, therefore, when the overall solute component exists at gas–liquid equilibrium, so do all of its molecular species and subspecies.

We may therefore write Eq. (3.5), where J denotes any of the descendant isomers of the dilute component. A straightforward extension [Eq. (3.6)] then yields a logarithmic concentration dependence for the partial free energy $G_J(M)$ of the molecular species and defines the standard partial free energy $G_J^0(M)$.

$$c_J/P_J = \mathcal{H}_J(M) \tag{3.5}$$

$$G_J(M) = G_J^0(M) + RT \ln c_J \tag{3.6a}$$

$$G_J^0(M) = G_J^0(g) - RT \ln \mathcal{H}_J(M) \tag{3.6b}$$

Although Eqs. (3.5)–(3.6) were derived with help from Detailed Balance, their actual validity requires only that J (in moles) be in equilibrium with J(g) and that the total solute concentration (in moles) be dilute, so that solute–solute interactions may be neglected. Equations (3.5)–(3.6) therefore apply even when the molecular species J is not at equilibrium with other species in the dilute solution—for instance, when J is a reactant in an ongoing rate process.

The assumption that solute–solute interactions are negligible in dilute solutions appears to contradict the fact that bimolecular, kinetically second-order reactions do take place in dilute solutions, even though such reactions require that the reacting molecules interact at short range. The dilute-solution model resolves this dilemma by letting the interaction define a transition-state species for the reaction. Molecules outside the transition state continue to interact negligibly. Because transition-state concentrations are stoichiometrically insignificant, transition states are included in composition trees only when the *rates* of the indicated reactions are of interest.

UNITS AND STANDARD STATES

This section might be called a medley of accounting principles for solution thermodynamics—commonly used formats for reporting data whose utilization facilitates public understanding. Some of the principles are obvious. For instance, one should avoid obscure units (such as degrees Baumé) and commercialisms (such as percent of minimum daily adult requirement per fluid ounce—Wow!). If such units *must* be used, they should be defined. If possible, one should use standard symbols recommended by a professional group such as the IUPAC [International Union, 1988]. One should clearly indicate the zero points of the energy or free-energy scales on which data are reported. And, in view of the growing practice of computer simulation, one should describe the underlying models and distinguish between modeled predictions and real-life observations.

Concentration Units

The preferred concentration units in chemical thermodynamics are based on molecular theory. For liquid solutions, three units are in common use:

- *For formal components*, the formula-weight fraction, the formality (formula weights of the component per kilogram of solvent), and the volume formality (formula weights of the component per liter of solution).
- *For molecular species*, the mole fraction, the molality (moles of the species per kilogram of solvent), and the molarity (moles of the species per liter of solution).

These units are defined and summarized in Table 3.1. For dilute solutions in a given solvent at constant T and P, they are practically proportional to each other. Conversion equations will be given below.

In closed systems, the formula-weight numbers of the components remain constant when other variables of state (T, P, and the α's) are changing. Formula-weight fractions and formalities are therefore convenient units for thermodynamic work when changes in T, P, and the α's are common. By the same token, the logically related mole fractions and molalities are convenient concentration units for molecular species. Volume formalities and molarities are good units for volumetric work and often for theoretical modeling, even though they vary inconveniently with T, P, and α. The inconvenience is not trivial, because the partial derivatives that allow for the variations clutter up the thermodynamic

Table 3.1. Summary of concentration units

Type	Symbol	Examples
Generalized concentration	c	c_1, c_2, c_A, c_U
Formula weight or mole number	n	n_1, n_2, n_A, n_U
Total formula weights		$\sum n_i$
Total moles		$\sum n_J = n_{total}$
Fraction	N	
Formula-weight fraction		$N_1 = n_1/\sum n_i$
Mole fraction		$N_A = n_A/n_{total}$
		$N_U = n_U/n_{total}$
Colligative fraction of solvent		$N_{1,\chi} = n_1/(n_1 + \sum n_{J,\text{solute}})$
Amount per kilogram of solvent	m	
Formal (f)[a]		$m_1 = 1000/\mathcal{L}_1; m_3/n_3 = m_1/n_1$
Molal (m)[b]		$m_A/n_A = m_1/n_1; m_U/n_U = m_1/n_1$
Amount per liter of solution	$\mathbf{c}$	
Volume formal (vf)[c]		$\mathbf{c}_2 = n_2/V$[e]
Molar (M)[d]		$\mathbf{c}_A = n_A/V; \mathbf{c}_U = n_U/V$

[a] Formula weights per kilogram of solvent.
[b] Moles per kilogram of solvent.
[c] Formula weights per liter of solution.
[d] Moles per liter of solution.
[e] V = volume of solution, in liters.

equations without adding essential insights—enough so that we shall rarely use volumetric units. The units of formality and molality offer the further advantage that concentration is always defined per kilogram of solvent, so that the formula-weight number n_1 of the solvent is fixed at $1000/\mathcal{L}_1$, where $\mathcal{L}_1$ is the formula weight.

For components and molecular species in the gas phase, current (1996) Standard Reference Tables still use the atmosphere as the unit of pressure, and we shall use this pressure unit throughout the book. However, the atmosphere is not part of the Système International (SI) of scientific units. The SI unit of pressure is the pascal (Pa), which equals 1 newton of force per square meter (Nm^{-2}). 1 atm equals $1.01325 \times 10^5 Nm^{-2}$. Another common pressure unit is the torr, defined as the pressure which produces a reading of 1 mm on a liquid mercury barometer at 0°C. 1 torr = 133.32 Pa = 1/760 atm.

Standard Partial Free Energies. Reference States and Standard States

Partial free energies such as G_i(M) and G_J(M) are functions of state, while the corresponding *standard* partial free energies, G_i^0(M) and G_J^0(M), are parameters of fit based on the dilute-solution model. These parameters characterize not a single state but the entire range of states in which $(G_i - RT \ln c_i)$ or $(G_J - RT \ln c_J)$ is independent of c_i or c_J. Perhaps through a quirk of tradition, this RANGE is called "the dilute reference STATE," but the appellation is not quite inappropriate. Except for G_i^0 being independent of c_i, it depends on the same variables of state as does G_i: It depends on the liquid solvent M, the nature of the dilute solute "i", and on T and P. It therefore makes sense to visualize a state in which $G_i = G_i^0$ and to call this state the *standard state*. The standard state may be identified with a real state in which $c_i = 1$, if the solute at unit concentration is still essentially dilute. Otherwise it is a hypothetical state. Analogous terminology is used for solutes in the low-pressure gas phase, except that c_i is now identified with the partial pressure P_i.

Physical Dimensions

Because $G_i = G_i^0 + RT \ln c_i$, G_i^0 must have the same dimensions as G_i, namely, energy per mole or energy per formula weight. RT has the dimensions of energy per mole; hence $\ln c_i$ (and c_i) are dimensionless. When concentration (c_i) is expressed as a fraction (e.g., $c_i = N_i$), c_i is indeed dimensionless. But when c_i is expressed on the m-scale or c-scale, c_i is dimensional, and $\ln c_i$ is a dimensional oddity. There are two stratagems to restore dimensional normalcy. (i) We may redefine the mathematical operation "$\ln x$" (where x is a dimensional variable such as c_i) so that x is first converted to its dimensionless magnitude, and the logarithm is then taken of the dimensionless magnitude. This requires no change in notation. Or (ii), we may explicitly replace x by the dimensionless ratio, $x/x^{\varnothing}$, where $x^{\varnothing}$ is a unit with the same dimensions as x, and then compute $\ln x/x^{\varnothing}$. This stratagem requires the substitution of $c_i/c_i^{\varnothing}$ for c_i in all logarithmic arguments involving dimensional concentrations. In this book we are employing stratagem (i).

Change of Concentration Units

Because the numerical value of c_i depends on the specific choice of concentration units, the actual value assigned to G_i^0 depends not only on the energy units and the zero point of the free-energy scale, but also on the concentration units.

To convert a set of standard partial free energies to a new concentration unit, let the generalized concentration c_i be expressed in two different units, x and y, which in dilute solutions are proportional to each other so that $y = \eta x$. Hence, in dilute solutions, $G_i = G_i^0(y) + RT \ln y = G_i^0(x) + RT \ln x$. The resulting dependence on concentration units is shown is Eqs. (3.7):

$$y = \eta x, \qquad G_i^0(y) = G_i^0(x) - RT \ln \eta \tag{3.7}$$

Particular solutions of Eqs. (3.7) for mole fraction, molality, and molarity are given in Eqs. (3.8), where ρ denotes the density of the solution:

$$G_{\mathrm{J}}^0(\text{molality scale}) = G_{\mathrm{J}}^0(\text{mole fraction scale}) - RT \ln(1000/\mathcal{L}_1) \tag{3.8a}$$

$$G_{\mathrm{J}}^0(\text{molarity scale}) = G_{\mathrm{J}}^0(\text{mole fraction scale}) - RT \ln(1000\rho/\mathcal{L}_1) \tag{3.8b}$$

One may often make the approximation that $\rho \simeq \rho_1$, the density of the solvent.

Medium Effects and Activity Coefficients

We have seen that when "i" is a dilute solute in a solvent medium M, $(G_i - RT \ln c_i)$ is independent of the concentration c_i and defines the standard partial free energy G_i^0. However, G_i^0 depends on the solvent medium through $\mathcal{H}_i(\mathrm{M})$, the Henry's law constant [Eq. (3.4b)], and we therefore write $G_i^0(\mathrm{M})$. The solvent medium need not be a pure liquid; it may be a liquid mixture, such as 1 f NaCl in water, or 20 wt% phenol in ethanol—indeed, any liquid in which the vapor pressure of the added dilute solute conforms to Henry's law.

Now allow the concentration c_i to increase. While c_i is in the dilute range, $(G_i - RT \ln c_i)$ remains constant at G_i^0. But as c_i becomes more concentrated, deviations from constancy grow in, and component "i" outgrows its status as a dilute solute. A typical behavior is shown in Fig. 3.4. We now have two options of accounting for the deviations. (i) We may say that "i" has outgrown its status, not only as a *dilute* solute, but altogether as a solute, so that "i" is now effectively a component of the solvent. (ii) We may say that "i" is still a solute,

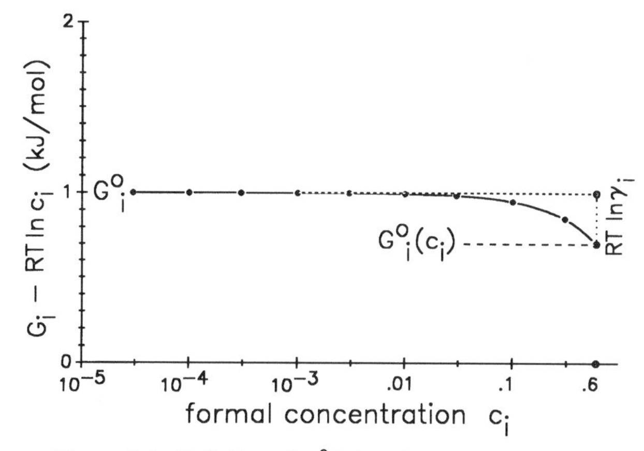

Figure 3.4. Definition of $G_i^0(c_i)$ and $RT \ln \gamma_i$. See text.

but c_i is high enough for solute–solute interaction to cause significant deviations from dilute laws. In case (i) we have a new solvent and may introduce a new standard partial free energy, $G_i^0(c_i)$, following the accepted definition: $G_i^0(c_i) \equiv (G_i - RT \ln c_i)$. In Fig. 3.4, the value of $G_i^0(c_i)$ at $c_i = 0.6$ formal is indicated by the dashed line.

In case (ii) we retain the standard partial free energy G_i^0 of the dilute solution, but introduce a correction term $RT \ln \gamma_i$, as indicated by the vertical dots in Fig. 3.4:

$$G_i = G_i^0 + RT \ln c_i + RT \ln \gamma_i \tag{3.8a}$$

or

$$G_i = G_i^0 + RT \ln a_i, \qquad a_i = c_i \gamma_i \tag{3.8b}$$

Approach (ii) is used when the deviation is small and $RT \ln \gamma_i$ may truly be regarded as a correction term. The form of $RT \ln \gamma_i$ is suggested by that of $RT \ln c_i$, so that Eq. (3.8a) can be rewritten in the form (3.8b), in which the *activity* a_i equals $c_i \gamma_i$. γ_i is the *activity coefficient*; it becomes unity in dilute solutions where the correction $RT \ln \gamma_i$ vanishes.

Equation (3.8b) creates a mathematical analogy between the solute concentration c_i in dilute solutions and the solute activity a_i in more concentrated solutions. That is, the equations derived for dilute solutions at constant T and P remain valid up to higher concentrations through the simple expedient of introducing activity in place of concentration.

For future reference, the equations related to partial free energies in dilute solutions are summarized in Table 3.2.

Table 3.2. Partial free energies in dilute solutions

1. The *i*th component[a] in a low-pressure gas mixture

$$P_i = PN_i = n_i RT/V, \qquad G_i = G_i^o(g) + RT \ln P_i$$

2. Molecular species J in a gas mixture

$$P_J = PN_J = n_J RT/V, \qquad G_J = G_J^o(g) + RT \ln P_J$$

3. The *i*th solute component[a] in dilute solution in a liquid solvent M

$$c_i/P_i = \mathcal{H}_i(M), \qquad G_i(M) = G_i^o(M) + RT \ln c_i$$
$$G_i^o(M) = G_i^o(g) - RT \ln \mathcal{H}_i(M)$$

Same, but nondilute: $G_i(M) = G_i^o(M) + RT \ln \gamma_i + RT \ln c_i$

4. Solute species J in dilute solution in a liquid solvent M

$$c_J/P_J = \mathcal{H}_J(M), \qquad G_J(M) = G_J^o(M) + RT \ln c_J$$
$$G_J^o(M) = G_J^o(g) - RT \ln \mathcal{H}_J(M)$$

Same, but nondilute: $G_J(M) = G_J^o(M) + RT \ln \gamma_J + RT \ln c_J$

5. Self-contained solvent in dilute solutions

$$N_{1,\chi} = n_1/(n_1 + \sum n_{J,\,\mathrm{solute}}), \qquad G_1 = G_1^o + RT \ln N_{1,\chi}$$

[a]Limited to components that are self-contained and isomeric with their descendant molecular species.

EQUILIBRIUM CONSTANTS IN DILUTE SOLUTIONS

In this section the role of standard partial free energies as marks of normalized stability will be illustrated by means of equilibrium constants. We shall begin with a simple 1 : 1 equilibrium, the symbolic isomerization $U \rightleftarrows V$ in dilute solution in a liquid solvent M. Since equilibrium depends on reactivity, U and V are molecular species, not components, and concentrations will be expressed in molal units. The partial free energies G_U and G_V are thus expressed in Eq. (3.9), in which G_U^0 and G_V^0 are the standard partial free energies.

$$G_U = G_U^0 + RT \ln m_U, \qquad G_V = G_V^0 + RT \ln m_V \tag{3.9}$$

To impose the condition that U exists in equilibrium with V at the given T and P, we apply the second Correspondence Theorem and specify that $\Delta_r G = 0$, where $\Delta_r G = G_V - G_U$. Introduction of (3.9) and minor rearrangement then yields Eq. (3.10a). Since T and P are fixed, $\Delta_r G^0$ and $\exp(-\Delta_r G^0/RT)$ are constants, with the latter [Eq. (3.10b)] defining the *equilibrium constant K*. Finally, Eq. (3.10c) relates the equilibrium constant, as measured by m_V/m_U at equilibrium, to $\Delta_r G^0$ and hence to $(G_V^0 - G_U^0)$, the difference in the standard partial free energies.

$$RT \ln (m_V/m_U) = -(G_V^0 - G_U^0) = -\Delta_r G^0 \tag{3.10a}$$

$$\frac{m_V}{m_U} = e^{-\Delta_r G^0/RT} = K \tag{3.10b}$$

$$\Delta_r G^0 = -RT \ln K \tag{3.10c}$$

Note that at equilibrium, $m_V = Km_U$. Hence only one of the concentrations is an independent variable; the other is constrained by the existence of equilibrium.

To generalize the above, we shall repeat the derivation for the more general symbolic reaction $a\text{A} + b\text{B} \rightleftarrows p\text{P} + q\text{Q}$ in the medium M. Again, T and P are constant, and A, B, P, and Q are molecular species at dilute concentrations. When chemical equilibrium exists, $\Delta_r G = pG_P + qG_Q - aG_A - bG_B = 0$. The concentration dependence of the partial free energies is again logarithmic; for example, $G_P = G_P^0 + RT \ln m_P$. On writing logarithmic expressions for the partial free energies and letting $\Delta_r G = 0$, we obtain Eq. (3.11a):

$$RT(p \ln m_P + q \ln m_Q - a \ln m_A - b \ln m_B) = -(pG_P^0 + qG_Q^0 - aG_A^0 - bG_B^0) = -\Delta_r G^0 \tag{3.11a}$$

Exponentiation then yields Eq. (3.11b). Since T and P are constant, $\Delta_r G^0$ is a constant, so that Eq. (3.11b) defines an equilibrium constant K. Note that the stoichiometric coefficients (a, b, p, q) appear as exponents in this expression.

$$\frac{m_P^p m_Q^q}{m_A^a m_B^b} = K = e^{-\Delta_r G^0/RT} \tag{3.11b}$$

$$\Delta_r G^0 = -RT \ln K \tag{3.11c}$$

Once again, the relative stability at equilibrium, as measured by the equilibrium constant, depends on the difference $\Delta_r G^0$ of the standard partial free energies.

THE SOLVENT IN DILUTE SOLUTIONS

This section will focus on the colligative nature of the partial free energy of the solvent in dilute solutions and, incidentally, introduce thermodynamic error tolerance. The following sections will extend the topics for the solvent in dilute solutions to solute-induced medium effects and to phenomena involving *two-component* solvents.

A colligative property is a property that depends on the *number* of molecular particles, but *not* on their atomic composition and molecular structure. The partial free energy of the solvent in dilute solutions is such a colligative property. We shall assume that the solvent consists of a single component (1) whose molecular formula, and corresponding molecular weight $\mathcal{L}_1$, are known in the gas phase or in the crystalline solid, but not necessarily in the liquid itself. Nonetheless, the molecular species in the liquid solvent will be stoichiometric multiples (or submultiples) of the known formula: If the gas formula or crystal formula is A, the species in the liquid will be A, or A_2, or A_n, or $A_{1/n}$, where n is an integer. We shall find that the ratios in which these species are present have no effect on the colligativity.

Given that the formula weight of the liquid solvent is $\mathcal{L}_1$, the solvent formality $m_1 = 1000/\mathcal{L}_1$. The partial free energy of the solvent component then is given by Eq. (3.12), where $\sum_J m_J$ is the *total* molality due to *all* solute species in the dilute solution.

$$G_1 - G_1^0 = -RT\left(\sum_J m_J\right)/m_1 \tag{3.12}$$

PROOF. The composition model (T-3.1) is elaborate enough to indicate the general scope of Eq. (3.12). The chemical formula assigned to the solvent component 1 is that of a monomeric species A, but the actual solvent is a self-contained equilibrium mixture consisting of A and its association products $A_2, A_3, \ldots, A_k$. The number of solute components and solute species is irrelevant, provided that the overall solution is dilute. It then follows from the first Correspondence Theorem that at equilibrium we obtain $G_1 = G_A$, and it follows from the second Correspondence Theorem that $G_A = (1/2)G_{A_2} = \cdots = (1/k)G_{A_k}$. Therefore, when equilibrium is maintained, $dG_1 = dG_A = (1/2)dG_{A_2} = \cdots = (1/k)dG_{A_k}$. Stoichiometry requires that $(n_A + 2n_{A_2} + \cdots + kn_{A_k}) = n_1$.

Composition Tree (T-3.1)

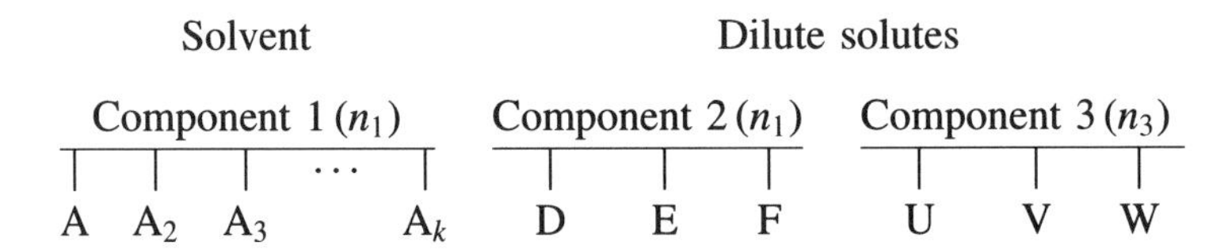

Since the Euler/Gibbs–Duhem equations apply to molecular species, it follows from Eq. (2.8d) that $n_A dG_A + n_{A_2}\ dG_{A_2} + \cdots + n_{A_k}\ dG_{A_k} = -\sum n_J\ dG_J$. When the changes $dG_A, dG_{A_2}, \ldots, dG_{A_k}$ take place with maintenance of equilibrium, we may substitute dG_1 and obtain

$$dG_A(n_A + 2n_{A_2} + \cdots + kn_{A_k}) = n_1 dG_1 = -\sum_J n_J\ dG_J$$

This equation is simplified by recalling that $m_1/n_1 = m_J/n_J$ and that $G_J = G_J^0 + RT \ln m_J$ for any solute species J. Hence $m_1\ dG_1 = -RT \sum_J dm_J$.

By definition, m_1, the number of *formula weights* of solvent per kilogram of solvent, is constant, independent of the m_J's. Integration of each term in the summation from 0 to m_J then yields Eq. (3.12). Q.E.D.

The same technique shows that Eq. (3.12) remains valid if A, $A_2, \ldots$ are mixtures of isomers rather than single species. Thermodynamic error tolerance enters because errors in the composition model for the self-contained solvent have no effect on the result, which is always Eq. (3.12).

The ratio $(\sum_J m_J)/m_1$ is a logical hybrid. $\sum_J m_J$ is the total solute molality in *moles* per kilogram of solvent, while m_1 is the solvent formality in *formula weights* per kilogram of solvent. Note that $(1-\sum_J m_J/m_1)$ is not the same as the *mole* fraction of the solvent, because the latter is sensitive to self-association or dissociation in the solvent component.

In applying Eq. (3.12) it will be convenient to define a fraction $N_{1,\chi} = m_1/(m_1 + \sum_J m_J)$, which will be called the *colligative fraction of the solvent.* Accordingly, $-\ln N_{1,\chi} = \ln(1 + \sum_J m_J/m_1)$, where $\sum_J m_J/m_1 \ll 1$. Expansion in power series and neglect of higher-order terms then yields $\sum_J m_J/m_1 = -\ln N_{1,\chi}$, with an error less than $(1/2)[\sum_J (m_J/m_1)]^2$, which is negligible when the solution is dilute. Equation (3.12) may therefore be written in the alternative logarithmic form (3.13):

$$G_1 - G_1^0 = RT \ln N_{1,\chi} \tag{3.13}$$

In Eq. (3.12), $(G_1 - G_1^0)$ is a property of the solvent component and is in principle measurable, since thermodynamic measurements operate on components. However, in dilute solutions the magnitude of $|G_1 - G_1^0|$ is very small—20 J/formula weight is a plausible upper limit. Yet during much of the history of chemistry, Eq. (3.12) provided virtually the only indisputable entry to *molar* solute concentrations, and much attention was paid to measuring colligative properties of the solvent, such as freezing-point lowering or vapor-pressure lowering, with meticulous precision. Applications of Eq. (3.12) provided molecular weights for literally thousands of new compounds and helped establish key chemical concepts, such as electrolytic dissociation in solvents of high dielectric constant, ionic aggregation in solvents of low dielectric constant, and the leveling effect of the solvent on acid–base equilibria.

Reaction with the Solvent

Equation (3.12) will detect reaction between solvent and solute only if there results a change in the total solute molality. For example, in anhydrous sulfuric acid (f.p. 10.5°C), methanol is converted to methylsulfuric acid and hydronium bisulfate:

$$CH_3OH + 2H_2SO_4 \rightarrow CH_3SO_3H + H_3O^+ + HSO_4^-$$

Accordingly, 1 formula weight of methanol produces a freezing-point lowering indicative of three moles of solute [Hammett, 1940].

On the other hand, reactions in which the total solute concentration remains unchanged are essentially undetectable, even though the reaction with solvent reduces the amount of unreacted solvent. In support of this statement we shall consider the detection of simple solvation complexes via high-precision measurements of $(G_1 - G_1^0)$. Our composition tree is (T-3.2). Here the dilute solute "S" reacts with h molecules of the solvent "A" to produce a genuine molecular complex $S \cdot hA$, where "genuine" means that one S and h A molecules form a united kinetic unit. (Solvation in which the molecules are neighbors without forming molecular complexes will be treated in Chapter 11.)

Composition Tree (T-3.2)

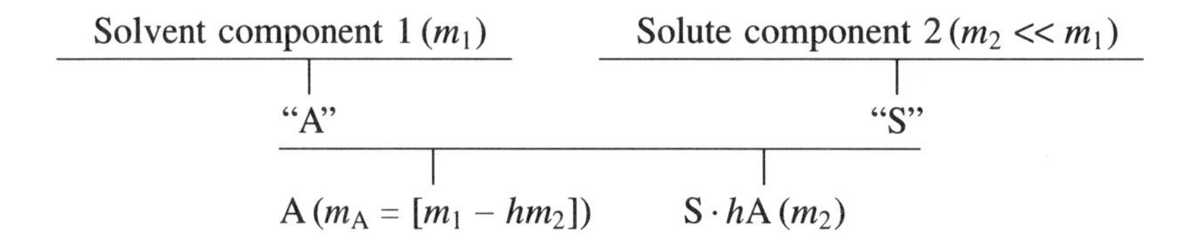

To prepare the system one might use 1 kg of solvent. Then $n_1 = m_1$ and $n_2 = m_2$. These formal concentrations define how the solution was prepared. By contrast, the molar concentrations define the amounts of the species that are actually present. Thus m_1 is fixed, while $m_A = m_1 - hm_{S \cdot hA}$. And, owing to the absence of uncomplexed S in (T-3.2), $m_{S \cdot hA} = m_2$.

In the derivation of Eq. (3.12) the key equation was $m_1\, dG_1 = -RT \sum_J dm_J$. This equation must now be modified in two ways. First, $\sum_J dm_J$ becomes $dm_{S \cdot hA} = dm_2$. Second, $m_1\, dG_1$ becomes $m_A\, dG_A$, where $m_A = [m_1 - hm_2]$, while dG_A at equilibrium continues to equal dG_1, by the first Correspondence Theorem. Thus

$$dG_1 = -RT\, dm_2/[m_1 - hm_2] = -(RT/m_1)\, dm_2[1 + hm_2/m_1 + O(hm_2/m_1)^2]$$

where the term of order $(hm_2/m_1)^2$ is small and may be neglected. Integration then yields Eq. (3.14):

$$G_1 - G_1^0 = -RT(m_2/m_1)[1 + hm_2/2m_1 + \cdots] \tag{3.14}$$

Letting $h = 8$, $m_2 = 0.1$, and $m_1 = 10$, we find that the effect of solvation-complex formation on $(G_1 - G_1^0)$—that is, the difference between Eqs. (3.12) and (3.14)—is 4%.

While a difference of 4% can be experimentally significant, solvation complex formation unfortunately is never the sole cause for deviations from Eq.

(3.12). Other deviations include those resulting from solute–solute (rather than solute–solvent) interactions. When $m_2 = 0.1$, such deviations amount to a few percent, similar to those due to solvation–complex formation, but their precise value is rarely known. Therefore, if we define a "measurement" as an operation whose result is well defined physically and whose error can be estimated objectively, then the solvation number h cannot be measured via $(G_1 - G_1^0)$.

All deviations from dilute-solution behavior are gathered up in the activity coefficient γ_i (defined in Fig. 3.4). The percentage deviation of γ_i from unity is about twice the percentage deviation of $(G_1 - G_1^0)$ from Eq. (3.12). Occasionally, but rarely, deviations other than those from solvation-complex formation can be predicted with adequate accuracy. The effect of solvation can then be estimated by difference. In this way, solvation numbers for ions in water were estimated by using the Debye–Hückel theory to represent the ion–ion interactions [Stokes and Robinson, 1948].

SOLUTE-INDUCED MEDIUM EFFECTS

The "structure" of a liquid includes not only the molecular species that are present, but also the nonbonded interactions—the intermolecular linkages and repulsions whose total effect defines the liquid cages available to the molecules in the medium. Facts about liquid structure can emerge from *solute-induced medium effects*—changes in structure-sensitive properties owing to the addition of a dilute solute. Solute-induced medium effects are of interest in solution thermodynamics because they add specific terms to the partial entropy, the partial heat capacity, and indeed to any property of a solution component that can be derived from the partial free energy. In dilute solutions the effects are often small, because the solvent is present in such great excess, but we shall describe examples in which the effects are quite significant.

Since the solution exists at equilibrium, the scope for solute-induced changes is restricted by thermodynamic constraints. We shall therefore digress briefly to consider the constraints and to prove that they leave freedom for solvent equilibria to be shifted by the addition of dilute solutes.

PROOF. It will be sufficient to consider solute-induced shifts in a 1 : 1 solvent equilibrium. The solvent (component 1) consists of two isomeric species A and B, to which a dilute solute (component 2) is added at formal concentration m_2. The formal solvent concentration is fixed at $m_1 = 1000/\mathcal{L}_1$. At equilibrium at constant T and P, the partial free energies of the solvent species are constrained by the Correspondence Theorems: $G_1 = G_A = G_B$, where G_1 follows Eq. (3.12). The molalities m_A and m_B are constrained by stoichiometry ($m_1 = m_A + m_B$) and by equilibrium ($m_A/m_B = K(M)$), but $K(M)$ is free to show medium effects.

That medium effects can in fact occur is documented by well-known propensity rules. For example, a strongly dipolar component added to the medium will shift the A $\rightleftarrows$ B equilibrium toward the more polar species; and an added component with a strong van der Waals functional group will shift it towards the species with the stronger complementary function. These freedoms are general and exist even if the added component is viewed as a solute. Thus, if we write the formal equations $G_A = G_A^0 + RT \ln m_A$ and $G_B = G_B^0 + RT \ln m_B$, and since $G_B - G_A = 0$, it follows that when a solute is added we obtain

$$(\partial/\partial m_2)(G_B^0 - G_A^0) + RT(\partial/\partial m_2)(\ln [m_B/m_A]) = 0$$

Since the ratio m_B/m_A of the solvent species can vary with m_2, the difference $(G_B^0 - G_A^0)$ of their *standard* partial free energies therefore can also vary with m_2. A shift in $(G_B^0 - G_A^0)$ with m_2 of course is tantamount to a shift in the equilibrium constant K with m_2. Q.E.D.

Use of Dielectric Data

To illustrate the reality of solute-induced medium effects we shall use two model systems: (i) a series of nonpolar and dipolar aprotic solutes in acetic acid and octanoic acid and (ii) the same series of aprotic solutes in 1-octanol. The measured property will be the dielectric constant. It is reasonable to assume that the aprotic solutes will behave according to Onsager's theory [Onsager, 1936], which treats the molecules as polarizable dipoles in a liquid continuum whose bulk properties are those of the real solution. The hydrogen-bonded solvents, on the other hand, are more likely to follow Kirkwood's theory [Kirkwood, 1939]—a development of Onsager's theory that allows for preferred relative orientations of the molecular dipoles. Our analysis will *not* require a detailed model for the solvent, but it *will* require that the solvent "structure" be easily perturbed.

The dielectric data will be analyzed as follows: Component 1 is the solvent and component 2 is the solute; ε denotes the dielectric constant of the solution, and ε_0 denotes that of the solvent; c denotes concentration in formula weights per milliliter. Both Onsager's and Kirkwood's theories then take the form (3.15):

$$\text{For the solution,} \quad (\varepsilon - 1)(2\varepsilon + 1)/9\varepsilon = c_1 P_1 + c_2 P_2 \tag{3.15a}$$

$$\text{For the pure solvent,} \quad (\varepsilon_0 - 1)(2\varepsilon_0 + 1)/9\varepsilon_0 = c_{1,0} P_{1,0} \tag{3.15b}$$

Here P_2 is the molar polarization of the solute, while P_1 and $P_{1,0}$ are the respective polarizations per formula weight of the solvent. P_2 will be predicted by Onsager's method [Grunwald and Pan, 1976], but the solvent polarizations are too uncertain to be predicted; they will be calculated from the dielectric data via Eqs. (3.15).

The difference between P_1 and $P_{1,0}$ will be identified with the solute-induced medium effect, which in dilute solutions varies linearly with c_2. We therefore

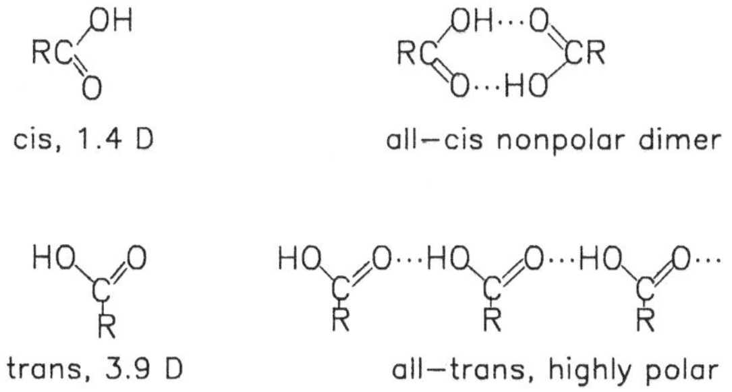

cis, 1.4 D

all–cis nonpolar dimer

trans, 3.9 D

all–trans, highly polar

Figure 3.5. *Cis–trans* conformations of carboxyl groups and topology of some of the linked structures.

write $(P_1 - P_{1,0}) = k_{1,2}c_2$, where the slope $k_{1,2}$ is characteristic of solvent and solute. Actually, a better approach is to define an apparent *solute* polarization $P_{2,\mathrm{app}}$ as shown below, and then subtract P_2. It can be shown that $P_{2,\mathrm{app}} - P_2 = k_{1,2}c_1$ and thus expresses the solute-induced medium effect in a form that is independent of solute concentration. Next we apply Onsager's theory in the form (3.16a)* and obtain the working equation (3.16b), which expresses the solute-induced medium effect (SIME) based on dielectric data, in (Debye units)^2.

$$P_{2,\mathrm{app}} = \left[(\varepsilon - 1)(\varepsilon + 2)/9\varepsilon - c_1 P_{1,0}\right]/c_2, \qquad P_{2,\mathrm{app}} - P_2 = k_{1,2}c_1$$

$$P_{2,\mathrm{app}} - P_2 = (4\pi L/9kT)\,(\mu_{2,\mathrm{app}}^2 - \mu_2^2)\,(n_2^2 + 2)^2/9 \tag{3.16a}$$

$$\mathrm{SIME} = (\mu_{2,\mathrm{app}}^2 - \mu_2^2) \times 10^{36} D^2 \tag{3.16b}$$

Carboxylic Acid Solvents

As shown in Fig. 3.5, the carboxyl group can exist in a *cis* conformation of low polarity and a *trans* conformation of high polarity. The OH hydrogen-bond donor group can link to the carbonyl oxygen or, less probably, to the hydroxyl oxygen: Each unshared electron pair provides a potential linkage site. Figure 3.5 shows also that the linked structures cover a wide range of polarity, from highly polar all-*trans* configurations to essentially nonpolar *cis* dimers, with unsystematic *cis–trans* sequences in between.

The following data for carboxylic acids and their esters [Riddick and Bunger, 1970] suggest that the polarity of the lower carboxylic acids is easily perturbed; c_1 is in formula wt/ml.

Formic acid:	ε 58,	c_1 0.0264	Methyl formate:	ε 8.5,	c_1 0.0162
Acetic acid:	ε 6.26,	c_1 0.0174	Methyl acetate:	ε 6.7,	c_1 0.0125
Butyric acid:	ε 2.95,	c_1 0.0108	Ethyl butyrate:	ε 5.0,	c_1 0.00752
Octanoic acid:	ε 2.46,	c_1 0.00629	Methyl oleate:	ε 3.2,	c_1 0.00293

*The factor $(n_2^2+2)/3$ accounts for the dipole enhancement by the reaction field; n_2 is the refractive index of the pure solute.

In normal dipolar liquids, ε decreases with the dipole concentration and hence with c_l, as exemplified by the data for the carboxylic esters. Relative to this norm, there is a marked decrease in ε between formic and acetic acid, and a sizable decrease between acetic and butyric acid, while the decrease between butyric and octanoic acid is unexceptional. It appears that as c_l decreases, the equilibrium among the carboxyl conformations and linked structures shifts from largely all-*trans* and highly polar in formic acid to largely all-*cis* and nonpolar dimer in octanoic acid. The linkage equilibria in acetic acid are in between. The steep slope of ε versus c_l at the point for acetic acid suggests a high sensitivity to perturbations and a high probability that solute-induced medium effects will be detectable. On the other hand, the dielectric constant of octanoic acid is near the nonpolar limit in the carboxylic acid series and the liquid "structure" should be difficult to perturb.

Results for SIME in acetic and octanoic acid are given in Table 3.3. In acetic acid most of the solute-induced medium effects are experimentally significant, while in octanoic acid they are within or just outside two standard deviations of error [Grunwald et al., 1976a,b].

Qualitatively, the solute-induced medium effect might be expected to depend on the polar character and size of the solute molecules. In fact, the empirical equation

$$\text{SIME} = (0.28 \pm 0.04)\mu_2^2 - (0.037 \pm 0.001)V_2$$

fits the data in acetic acid with an error that is 1.2 times the experimental error. As expected, the addition of solute dipoles (as measured by μ_2^2) increases SIME, while the dilution of solvent dipoles (as measured by V_2) decreases it beyond the normal effect due to the decrease in c_l. [The latter is included in Eq. (3.15).]

1-Octanol Solvent

A hydrogen bond between a donor, AH, and an acceptor, B, may be a three-center bond that produces a distinct molecular complex [Pimentel and McClellan, 1960], or it may be a nonbonded linkage of above-average van der Waals strength whose presence leaves the original molecules perturbed but essentially intact. A necessary (but not sufficient) condition for three-center bond formation is that the sums of the standard free energies for the reactants (A—H + B) and for the Brønsted conjugate ions (A^- + HB^+) be nearly equal [Kreevoy and Chang, 1976; Lindemann and Zundel, 1972, 1977; Zundel and Eckert, 1989]. The formation of a molecular complex can be recognized because the ultraviolet and infrared spectra differ markedly from simple superpositions of separate (A—H + B) and (A^- + HB^+) spectra, and the energy of dissociation to ligands is substantial, approaching that of a covalent single bond. Examples

Table 3.3. Solute-induced medium effects from dielectric data

Solute (V_2)[a]	$\mu_2(D)$	μ_2^2	$\mu_{2,\,app}^2$	SIME[b]
		Solvent: Acetic Acid (ε_0 6.265, 25°C)		
$PhNO_2$ (102.7)	3.93	15.4	15.6	0.2 ± 0.8
PhCN (103.1)	3.93	15.4	16.5	1.1 ± 0.8
PhCl (102.2)	1.58	2.50	−0.9	−3.4 ± 0.3
Benzene (89.4)	0	0	−3.1	−3.1 ± 0.3
p-(*t*-Bu)$_2$Ph (221)	0	0	−7.3	−7.3 ± 0.4
CCl_4 (97.1)	0	0	−4.7	−4.7 ± 0.3
		Solvent: Octanoic Acid (ε_0 2.46, 25°C)		
PhCN (103.1)	3.93	15.4	17.3	1.9 ± 0.8
PhCl (102.2)	1.58	2.50	2.9	0.4 ± 0.3
Benzene (89.4)	0	0	−0.2	−0.2 ± 0.1
CCl_4 (97.1)	0	0	0.15	0.15 ± 0.1
		Solvent: 1-Octanol (ε_0 10.01, 25°C)		
$PhNO_2$ (102.7)	3.93	15.4	11.4	−4.0 ± 0.8
PhCl (102.2)	1.58	2.50	−1.7	−4.2 ± 0.3
Benzene (89.4)	0	0	−4.5	−4.5 ± 0.3
p-(*t*-Bu)$_2$Ph (221)	0	0	−9.9	−9.9 ± 0.5

[a] V_2 in milliliters per formula weight. $PhNO_2$ = nitrobenzene; PhCN = benzonitrile; PhCl = chlorobenzene; *p*-(*t*-Bu)$_2$Ph = 1,4-di-*t*-butylbenzene; μ_2 measured in gas phase or in dilute solution in benzene.

[b] SIME is the solute-induced medium effect, $\mu_{2,\,app}^2 - \mu_2^2$, in (Debye units)^2.

Source: T.-P. I and Grunwald [1976] and Grunwald et al. [1976a, b].

include the ultraviolet (UV) spectrum of the pyridine-*N*-oxide··dichloracetic acid complex in sulfolane [Kreevoy and Chang, 1976], the infrared (IR) spectrum of liquid imidazole–acetic acid mixtures (which shows broad bands at 2950, 2550, and 1970 cm^{-1} attributed to O · H · N groups with high proton polarizabilities [Zundel and Eckert, 1989]), and the 160-kJ/mol dissociation energy for $F \cdot H \cdot F^-$ going to F—H + F$^-$ in the gas phase [Larson and McMahon, 1983].

On the other hand, when the hydrogen "bond" is a van der Waals linkage, the electron distributions of the original molecules are mildly perturbed, but the linked molecules retain their status as independent entities. This description fits the hydrogen-bonding in liquid aliphatic alcohol solvents, where there is no doubt that hydrogen bonds are formed: The boiling points, viscosities, IR and nuclear magnetic resonance (NMR) spectra, and dielectric constants all show it.

We are especially interested in 1-octanol, the solvent for which Table 3.3 lists SIMEs. The dielectric constant of 1-octanol deviates from Onsager's theory, but, in common with that of other straight-chain alcohols, it comes close to fitting a linkage model in which linear O—H··O hydrogen bonds oligomerize to form ··O—H··(O—H)$_n$·· molecular chains. The linked molecules retain the dipole moments of the unlinked molecules. Equally important, they retain a high degree of dynamic independence, rotating freely about the linear O—H··O hydrogen-bond axes [Oster and Kirkwood, 1943].

A linkage model with free rotation about the hydrogen-bond axis, in which the hydrogen-bond acceptor is a tetrahedral or trigonal unshared electron pair, also fits the measured dipole moments of 1 : 1 complexes between 1-octanol and a series of ligands, including acetone, benzaldehyde, pyridine, chloroform, and methylisobutyl ketone [Pan and Grunwald, 1976].

To assess the sensitivity of the dielectric constant to changes in the molecular environment, it is instructive to consider the dielectric constants of a series of isomeric octanols. The following data are for 25°C [Dannhauser, 1968]. The dipole concentration c_1 in the pure liquids is nearly uniform at 0.0063 ± 0.0001 formula wt/ml. The dielectric constants vary by more than a factor of three, showing great sensitivity.

1 – Octanol:	ε 10.0	2 – Methyl – 3 – heptanol:	ε 3.3
2 – Octanol:	ε 7.8	3 – Methyl – 3 – heptanol:	ε 3.1
3 – Octanol:	ε 5.3	4 – Methyl – 3 – heptanol:	ε 3.4
4 – Octanol:	ε 4.4	5 – Methyl – 3 – heptanol:	ε 3.9
		6 – Methyl – 3 – heptanol:	ε 4.9

If we write the structural formula of the alcohol in the form $R_1R_2R_3COH$, the data suggest that the substitution of an alkyl group for H at the carbinol carbon lowers the dielectric constant—presumably because it forces the linked chains to shift toward less polar or nonpolar cyclical or helical configurations.

Returning to Table 3.3, the SIMEs in 1-octanol are clearly significant for all solutes. They are reproduced by the empirical equation

$$\text{SIME} = (0.07 \pm 0.05)\mu_2^2 - (0.046 \pm 0.002)V_2$$

with an error of fit that is 0.7 times the experimental error. The data show only marginal dependence on the solute's dipole moment, but there is a marked dependence on V_2. The dominant effect appears to be a nonspecific response to dilution of the solvent dipoles by the added solute—a response that exceeds the effect of the decrease in c_1 allowed for by Eq. (3.15). The extraordinary sensitivity to changes in the covalent environment of the carbinol group, which emerges from the dielectric constants of the octanol isomers, evidently does not extend to changes in the nonbonded environment.

TWO-COMPONENT SOLVENTS

Mixed solvents offer a blend of the solvating actions of the individual components. This is especially useful when the solvating actions are distinct and complementary—when, in the jargon of "like dissolves like," the components are "unlike." In consequence, the thermodynamics of mixed solvent systems is often intricate, with complicated, even messy, dependences on composition, temperature and pressure.

The basic laws and theorems are of course unchanged. Henry's law applies to dilute solutes in any solvent of fixed composition, whether mixed or simple. The Euler equation (2.7) and the Gibbs–Duhem equation (2.8d) apply to any homogeneous phase. These bedrocks, and others, will be basic to the following discussion of two-component solvents.

Two numbering systems for the components are in common use. When the solvent components enjoy equal logical status, they are numbered 1 and 2, and the dilute solutes begin with 3. For a single solute, this system is outlined in composition tree (T-3.3). Molecular species are not shown, because we shall consider components only.

Composition Tree (T-3.3)

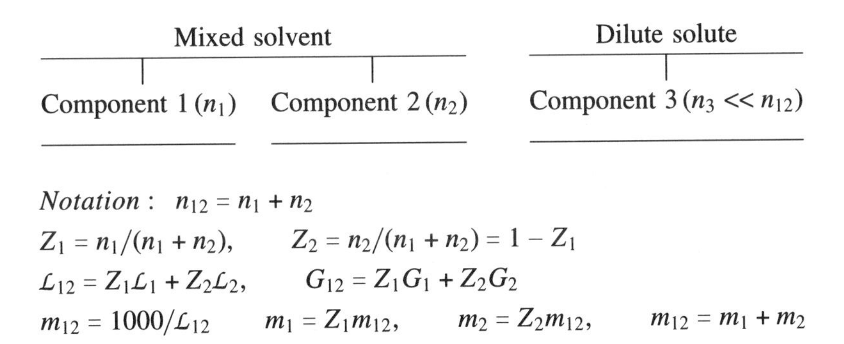

An alternative numbering system is used when one solvent component is the nominal solvent and the other is the cosolvent. The nominal solvent now is 1, the cosolvent is 3, and the solute is 2. This system is often used in biochemistry, where component 1 is likely to be water, the solute component 2 is likely to be a protein or other biochemical of high molecular weight, and the cosolvent 3 serves a specific purpose, such as changing the water "structure" or providing an ionic strength. The biochemical cosolvent might be urea, alcohol, or KCl.

We shall employ the numbering system and related notation shown in (T-3.3). The notation is designed to treat the solvent as an entity, rather than as a pair of components. Thus n_{12}, the total number of formula weights, is analogous to n_1 for a single-component solvent. Similarly, the formula weight $\mathcal{L}_{12}$, the partial free energy G_{12}, and the concentration m_{12} are analogous to $\mathcal{L}_1$, G_1, and m_1 for a single-component solvent. The solvent composition is specified by the formula-weight ratio Z_1 and its complement, $Z_2 = 1 - Z_1$.

The basic Euler equation for the Gibbs free energy,

$$G = n_1 G_1 + n_2 G_2 + n_3 G_3$$

expresses G as a function of $\{n_1, n_2, n_3\}$ at constant T and P. G_1, G_2, and G_3 follow the usual definitions; for example, $G_1 = (\partial G/\partial n_1)_{T,P,n_2,n_3}$. To obtain the desired relationships, we will transform the composition variables from (n_1, n_2, n_3) to (n_{12}, Z_1, and n_3). This yields Eqs. (3.17):

***T* and *P* are constant in Eqs. (3.17)**

$$(\partial G/\partial n_{12})_{Z_1,n_3} = Z_1 G_1 + Z_2 G_2 = G_{12} \tag{3.17a}$$
$$(\partial G/\partial Z_1)_{n_{12},n_3} = n_{12}(G_1 - G_2) \tag{3.17b}$$
$$(\partial G/\partial n_3)_{n_{12},Z_1} = G_3 \tag{3.17c}$$

PROOF. It will be sufficient to prove Eq. (3.17a). We begin with $G(n_1, n_2, n_3)$ and use the shortcut recipe preceding Eq. (2.11). Thus,

$$\left(\frac{\partial G}{\partial n_{12}}\right)_{Z_1,n_3} = \left(\frac{\partial G}{\partial n_1}\right)_{n_2,n_3} \left(\frac{\partial n_1}{\partial n_{12}}\right)_{Z_1,n_3} + \left(\frac{\partial G}{\partial n_2}\right)_{n_1,n_3} \left(\frac{\partial n_2}{\partial n_{12}}\right)_{Z_1,n_3}$$
$$+ \left(\frac{\partial G}{\partial n_3}\right)_{n_1,n_2} \left(\frac{\partial n_3}{\partial n_{12}}\right)_{Z_1,n_3}$$

Since $n_1 = Z_1 n_{12}$ (see T-3.3), $\partial n_1/\partial n_{12}$ at constant Z_1 equals Z_1. Similarly, $(\partial n_2/\partial n_{12})_{Z_1}$ equals Z_2; and $(\partial n_3/\partial n_{12})_{n_3}$ vanishes. Q.E.D.

A Two-Component Solvent with Z_1 Constant

In view of Eq. (3.17a), the Gibbs–Duhem equation, $n_1\, dG_1 + n_2\, dG_2 = -n_3\, dG_3$, may be rewritten in the form (3.18):

$$n_{12}(Z_1\, dG_1 + Z_2\, dG_2) = n_{12}\, dG_{12} = -n_3\, dG_3 \tag{3.18}$$

Comparison with the equation $n_1\, dG_1 = -n_3\, dG_3$, for solute 3 in a single-component solvent 1 shows mathematical analogy, with $n_{12} \leftrightarrow n_1$ and $G_{12} \leftrightarrow G_1$. Given that G_1 for a single-component solvent is colligative [Eqs. (3.12), (3.13)], it follows from mathematical analogy that G_{12} for a two-component solvent at constant Z_1 is also colligative, so that two-component solvents can be used to measure molecular weights [Inone and Timasheff, 1968].

Even more important, the presence of two components in the solvent (rather than only one) adds a mathematical degree of freedom whose physical interpretation can elucidate the solvating actions by the solvent components. To explore this, let us rewrite Eq. (3.18) specifically for 1 kg of the mixed solvent; that is, $n_1\mathcal{L}_1 + n_2\mathcal{L}_2 = 1000g$. Then $n_{12} \rightarrow m_{12} = 1000/\mathcal{L}_{12}$, and $n_3 \rightarrow m_3$. Next, we relate G_1 and G_2 to the partial pressures P_1 and P_2 of the solvent components in the vapor in equilibrium with the solutions; that is, $G_1 = G_1(g) = G_1^0(g)+RT \ln P_1$; and $G_2 = G_2(g) = G_2^0(g)+RT \ln P_2$. Finally, we assume that the dilute solute obeys Henry's law so that $G_3 = G_3^0(Z_1)+RT \ln m_3$, where $G_3^0(Z_1)$ is the *standard* partial free energy of the dilute solute at the given Z_1. Validity of Henry's law implies that the *formal* concentration m_3 equals the *molar* concentration, which in turn conveys the analysis into the domain of molecular species.

On this basis, substitution in Eq. (3.18) and integration at constant Z_1 gives the following result:

$$RT\left[Z_1 \int_i^f d\ln P_1 + Z_2 \int_i^f d\ln P_2\right] = -(RT/m_{12}) \int_0^{m_3} dm_3 \quad (3.19a)$$

$$\therefore \quad Z_1\delta \ln P_1[Z_1] + Z_2\delta \ln P_2[Z_1] = -m_3/m_{12} \quad (3.19b)$$

In Eq. (3.19a), the integration is from the initial state $(i; m_3 = 0)$ to the final state $(f; m_3)$. In Eq. (3.19b), $\delta \ln P_1[Z_1] = \ln\left[P_1[Z_1, m_3]/P_1[Z_1]\right]$, and $\delta \ln P_2[Z_1] = \ln\left[P_2[Z_1, m_3]/P_2[Z_1]\right]$, and $P_1[Z_1]$ and $P_2[Z_1]$ are the partial pressures of 1 and 2 at $m_3 = 0$.

Although the changes in P_1 and P_2 are not independent—they are constrained by Eq. (3.19b)—there remains a degree of freedom. The right-hand side of Eq. (3.19b) is negative and close to zero (since $m_3 \ll m_{12}$). Therefore Eq. (3.19b) may be rewritten in the approximate form $Z_1\delta \ln P_1[Z_1] \approx -Z_2\delta \ln P_2[Z_1]$, which shows that P_1 is free to increase or decrease by any amount, provided that P_2 varies in a nominally complementary fashion.

Some experimental results for ionic and molecular solutes in 50 wt% *p*-dioxane/water are listed in Table 3.4 [Grunwald et al., 1960]. *p*-Dioxane,

CH_2CH_2
O　　O
CH_2CH_2

is a lipophilic solvent whose dielectric constant ($\varepsilon = 2.21$ at 25°C) is similar to that of benzene but which, unlike benzene, is completely miscible with water. The values of $\delta \ln P_1/m_3$ and $\delta \ln P_2/m_3$ in Table 3.4 are of opposite sign, except for pentaerythritol, $C(CH_2OH)_4$, where both are near zero and negative. $\delta \ln P_1/m_3$ for the water solvent-component is negative for hydrophilic electrolytes and positive for hydrophobic naphthalene.

Table 3.4. Solvation effects in a two-component solvent consisting of water (1, 50 wt %) and *p*-dioxane (2, 50 wt%) at 25°C

Data for solvent: Z_1 0.8302, Z_2 0.1698, $\mathcal{L}_{12}$ 29.92, $(\mathcal{L}_1 - \mathcal{L}_2)/\mathcal{L}_{12}$ −2.097. m_{12} = 33.43 formula wt/kg of mixed solvent.

Solute	m_3	$\delta \ln P_1/m_3$	$\delta \ln P_2/m_3$	$\partial G_3^0/\partial Z_1$ (kJ/formula weight)
Na^+OH^-	0.0242	−0.189	+0.573	−68.3[a]
Na^+Cl^-	0.0256	−0.164	+0.461	−57.0[a]
H^+Cl^-	0.0247	−0.111	+0.193	−30.4[a]
$C(CH_2OH)_4$	0.050	−0.024	−0.057	−2.5[a]
Naphthalene	0.029	+0.077	−0.550	+46.7[a]
$Na^+BPh_4^-$	0.0493	+0.213	−0.688	+59.7[a]

[a]For electrolytes, $\partial G_3^0/\partial Z_1 = \partial G^0/\partial Z_1[\text{cation}] + \partial G^0/\partial Z_1[\text{anion}]$. The results given in this table (in contrast to those tabulated by Grunwald et al. [1960]) have not been corrected for interionic effects.

Sources: Grunwald and Bacarella [1958], Grunwald and Baughman [1960], and Grunwald et al. [1960].

Change of Z_1

Generalizing from the examples in Table 3.4, hydrophilic solutes in *p*-dioxane/water solvents have a negative $\delta \ln P_1/m_3$ and a positive $\delta \ln P_2/m_3$; and hydrophobic solutes have a positive $\delta \ln P_1/m_3$ and a negative $\delta \ln P_2/m_3$. To show the thermodynamic origin of this inference, we need to derive an expression for the derivative, $\partial G_3^0/\partial Z_1$, of the standard partial free energy of the solute with respect to Z_1. Since G_3^0 is a normalized measure of stability for the solvated solute, a negative value for $\partial G_3^0/\partial Z_1$ proves that the solvated solute becomes more stable as the mixed solvent becomes richer in water.

The derivation of $\partial G_3^0/\partial Z_1$ begins with the identity, $\partial^2 G/\partial Z_1 \partial n_3 = \partial^2 G/\partial n_3 \partial Z_1$, both partial derivatives being at constant (T, P, n_{12}). Operationally, one differentiates Eq. (3.17b) with respect to n_3 and Eq. (3.17c) with respect to Z_1. The result is Eq. (3.20):

$$\frac{1000RT}{\mathcal{L}_{12}}\left(\frac{\partial \ln[P_1/P_2]}{\partial m_3}\right)_{Z_1} = \frac{\partial G_3^0}{\partial Z_1} - RT\frac{\mathcal{L}_1 - \mathcal{L}_2}{\mathcal{L}_{12}} \tag{3.20}$$

PROOF. (i) Left-hand side of (3.20):

$$\begin{aligned}(\partial n_{12}[G_1 - G_2]/\partial n_3)_{Z_1,n_{12}} &= n_{12}(\partial[G_1 - G_2]/\partial n_3)_{Z_1,n_{12}} \\ &= m_{12}(\partial[G_1 - G_2]/\partial m_3)_{Z_1} \\ &\qquad (\text{at constant } Z_1, n_3/n_{12} = m_3/m_{12}) \\ &= (1000RT/\mathcal{L}_{12})(\partial \ln[P_1/P_2]/\partial m_3)_{Z_1}\end{aligned}$$

(ii) Right-hand side of (3.20):

$$(\partial G_3/\partial Z_1)_{n_{12},n_3} = (\partial[G_3^0 + RT \ln m_3]/\partial Z_1)_{n_{12},n_3}$$
$$m_3 = (1000/\mathcal{L}_{12})(n_3/n_{12}); \quad \ln m_3 = \ln(1000n_3/n_{12}) - \ln \mathcal{L}_{12}$$

G_3^0 and $\mathcal{L}_{12}$ are functions of Z_1, independent of n_3, n_{12}:

$$\therefore \quad (\partial G_3/\partial Z_1)_{n_{12},n_3} = \partial G_3^0/\partial Z_1 - RT[\mathcal{L}_1 - \mathcal{L}_2]/\mathcal{L}_{12}$$

Comment. The term $-RT[\mathcal{L}_1 - \mathcal{L}_2]/\mathcal{L}_{12}$ in (ii) corrects for the Z_1-dependence of m_3. Note the technique used in deriving it! One expresses m_3 as a function of the ongoing variables (n_{12}, Z_1, n_3) and substitutes the result in the expression to be differentiated. Q.E.D.

The final column in Table 3.4 lists values for $\partial G_3^0/\partial Z_1$ according to Eq. (3.20), based on the data in the two preceding columns. As expected, $\partial G_3^0/\partial Z_1$ is negative for the hydrophilic electrolytes and positive for naphthalene. But note the magnitudes! They are typically well over 10 kJ/mol, indicating a considerable "unlikeness" in the solvent actions of water and *p*-dioxane. Note also that the hydrophobic effects are of comparable magnitude to the hydrophilic effects, in spite (for ionic solutes) of the hydrophilic increase in the dielectric constant. Indeed, by the evidence of $Na^+BPh_4^-$ versus Na^+Cl^-, the hydrophobic effects can swamp the dielectric-constant effect for ions with extensive hydrophobic surfaces, such as BPh_4^-.

Solvation Numbers

For a dilute solute in a *one*-component solvent, we found that the change in G_1 of the solvent component *cannot* measure the solvation number of the solute per se, but that it *can* yield the solvation number as a parameter in a more comprehensive theory. Similar remarks apply to changes in G_1 and G_2 in two-component solvents. The measured result for $\partial G_3^0/\partial Z_1$ is clearly relevant to solvation, but it includes *all* relevant interactions. Specific solvation numbers can evolve only as parameters in an explicit theory. It should be stressed that this conclusion is independent of the choice of the composition variables (n_{12}, Z_1, n_3) in the original Eq. (3.17). Mathematical transformation to other independent variables will generate equations with a different look to them, but mathematics alone cannot change physical content.

On the other hand, transformation of variables can generate equations that are more congenial to particular models. One model, the *endostatic* model (*static* unchanging, *endo* on the inside [Grunwald and Effio, 1974]), lets us add a small amount δm_3 of the solute to the intial solution, accompanied by model-predicted amounts δm_1 and δm_2 of the solvent components chosen so that the solution environments remain constant. If only short-range interactions are significant, the partial derivatives

$$(\partial m_1/\partial m_3)_{\text{endostatic}} \quad \text{and} \quad (\partial m_2/\partial m_3)_{\text{endostatic}}$$

measure the respective solvation numbers per formula weight of the solute.

Is the endostatic model workable? A single phase with three components at constant T and P is defined by specifying 2 independent variables, according to the phase rule. When we write $(\partial m_1/\partial m_3)_{\text{endostatic}}$, we specify that m_1 is the dependent variable and m_3 is the active independent variable. This leaves a choice of one independent variable, which must be chosen so that its constancy fulfills endostaticity.

A credible choice for the variable is $[G_1 - G_2]$. In a study of the solubility of nonelectrolytes in p-dioxane/water, Grunwald and Butler [1960] found that salt effects at constant Z_1 follow the reverse pattern of salt effects in water; but at constant $[G_1 - G_2]$, the patterns become parallel. The result suggests that two-component solvents resemble one-component solvents if $[G_1 - G_2]$ remains constant as solute is added—that is, if the escaping tendencies of the solvent components (as measured by P_1 and P_2) remain in a constant ratio. Constancy of $[G_1 - G_2]$ promotes endostaticity because, in the absence of solute-induced medium effects, solvation in one-component solvents is necessarily endostatic.* Approximate solvation numbers might therefore be obtained by measuring the partial derivatives

$$(\partial m_1/\partial m_3)_{[G_1 - G_2]} \quad \text{and} \quad (\partial m_2/\partial m_3)_{[G_1 - G_2]}$$

As the number of components increases, the number of inactive variables that must be kept constant increases as well. Thus a phase with four components at constant T and P leaves a choice of two variables whose constancy might promote endostaticity. In fact, the partial derivative

$$(\partial m_1/\partial m_3)_{G_1, G_2}$$

has been proposed to estimate changes in hydration numbers accompanying protein processes in water/cosolvent media, where 1 and 2 are the solvent components [Timasheff, 1993]. The composition variables here are transformed from the traditional $\{n_1, n_2, n_3, n_4\}$ to $(n_1 + n_2), m_3, G_1, G_2\}$. Since $G_1 = \partial G/\partial n_1$, and $G_2 = \partial G/\partial n_2$, the transformations $n_1 \leftrightarrow G_1$ and $n_2 \leftrightarrow G_2$ are Legendre transformations. In a far-reaching review, Alberty [1994] has shown that Legendre transformations of the type $n_i \leftrightarrow G_i$ sometimes give keener insights into systems of molecular species. In the present case the separate constancy of G_1

*Unfortunately, a subsequent attempt to generalize the endostatic model [Grunwald and Effio, 1974] was flawed [Bertrand and Fagley, 1976].

and G_2 is an improvement over constancy of $[G_1 - G_2]$, but, as will be shown in Chapter 11, even this level of control does not vouchsafe constancy of the molecular environments.

The transfer of solvent molecules from the bulk of the solvent into the solvation shells of solutes is called *solvent reorganization*. It is an integral part of the thermodynamics of solvation and, as shown in Chapter 11, contributes to the measured values of ΔH^0, ΔS^0, and ΔC_P^0. Solvent reorganization can be especially important in processes involving biochemical macromolecules (e.g., proteins or nucleotides), partly because the molecules are built on a grander scale, but largely because the concept of a *localized* reaction zone fails in an environment whose pieces are tied together by linkage bonds. Linkage, it will be recalled, is the coupling of covalently bound units by van der Waals interactions of above-average strength. Linkage bonds are perturbable, as shown for hydrogen-bonding solvents by the data in Table 3.3. When the biomolecule binds a substrate at an active site, the linked structure yields to accommodate the substrate, using a mechanism in which some linkage bonds break or form or deform, and some solvent molecules move into or out of the region affected by the perturbation. The motion of solvent molecules of course is solvent reorganization. At this stage we know little about its quantitative aspects, except that the contributions to ΔH^0, ΔS^0, and ΔC_P^0 can be large. Calibrating facts are needed, and any respectable appraisal of changes in solvation numbers, even though approximate, can be useful. Two-component solvents are congenial to such appraisals, and action is simplified by the availability of the Wyman linkage relations, a remarkable catalog of derived equations that cover a suitable range of components, stoichiometries, interaction mechanisms, and thermodynamic scenarios [Wyman, 1964; Wyman and Gill, 1990].

We will describe one such appraisal, for the solvent reorganization accompanying oxygen binding by hemoglobin [Colombo et al., 1992]. The studied system consists of four components: water (1), a cosolvent (2), hemoglobin (Hb), and oxygen (O_2). Hemoglobin is a linked tetramer made up of four nearly equivalent subunits. [Lippard and Berg, 1994; Colombo et al., 1992] These are linked either in a *tense* form (T) or in a *relaxed* form (R). In the absence of bound oxygen the favored form is T. When each subunit binds one molecule of O_2, the favored form is R. In the T $\rightarrow$ R transition, several linkages come apart, namely, electrostatic "salt bridges" between CO_2^- groups and cationic groups such as imidazolium in histidine side chains. It was assumed by Colombo et al. [1992]* that water molecules, but no molecules of cosolvent, move onto the newly exposed surface. On that basis, the results for oxygen uptake indicate that the change in hydration number, $[\Delta_{T \rightarrow R}(\partial n_1/\partial n_{Hb})]$, is near 60 water molecules per formula weight of hemoglobin. The neglectability of solvation by the cosolvent was tested by changing the cosolvent; the effect was negligible. The sug-

*A second model assumption is that the derivative $(\partial n_2/\partial G_{O_2})_{G_1, n_{Hb}}$ equals 0. This may be unrealistic because it deprives the model system of the mixed solvent properties illustrated in Table 3.4.

gested reorganization of 60 water molecules is an order-of-magnitude greater than the likely solvent reorganization in reactions of ordinary-sized molecules.

STABILITY THEOREM

When consistent concentration units are used, the standard partial free energy is a normalized index of stability at equilibrium. In the case of molecular species, we saw, for example, that the equilibrium constant K for a reaction is related to ΔG^0, the standard partial free energy of the products *minus* that of the reactants, through the equation $\Delta G^0 = -RT \ln K$.

The Stability Theorem relates the stability of a molecular species to that of its parent *component*. It may be stated as follows:

Stability Theorem. The standard partial free energy G_i^0 of any ith formal component is related to the standard partial free energy G_J^0 of any of its isomeric molecular species or subspecies J by

$$G_i^0 = G_J^0 + RT \ln \alpha_J \tag{3.21}$$

where α_J is the stoichiometric fraction of J in the ith component *at equilibrium*. Here $G_i^0 = G_i - RT \ln c_i$ and $G_J^0 = G_J - RT \ln c_J$, and ci (the formula-weight concentration) and c_J (the molecular-species concentration) are expressed in consistent units.

PROOF. At equilibrium $G_J = G_i$ according to the first Correspondence Theorem, and $c_J = \alpha_J c_i$ by definition of α_J. Hence $[G_i^0 - G_J^0] = -RT \ln(c_i/c_J) = RT \ln \alpha_J$. Q.E.D.

Since c_i cannot be less than c_J, we have $\alpha_J \leq 1$ and $G_i^0 \leq G_J^0$. Moreover, since formal components can always be modeled as mixtures of isomeric molecular species or subspecies, the "less than" inequality applies. That is, the standard partial free energy of any formal component is smaller, in the algebraic sense, than that of any of its descendant molecular species or subspecies. Since standard partial free energies measure relative stability, it follows that *a formal component is more stable than any of its isomeric descendant molecular species or subspecies.*

Furthermore, the Stability Theorem expresses the fact that at equilibrium, unstable molecular species are present at a low mole fraction. Figure 3.6 shows a plot of α_J versus $[G_J^0 - G_i^0]$ on a semilogarithmic scale. If we let 0.1–1% be the minimum for stoichiometric significance, then a purely stoichiometric model may omit all species with $[G_J^0 - G_i^0] > 12$–17 kJ/mol. Molecular species with $[G_J^0 - G_i^0]$ in the range 17–150 kJ/mol are stoichiometrically negligible but may have significant reactivity owing to their high free energy. Such species also are likely to have interesting physical and chemical properties. Above ~ 160 kJ/mol,

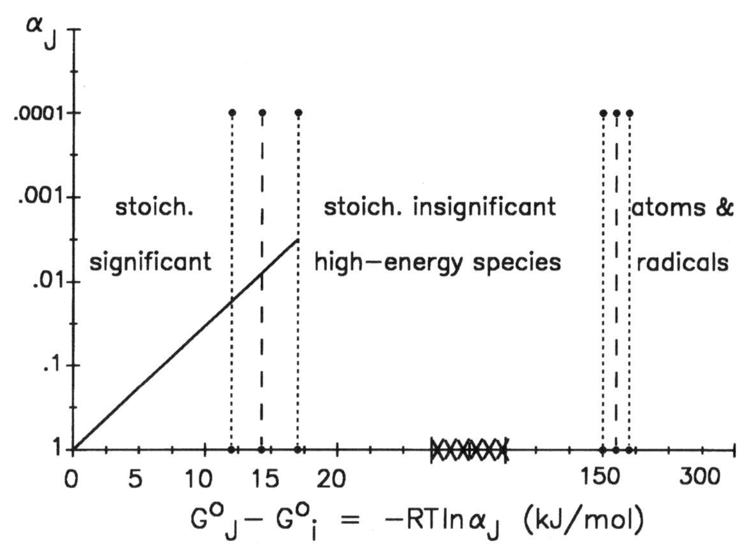

Figure 3.6. A stoichiometrist's view of the Stability Theorem.

G_J^0 is high enough to permit molecular dissociation. At equilibrium in systems of ordinary temperatures and volumes, species in this range average less than one molecule for the entire system and are significant only under chemical nonequilibrium or extreme physical conditions.

The nominal 15-kJ/mol limit for stoichiometric significance gives a clue as to what sorts of isomers will be found in purely stoichiometric equilibrium models. One would rarely find two structural isomers with unlike neighbor atoms (such as CH_3SSCH_3 versus $HSCH_2CH_2SH$), because differences in covalent bond energies typically exceed 15 kJ/mol. One might find skeletal isomers—structural isomers with the same bonds in different arrangements—but it would not be a safe guess because deviations from constancy of bond energies are of the order of 10 kJ/mol. The isomerism most commonly found in stoichiometric models is based on nonbonded interactions—conformational, geometric, and environmental isomerism. The intramolecular interactions that are responsible for geometric and conformational isomerism are often known and understood [Eliel, 1965; Riddell, 1980], while the liquid-cage effects that produce environmental isomerism are more elusive. We shall take a close look at environmental isomers in Chapter 10.

STATISTICAL FORM OF THE STABILITY THEOREM

We shall now derive a corollary of the Stability Theorem which is useful because it builds a bridge to statistical thermodynamics. The Stability Theorem applies to a formal component and those of its molecular species and subspecies that are isomeric with the component, but it does not forbid the presence of species in which the component is associated or dissociated. In the statistical form of the Stability Theorem, the component must consist *entirely* of isomeric species and subspecies. These isomeric "species" may be as dainty as the populations of individual energy levels.

The composition tree for the statistical form of the Stability Theorem is (T-3.4). Note that the sum of α_V over all isomeric species and subspecies of component i equals 1.

Composition Tree (T-3.4)

Components: $1, 2, \ldots, i, \ldots$

$\mathrm{T}(\alpha_\mathrm{T})$ $\mathrm{U}(\alpha_\mathrm{U})$ $\cdots$ $\mathrm{V}(\alpha_\mathrm{V})$ $\cdots$ $\mathrm{Z}(\alpha_\mathrm{Z})$

$$\alpha_\mathrm{T} + \alpha_\mathrm{U} + \cdots + \alpha_\mathrm{Z} = \sum_{V=T}^{Z} \alpha_\mathrm{V} = 1$$

The statistical form of the Stability Theorem is as follows:

Statistical Form. The standard partial free energy of any ith formal component consisting solely of isomeric molecular species and subspecies is equal to the average standard partial free energy of these species and subspecies, *plus* a term for the mixing of the species to form the component:

$$G_i^0 = \sum_{V=T}^{Z} \alpha_\mathrm{V} G_\mathrm{V}^0 + RT \sum_{V=T}^{Z} \alpha_\mathrm{V} \ln \alpha_\mathrm{V} \tag{3.22}$$

Average G^0 + Mixing term

The mixing term equals the free energy of mixing if the component is a dilute solute or if mixing follows the ideal-solution model.

PROOF OF EQ. (3.22). Apply the Stability Theorem (3.21) to each species, multiply by the corresponding α, and add

$$\alpha_T G_i^0 = \alpha_\mathrm{T} G_\mathrm{T}^0 + RT\,\alpha_\mathrm{T} \ln \alpha_\mathrm{T}$$

$$\alpha_U G_i^0 = \alpha_\mathrm{U} G_\mathrm{U}^0 + RT\,\alpha_\mathrm{U} \ln \alpha_\mathrm{U}$$

$$\vdots$$

$$\alpha_Z G_i^0 = \alpha_\mathrm{Z} G_\mathrm{Z}^0 + RT\,\alpha_\mathrm{Z} \ln \alpha_\mathrm{Z}$$

$$\therefore\quad G_i^0 \sum \alpha_\mathrm{V} = \sum \alpha_V G_\mathrm{V}^0 + RT \sum \alpha_\mathrm{V} \ln \alpha_\mathrm{V}$$

Q.E.D.

Because mole fractions decrease exponentially with G_V^0, species with high standard partial free energies carry little weight in Eq. (3.22). But if there are many such species, their cumulative contribution to the mixing term may become significant, since $-(\alpha_V \ln \alpha_V)$ approaches zero more slowly than does α_V alone. A plot of $-(\alpha \ln \alpha)$ versus α is shown in Fig. 3.7.

Consistency with Statistical Thermodynamics

In Eq. (3.22) the component is dissected into distinguishable molecular species and subspecies. In statistical thermodynamics the same component would be dissected into the populations of distinguishable quantum states. The two operations differ only in degree—the degree of fineness of the dissection. Since both accord with the same macroscopic thermodynamics, we may assume that consistency exists. Nonetheless it is worthwhile to examine the conditions for consistency, because they clarify the constraints and freedoms surrounding the concept of molecular species.

Statistical thermodynamics is more general than the thermodynamics of molecular species. The "molecules" of chemistry that make up the molecular species act as individuals; while the physical "particles" that make up the statistical–mechanical ensembles range from independent particles of molecular size to arbitrarily complex aggregates of interacting particles.

Do the individually acting "molecules" of chemistry correspond to independent molecular particles? For molecules in a dilute gas, where the intermolecular interactions are negligible, the answer is clearly "yes." But for molecules in a liquid, a dense gas, or on a surface the answer is not so obvious, because the molecules interact with the molecules of their environments. In principle, these interactions interfere with freedom of individual molecular action. The only way in which such molecules can retain their freedom is that the poten-

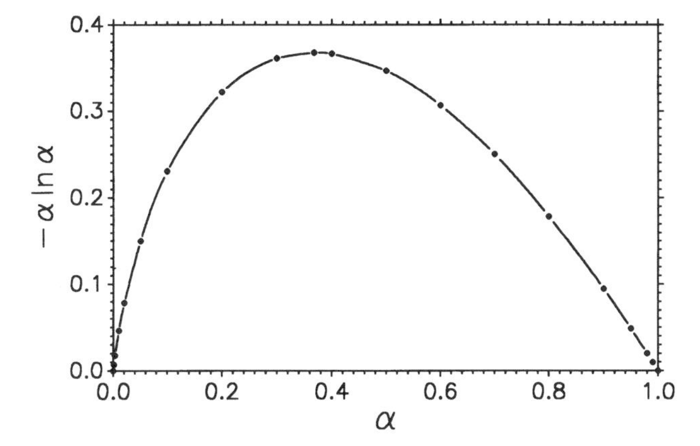

Figure 3.7. Plot of $-(\alpha \ln \alpha)$ versus α.

tial energy associated with interaction with the environment becomes part of the potential energy of the molecule itself; that is, as the molecule diffuses through the medium, the potential energy of interaction must accompany it. A straightforward mechanism is to model the medium as a continuum. An alternative mechanism (and one that does not overlook the molecular nature of the medium) employs the liquid-cage model derived from early computer simulations by Rabinowitch and Wood [1936]. We shall examine the cage model in Chapter 9.

In crystalline solids, chemical reactions and physical processes often require cooperative actions by the molecules, and the concept of molecular species consisting of molecules acting individually tends to break down.

INDICATION OF CONSISTENCY. It will be sufficient if we rewrite Eq. (3.22) in the language of the statistical thermodynamics of independent molecular particles and show that the left-hand side remains equal to the right-hand side.

Let $RT = LkT$; let $\epsilon_a, \ldots, \epsilon_j, \ldots$ denote the energy values of the quantum-mechanical energy states (relative to a chosen zero of energy); let $\alpha_a, \ldots, \alpha_j, \ldots$ denote the fractional populations of the respective states at statistical equilibrium at constant T and P; and let the molecular partition function $z = \sum e^{-\epsilon_j/kT}$. In this notation, $\alpha_j = e^{-\epsilon_j/kT}/z$.

The molecules of the formal component may be viewed as localized or nonlocalized. If localized, the left-hand side of Eq. (3.22) becomes $G^0 = -LkT \ln z$. On the right-hand side, the molecular species T...V...Z become the quantum-mechanical energy states $a \ldots j \ldots$: Thus G_V^0 becomes $L\epsilon_j$, and α_j is given above. The right-hand side of Eq. (3.22) then becomes

$$L \sum \frac{\epsilon_j e^{-\epsilon_j/kT}}{z} - LkT \sum \frac{e^{-\epsilon_j/kT}}{z} \left[\frac{\epsilon_j}{kT} + \ln z \right] = -LkT \ln z$$

The left-hand side of Eq. (3.22) thus remains equal to the right-hand side.

Q.E.D.

If the molecules follow the model of nonlocalized particles, the term $RT \ln L!$ must be subtracted from the left in the above, and $RT \sum (\alpha_j \ln L!) = RT \ln L!$ must be subtracted from the right. The equation therefore remains in balance. Q.E.D.

Having shown that consistency exists regardless of whether the individual molecules are localized or nonlocalized, let us now consider distinguishability. As stated in Chapter 1, there are two kinds of distinguishability: real and ideal. Real distinguishability is observable. Ideal distinguishability is predicted by primary theory, usually quantum mechanics. Statistical thermodynamics accepts ideal distinguishability as sufficient—even if the capability to observe the distinction does not currently exist.

Chemistry cannot afford to be so permissive. Chemical reactions must be real and observable, which means that reactants and products, and hence molecular species in general, must have real distinguishability.

On the other hand, as molecular species become divided into subspecies and sub-subspecies, eventually a point is reached where real distinguishability comes to an end and ideal distinguishability alone remains. In such cases we shall follow the example of statistical thermodynamics and accept ideal distinguishability as sufficient. That is, subspecies and sub-subspecies with ideal distinguishability may be included in composition trees. This degree of freedom will be essential in the thermodynamic treatment of solvation in Chapter 11.

4

DISPLACEMENTS FROM EQUILIBRIUM AND ERROR TOLERANCE

Because its laws are exact, thermodynamics is a rigorous and exacting science. But it is also a friendly science, because it has built-in error tolerance. To appreciate error tolerance, consider a continuous function $f(x, y)$ and a small error δx at the point (x_0, y_0). The resulting error $\delta f \simeq (\partial f/\partial x)_{x_0, y_0} \delta x$. If f is at a maximum or minimum, $(\partial f/\partial x)_{x_0, y_0} = 0$ and errors in x cause virtually no errors in f.

There is error tolerance in thermodynamics because the free energy G, for a closed system at constant T and P, reaches a minimum whenever equilibrium is reached. Operational equilibrium exists, it will be recalled, when the properties of the system become stable on the observational time scale and when this stable limit is independent of the manner in which the system was prepared. Moreover, as shown in Chapter 2 [composition tree (T-2.3)], each independent progress variable α_i satisfies the condition $\partial G/\partial \alpha_i = 0$ independently, so that error tolerance exists not only for the global system, but individually for each reaction that has reached equilibrium. At constant T and P, this kind of error tolerance applies specifically to the Gibbs free energy G. It does not apply to the enthalpy H (for example), nor does it apply to any function that lacks a minimum when equilibrium is reached at constant T and P.

As an example of error tolerance, Fig. 4.1 plots G/n_1 versus α for the dimerization of nitrogen dioxide in the gas phase at 298 K:

$$\underset{(1-\alpha)n_1}{1\,NO_2} \rightleftarrows \underset{\alpha n_1/2}{\tfrac{1}{2}\,N_2O_4}$$

As reaction goes to equilibrium, G/n_1 reaches a minimum of -2.85 kJ as α goes

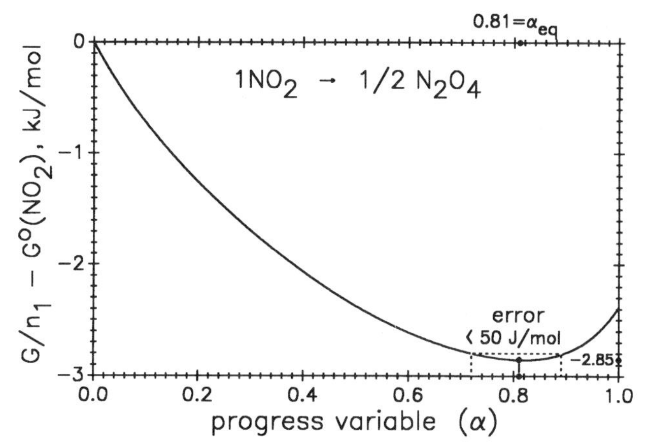

Figure 4.1. G/n_1 for the dimerization of nitrogen dioxide at 298 K and a constant pressure of 1 atm. $\Delta G^0 = -2.38$ kJ. $P_{NO_2} = (1-\alpha)P/(1-\frac{1}{2}\alpha)$. $G/n_1 = G^0_{NO_2} + \alpha\Delta_r G^0 + RT[(1-\frac{1}{2}\alpha)\ln P + (1-\alpha)\ln(1-\alpha) + (\frac{1}{2}\alpha)\ln(\frac{1}{2}\alpha) - (1-\frac{1}{2}\alpha)\ln(1-\frac{1}{2}\alpha)]$. (Calculated from data in JANAF [1971].)

to its equilibrium value of 0.81. However, allowing for a typical experimental error of 50 J/mol in G/n_1, the error in α at equilibrium may be as large as ± 0.08 before the resulting error in G/n_1 can be detected.

In this chapter we shall examine error tolerance for deviations from equilibrium in chemical reactions. We shall find that error tolerance encompasses not only the free energy G of the system, but also the partial free energy of any ith component, $G_i = \partial G/\partial n_i$, and the corresponding standard partial free energy, G_i^0. That is, at equilibrium in a closed system at constant T and P, $\partial G/\partial\alpha$, $\partial G_i/\partial\alpha$, and $\partial G_i^0/\partial\alpha$ all vanish, for any progress variable α. The error tolerance for G_i and G_i^0 is distinct from that for G and must be proved, because error tolerance is not transmitted automatically from a function to its derivatives. For example, the volume $(\partial G/\partial P)_T$ and the partial entropy of any ith component $-(\partial G_i/\partial T)_P$ in a closed system are derivatives of the free energy G, but these properties do not reach an extremum at chemical equilibrium.

Let us be clear! We are not offering error tolerance to excuse sloppy experiments—that would be unfair to the professionals whose measurements are setting ever higher standards. Rather, we are offering error tolerance as a fact to be reckoned with, not only for interpreting experimental results, but also for judging that elusive ingredient of "verity" in scientific models. Some models look good because the properties they predict are protected by error tolerance, even though the models themselves omit known complexities. This is especially true for the kind of modeling that relies on chemical analogies, whose successful products include the popular linear free energy relationships [Hammett, 1937; Leffler and Grunwald, 1989]. We shall return to the contribution of error tolerance to the success of linear free energy relationships in Chapter 5.

CHEMICAL SYSTEMS OFF EQUILIBRIUM

When it comes to equilibrium, all chemical systems are hybrids: Some reactions exist at operational equilibrium, while others are mired or "stuck" off equilibrium. The difference is one of time scale. Reactants and products appear to be at equilibrium when the time between observations is *long* compared to the relaxation time τ for return to equilibrium, while a reaction appears to be mired or "stuck" off equilibrium when the time between observations is *short* compared to τ. In the latter case the reactants and products remain displaced from equilibrium for many observations and, in effect, are separate components.

Since the difference between at equilibrium and off equilibrium is one of time scale, it is *not* a difference of kind. This is important, because the laws of thermodynamics do not heed the speed at which we make our measurements. Accordingly, the same thermodynamic framework applies to reactions at equilibrium and off equilibrium—the same basic laws, the same state variables and state functions—except that at equilibrium $\partial G/\partial \alpha$ equals zero.

Nomenclature

For describing chemical reactions off equilibrium, two systems of nomenclature are in use. One nomenclature uses the terms *stress* and *strain*, which have a mechanical flavor [Harris and Hemmerling, 1955]. For instance, biochemists talk about osmotic stress, and textbooks dealing with the Principle of Le Châtelier describe the relief of molar stress introduced by sudden changes in external conditions. The other nomenclature uses the term *affinity* to denote essentially the negative of molar stress. This term was used by De Donder and his school in their classic analysis of chemical systems moving toward equilibrium [De Donder, 1922; Prigogine and Defay, 1954]. The choice of one nomenclature over the other is largely a matter of taste. We shall use the terms *molar stress* and *molar strain*, partly because the mechanical analogy is useful, and partly because the term *affinity* has other meanings in close-enough contexts to be ambiguous.*

Definition of Molar Strain and Molar Stress

In a generalized displacement or distortion from equilibrium, the *strain* is the amount of the distortion, and the *stress* is the active force, defined as the derivative of the work of distortion with respect to the strain [Harris and Hemmerling, 1955]. Thus, in the displacement of a chemical reaction from equilibrium in a closed system, the molar strain is the amount, $\alpha - \alpha_{eq}$, by which the progress variable α is displaced from equilibrium. (The term "molar" here means "relating to molecular species.") The work of distortion at constant T

***Webster's Dictionary* defines "chemical affinity" as "an attractive force exerted in different degrees between atoms, which causes them to enter into and remain in combination."

and P is $G(\alpha) - G(\alpha_{eq})$, and the molar stress is its derivative with respect to the strain:

$$\text{Molar stress} = [\partial[G(\alpha) - G(\alpha_{eq})]/\partial(\alpha - \alpha_{eq})] = (\partial G/\partial\alpha), \text{at constant } T, P, \{n\}$$

The molar stress simplifies to $\partial G/\partial\alpha$ because α_{eq} and $G(\alpha_{eq})$ are constant in a closed system at constant T and P.* The molar strain, work of distortion, and molar stress all vanish at equilibrium; in particular, the stress $(\partial G/\partial\alpha)$ vanishes as a consequence of the Second Law of Thermodynamics.

Properties of Molar Stress

Because of the importance of molar stress, we shall simplify the notation and let $(\partial G/\partial\alpha)_{T,P,\{n\}} = y$. Since G and α are functions of state, y is a function of state. The equation $y = \partial G/\partial\alpha$ puts y on a logical par with such functions of state as the volume $V = \partial G/\partial P$ or the entropy $S = -\partial G/\partial T$. In closed chemical systems not at constant T and P, the molar stress y is related to other thermodynamic functions. The relations of y to the Helmholtz free energy A, the enthalpy H, and the energy U are given in Eq. (4.1).

Alternative expressions for y; $\{n\}$ = constant

$$y = \left(\frac{\partial G}{\partial\alpha}\right)_{T,P} = \left(\frac{\partial A}{\partial\alpha}\right)_{T,V} = \left(\frac{\partial H}{\partial\alpha}\right)_{S,P} = \left(\frac{\partial U}{\partial\alpha}\right)_{S,V} \tag{4.1}$$

These relations are general for chemical systems.

PROOF. Because similar mathematics is used in all cases, it will be sufficient to prove that $(\partial A/\partial\alpha)_{T,V} = (\partial U/\partial\alpha)_{S,V}$:

$$A = U - TS, \qquad dU = T\,dS - P\,dV, \qquad \{n\} = \text{constant}$$

$$\left(\frac{\partial A}{\partial\alpha}\right)_{T,V} = \left(\frac{\partial U}{\partial\alpha}\right)_{T,V} - T\left(\frac{\partial S}{\partial\alpha}\right)_{T,V}$$

$$\left(\frac{\partial U}{\partial\alpha}\right)_{T,V} = \left(\frac{\partial U}{\partial\alpha}\right)_{S,V} + \left(\frac{\partial U}{\partial S}\right)_{\alpha,V}\left(\frac{\partial S}{\partial\alpha}\right)_{T,V}; \qquad \left(\frac{\partial U}{\partial S}\right)_{\alpha,V} = T$$

$$\therefore \left(\frac{\partial A}{\partial\alpha}\right)_{T,V} = \left(\frac{\partial U}{\partial\alpha}\right)_{S,V} \qquad \text{Q.E.D.}$$

*In the terminology of De Donder [1922] and Prigogine and Defay [1954], reaction progress is measured by n_{product}, and the affinity is $-(\partial G/\partial n_{\text{product}})_{n_1,n_2,\ldots,T,P}$, which is an intensive variable. Except for the use of n_{product} in place of α, the affinity is therefore the negative of the molar stress. In common with the molar stress, the affinity is zero at equilibrium.

The chemical statement of the Second Law, reviewed in Chapter 2, implies that as equilibrium is reached in a closed system, G goes to a minimum at constant T and P. Similar proofs show that A in a closed system goes to a minimum as equilibrium is reached at constant T and V; H goes to a minimum at constant S and P; and U goes to a minimum at constant S and V. Thus, according to Eq. (4.1), the molar stress y goes to zero at equilibrium not only at constant T and P, but under virtually any combination of external conditions—at constant T and P, at constant T and V, at constant S and P, and at constant S and V. Moreover, *y remains constant at zero* in any change of state in which chemical equilibrium is maintained, under any of these external conditions. This certainty, that at equilibrium $y = 0$, makes the molar stress y a uniquely useful indicator for the existence and maintenance of chemical equilibrium. We shall use it often.

STRESS–STRAIN RELATIONSHIPS IN DILUTE SOLUTIONS

To illustrate the calculation of molar stress as a function of molar strain, we will use dilute solutions at constant T and P, exemplified by the composition tree (T-4.1). The solvent is a single molecular species, but the solute consists of the isomeric species U and V. The stress–strain relationships involve departures of U and V from equilibrium.

Composition Tree (T-4.1)

Solvent $(1, n_1, m_1)$	Dilute solute $(2, n_2, m_2)$	
$\mathrm{A}\,(n_\mathrm{A}, m_\mathrm{A})$	$\mathrm{U}\,(n_\mathrm{U}, m_\mathrm{U})$	$\mathrm{V}\,(n_\mathrm{V}, m_\mathrm{V})$
$m_\mathrm{A} = m_1$	$m_\mathrm{U} = (1-\alpha)\,m_2$	$m_\mathrm{V} = \alpha m_2$

T and P are constant; the V/U equilibrium constant K equals $\alpha_\mathrm{eq}/(1-\alpha_\mathrm{eq})$

Except for the trivial change in symbols for molecular species, composition tree (T-4.1) is equivalent to (T-2.1). The representation of the free energy G by Eq. (4.2) thus requires no comment. Equation (4.3) follows the format for dilute solutes (Table 3.2), but it raises the question whether the *standard* partial free energies, G^0_U and G^0_V, remain constant in departures from equilibrium. The answer is "yes," because standard partial free energies are independent of concentration in dilute solutions and, fundamentally, the difference between chemical systems at equilibrium and off-equilibrium is not a difference of kind.

T, P = constant in Eqs. (4.2)–(4.4)

$$G = n_1 G_\mathrm{A} + n_2(1-\alpha)G_\mathrm{U} + n_2\alpha G_\mathrm{V} \tag{4.2}$$

$$G_\mathrm{U} = G^0_\mathrm{U} + RT\ \ln\ m_\mathrm{U} = G^0_\mathrm{U} + RT\ \ln(1-\alpha)m_2$$

$$G_V = G_V^0 + RT \ln m_V = G_V^0 + RT \ln \alpha m_2 \tag{4.3}$$

Recalling that the Euler and Gibbs–Duhem equations apply to homogeneous phases regardless of the existence of chemical equilibrium, Eqs. (4.4) for the molar stress then follow in a straightforward way.

$$\begin{aligned}(\partial G/\partial \alpha)_{n_1,n_2} &= y = n_2(G_V - G_U) \\ &= n_2[G_V^0 - G_U^0 + RT \ln(\alpha/[1-\alpha])] & (4.4a) \\ y/n_2 &= RT[\ln(\alpha/[1-\alpha]) - \ln K] & (4.4b) \\ y/n_2 &= RT[\ln(\alpha/[1-\alpha]) - \ln(\alpha_{eq}/[1-\alpha_{eq}])] & (4.4c)\end{aligned}$$

The second and third derivatives of G with respect to α at constant T and P follow by straightforward mathematics and are given by Eqs. (4.5) and (4.6). Note that at constant $\{n_1, n_2, T, P\}$, both α_{eq} and K are constant.

$$\left(\frac{\partial y}{\partial \alpha}\right)_{T,P,n_1,n_2} = \left(\frac{\partial^2 G}{\partial \alpha^2}\right)_{T,P,n_1,n_2} = \frac{n_2 RT}{\alpha(1-\alpha)} \tag{4.5}$$

$$\left(\frac{\partial^2 y}{\partial \alpha^2}\right)_{T,P,n_1,n_2} = \left(\frac{\partial^3 G}{\partial \alpha^3}\right)_{T,P,n_1,n_2} = \frac{n_2 RT(1-2\alpha)}{\alpha^2(1-\alpha)^2} \tag{4.6}$$

Figure 4.2 plots y/n_2, calculated from Eq. (4.4c), versus the absolute and fractional strain, letting $\alpha_{eq} = 0.1$, or 0.5, respectively. [The fractional strain $(\alpha - \alpha_{eq})/\alpha_{eq}$ is a useful index when α_{eq} is small or when α is not normalized from 0 to 1.] Note the magnitudes. Note also that the molar stress is nearly proportional to the molar strain, up to fractional strains of at least 0.2. This near-proportionality is rationalized by developing y/n_2 as a power-series expansion in $\alpha - \alpha_{eq}$ about the equilibrium point [Eq. (4.7)]; the slope is $RT/[\alpha_{eq}(1 - \alpha_{eq})]$.

$$\frac{y}{n_2} = \frac{RT(\alpha - \alpha_{eq})}{\alpha_{eq}(1-\alpha_{eq})}\left[1 - \frac{(\alpha - \alpha_{eq})(1 - 2\alpha_{eq})}{2\alpha_{eq}(1-\alpha_{eq})} + \cdots\right] \tag{4.7}$$

SELF-CONTAINED SETS OF ISOMERS

A component or molecular species is *self-contained*, on the observational time scale, if it is chemically inert toward all other components and toward all molecular species not contained within itself. For example, in (T-4.1) the dilute solute 2 is a self-contained component consisting of the isomers U and V. The solvent A is a single unreactive molecular species. But if appropriate, A may be subdivided into subspecies that may be as fine-grained as the populations of

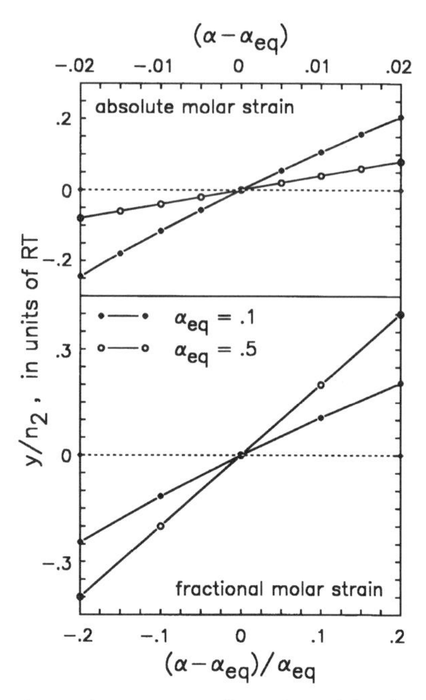

Figure 4.2. Plots of y/n_2, the molar stress per formula weight of solute, versus the absolute (*top*) and fractional molar strain. [Calculations are based on Eq. (4.4c).]

individual energy states. When the subdivision is coarse, as for solute 2 and its two isomers, one can usually find out experimentally whether isomer equilibrium exists. On the other hand, when the subdivision is fine, some of the subspecies may only have ideal (rather than real) distinguishability, and their equilibrium status cannot be observed.

Yet fine subdivisions—unto the populations of individual energy states—are important. They are the be-all and end-all of statistical thermodynamics and, through theorems of detailed balance and microscopic reversibility, exist at equilibrium whenever the macroscopic system is at equilibrium. In chemical kinetics, where the macroscopic system is off equilibrium, the key species are mechanically unstable transition states whose energy-state populations, owing to the overall imbalance, are more or less off equilibrium.

In this section we shall calculate the work of distortion—that is, the increase in the Gibbs free energy at constant T and P when the isomeric species or subspecies or energy-state populations, in a self-contained component or molecular species, deviate from equilibrium. Thermodynamic error tolerance will be significant. The relevant branch of the composition tree is (T-4.2). The self-contained parent, P, is a component or molecular species. P is subdivided into an arbitrary number of molecular species or subspecies a, b, ..., j, ..., z. The progress variables are α at equilibrium and α' off equilibrium. The stoichiometric constraints are given in Eqs. (4.8).

Composition Tree (T-4.2)

Self-contained parent (P)

At equil.:	$a(\alpha_a)$	$b(\alpha_b)$	$\cdots$	$j(\alpha_j)$	$\cdots$	$z(\alpha_z)$
Off equil.:	$a(\alpha'_a)$	$b(\alpha'_b)$	$\cdots$	$j(\alpha'_j)$	$\cdots$	$z(\alpha'_z)$

Absolute strain: $\rho_j = \alpha'_j - \alpha_j$

Relative strain: $\sigma_j = \rho_j/\alpha_j$

Stoichiometric constraints: $$\sum \alpha_j = 1, \quad \sum \alpha'_j = 1, \quad \sum \rho_j = 0 \tag{4.8}$$

The calculation of the work of distortion per mole of parent P, when the progress variables are strained from α to α', will be based on the Stability Theorem (4.9a) and its statistical form (4.9b). At equilibrium, the α's are such that G_P^0 is at a minimum. But as α is strained to α', substitution of the strained value yields the distorted free energy, G_P^0 (strain), which is greater than G_P^0 [Eq. (4.9c)].

Stability Theorem, at equil. : $$G_P^0 = G_j^0 + RT \ln \alpha_j \quad \text{for any j} \tag{4.9a}$$

Statistical form, at equil. : $$G_P^0 = \sum \alpha_j G_j^0 + RT \sum \alpha_j \ln \alpha_j \tag{4.9b}$$

$$G_P^0(\text{strain}) = \sum \alpha'_j G_j^0 + RT \sum \alpha'_j \ln \alpha'_j \tag{4.9c}$$

The work of distortion per mole, W(strain), is the difference between G_P^0(strain) and G_P^0, as given in Eqs. (4.9). The resulting expression then is cast in a useful form by introducing the variables ρ_j and σ_j defined in (T-4.2), and noting that $\sum \rho_j = 0$.

$$\begin{aligned} W(\text{strain}) &= G_P^0(\text{strain}) - G_P^0 \\ &= \sum \rho_j(G_j^0 + RT \ln \alpha_j) + RT \sum \alpha'_j \ln(\alpha'_j/\alpha_j) \\ &= G_P^0 \sum \rho_j + RT \sum \alpha_j(1 + \sigma_j) \ln(1 + \sigma_j) \\ &= RT/2 \sum \alpha_j \sigma_j^2 - RT/6 \sum \alpha_j \sigma_j^3 + \cdots \end{aligned} \tag{4.10}$$

Note that $\sum \alpha_j \sigma_j^2$ is the α_j-weighted mean square of the fractional strain. This average includes all species and subspecies in the parent P. However, relatively unstable subspecies carry little weight since α_j decreases exponentially with increasing G_j^0.

In the following it will be convenient to introduce the root-mean-square fractional strain, σ_{rms}, defined by

$$\sigma_{\mathrm{rms}}^2 = \sum \alpha_j \sigma_j^2$$

When $\sigma_{\mathrm{rms}} < 0.3$, the cubic and higher-order terms in Eq. (4.10) amount to less than 10% of W(strain) and will be neglected, to give the working approximation (4.11):

$$W(\text{strain}) \approx (RT/2)\sigma_{\mathrm{rms}}^2, \qquad \sigma_{\mathrm{rms}} < 0.3 \tag{4.11}$$

Equation (4.11) expresses error tolerance because W(strain) is proportional to the *square* of a *fractional* strain that normally, unless deliberately imposed, would be small compared to 1. Equation (4.11) is an adequate approximation when $\sigma_{\mathrm{rms}} < 0.3$, which represents quite a large inadvertent distortion from equilibrium. Yet when $\sigma_{\mathrm{rms}} = 0.3$, the resulting strain energy W(strain) at ordinary temperatures is only 100 J/mol, and $\exp[-W(\text{strain})/RT]$ is 0.05. If 100 J/mol were the error, from whatever cause, in the energy of a reaction product or transition state, the resulting error in the equilibrium or rate constant would be less than 5%, which is tolerable for most purposes.

A Numerical Example

To illustrate the molar strain tolerance of a self-contained component, we shall consider a real equilibrium. 1,2-Dimethylhexahydropyridazine (DHP) can consist of three conformational isomers, shown in (4.12), because the individual N-methyl groups can be either equatorial (e) or axial (a).

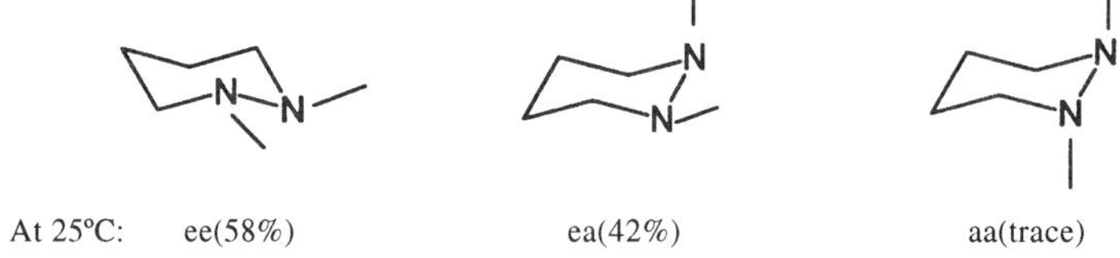

At 25°C: ee(58%) ea(42%) aa(trace) (4.12)

On the basis of nuclear magnetic resonance experiments in acetone-d_6, the equilibrium mixture at 25°C consists of 58% ee, 42% ea, and a trace of aa, too small to be seen in the spectrum but apparent as an intermediate in the ee $\rightleftarrows$ ea interconversion. The half-time for ee $\rightleftarrows$ ea interconversion is 2 μs at 25°C [Nelsen and Weisman, 1976]. The presence of the trace of aa will be neglected in the following calculations.

In Table 4.1 we are mentally stopping the interconversion at various off-equilibrium values of the progress variable α'_{ee}. The calculations of σ_{rms}, W(strain), and percent error (defined below) span a broad range of α'_{ee}, enough to surpass even overestimates of the experimental error. The column labeled "percent

Table 4.1. Strain properties in the ee $\rightleftarrows$ ea isomerization of 1,2-dimethylhexahydropyridazine if ee and ea were off equilibrium

α'_{ee}	σ_{rms}	W(strain)[a]	Percent Error[b]
0.8	0.447	248	−10.3
0.7	0.241	72	−3.1
0.58[c]	0.000	0	0.0
0.5	0.161	32	−1.3
0.4	0.366	166	−6.3
0.3	0.566	396	−14.8

[a]J/mol at 298 K [Eq. (4.11)].
[b]See text.
[c]Equilibrium, α_{ee}.

error" refers to the error that would be made in an equilibrium constant involving DHP as a reaction product—if the given molar strain were neglected! An example might be the acid dissociation of $(DH_2P)^+$:

$$(DH_2P)^+ = H^+ + \alpha_{ee}DHP(ee) + \alpha_{ea}DHP(ea)$$

The calculations in Table 4.1 show that the strain in α_{ee} may be as high as 0.15, or 30% of α_{ee} at equilibrium, without exceeding a 5% error tolerance. The corresponding relative strain for the ee and ea isomers is about 0.3. The very magnitude of these figures allows us to conclude that molar strains arising inadvertently due to incomplete equilibration, or molar strains arising naturally in chemical kinetics, are normally quite tolerable.

THE TOLERANCE THEOREM

The Tolerance Theorem deals with the effect of departures from equilibrium on the partial free energies of the components. It may be stated as follows:

Tolerance Theorem. For any ith component in a closed homogeneous phase at constant T and P and for any independent reaction in the phase, the partial free energy G_i is insensitive to deviations of the progress variable α and the molar stress y from equilibrium, as stated in Eq. (4.13):

$$\text{Strain tolerance:} \quad (\partial G_i/\partial \alpha)_{T,P,\{n\},y=0} = 0 \tag{4.13a}$$

$$\text{Stress tolerance:} \quad (\partial G_i/\partial y)_{T,P,\{n\},y=0} = 0 \tag{4.13b}$$

Note that the reaction is at equilibrium when the partial derivatives vanish. The reaction may be of any type—an interconversion of isomers within the ith component, a reaction of the ith component with other components, or a reaction not involving the ith component. The simultaneous presence of other reactions does not matter.

Error tolerance with respect to chemical reactions implies that chemical complexities—the richness of chemistry intrinsic to components—may often be neglected in free-energy formulations. Traditional chemical thermodynamics does just that, because it strives to be a macroscopic science and thus curtails the intrusion of molecular theory. But even though this tactic is protected by the Tolerance Theorem for partial free energies, it is risky for unprotected partial properties—for example, for partial entropies and enthalpies of formal components. Typically, chemical phenomena, such as molar shifts and changes in neighbor environments, then make significant contributions.

PROOF OF EQ. (4.13). The trick in proving theorems for functions of many variables is to identify the active variables, create a memo of the inactive variables, and then proceed with a notation based solely on the active variables. The following proof of the Tolerance Theorem will consist of three parts. In the first part we consider a single reaction with progress variable α and molar stress y, in a homogeneous phase at constant T and P. The proof here will be graphical. In the second part the graphical proof becomes analytical, and in the third part the analytical proof for a single reaction is generalized. Throughout the proof we assume the presence of an arbitrary number of components, among which we single out component 2 for action. That is, we derive error tolerance for the partial free energy G_2. Since $G_2 = \partial G/\partial n_2$, one of the active variables evidently is n_2. When there is a single reaction, either α or y may serve as the second active variable. There are no others.

Graphical Proof. This proof is based on the mathematical identity $(\partial^2 G/\partial n_2 \partial\alpha)_{\text{eq}} = (\partial^2 G/\partial\alpha\partial n_2)_{\text{eq}}$. The left-hand side of Eq. (4.13a) is therefore equal to $(\partial y/\partial n_2)_{\text{eq}}$, evaluated at constant $\{T, P, (n \neq n_2)\}$. Now consider a homogeneous phase at constant $\{T, P, (n \neq n_2)\}$ and allow n_2 to vary. As the phase reaches equilibrium, $y \to 0$, and this consequence of the Second Law is true *at any composition*. The resulting plot of $(\partial G/\partial\alpha)_{\text{eq}}$ versus n_2 is shown in Fig. 4.3. It is obvious that $(\partial y/\partial n_2)_{\text{eq}}$, and hence $(\partial G_2/\partial\alpha)_{\text{eq}}$, must be zero. Q.E.D.

Analytical Proof. The analytical proof is worth giving because it shows the breadth of the Tolerance Theorem, even though the procedure involves a Legendre transformation (see Chapter 2) and is slightly circuitous. It consists of eight lines and commentary. Line 1 records the inactive variables; the active variables again are n_2 and either the progress variable α or the molar

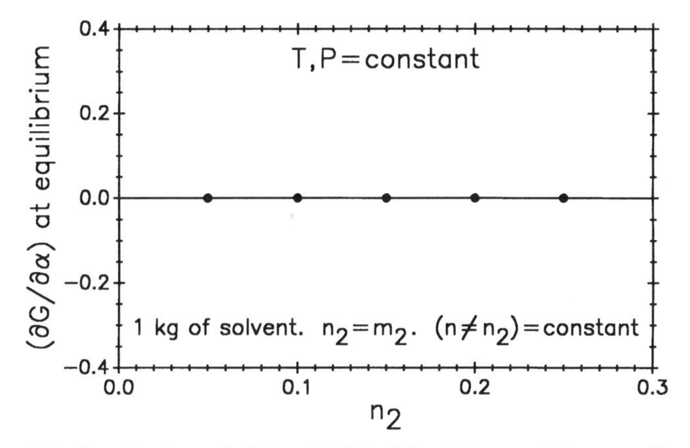

Figure 4.3. Graphical proof of Eq. (4.13a) of the Tolerance Theorem. See text.

stress y. $\{n\}$ denotes the full set of formula-weight numbers, and $\{n \neq n_2\}$ denotes the full set except for n_2.

Line 2 expresses G as a function of n_2 and α. Line 3 rewrites line 2 in terms of the molar stress y and introduces a function $(G_2)_\alpha$, whose definition is given.

1. Memo: T, P, and $\{n \neq n_2\}$ are constant throughout.
2. $dG = (\partial G/\partial n_2)_\alpha \, dn_2 + (\partial G/\partial \alpha)_{n_2} \, d\alpha$.
3. $dG = (G_2)_\alpha \, dn_2 + y \, d\alpha; \qquad (G_2)_\alpha = (\partial G/\partial n_2)_\alpha$.

Line 4 transforms variables to (n_2, y) and uses the shortcut recipe, described before Eq. (2.11), to evaluate the function $(G_2)_y = (\partial G/\partial n_2)_y$.

Line 5 differentiates line 4 with respect to y, thus entering the domain of Eq. (4.13).

4. $$(G_2)_y = \left(\frac{\partial G}{\partial n_2}\right)_y = (G_2)_\alpha + y\left(\frac{\partial \alpha}{\partial n_2}\right)_y.$$

5. $$\left(\frac{\partial (G_2)_y}{\partial y}\right)_{n_2} = \left(\frac{\partial (G_2)_\alpha}{\partial y}\right)_{n_2} + \left(\frac{\partial \alpha}{\partial n_2}\right)_y + y\left(\frac{\partial^2 \alpha}{\partial n_2 \partial y}\right).$$

Unfortunately, line 5 expresses $\partial (G_2)_y/\partial y$ as a sum of three terms, of which only the third is proportional to y and vanishes at equilibrium. But fortunately, the first two terms cancel. To prove this, we introduce a Legendre transformation.

Let $J = G - y\alpha$. This operation is similar to the familiar operation, $G = A + PV$, since $y = \partial G/\partial \alpha$, and $P = -\partial A/\partial V$. Since G, y, and α are functions of state, J is a function of state. Hence $dJ = dG - y \, d\alpha - \alpha \, dy$. Now substitute line 3: $dJ = (G_2)_\alpha\, dn_2 + y\, d\alpha - y\, d\alpha - \alpha\, dy = (G_2)_\alpha\, dn_2 - \alpha\, dy$. It follows that

$$(G_2)_\alpha = (\partial J/\partial n_2)_y \quad \text{and} \quad \alpha = -(\partial J/\partial y)_{n_2}$$

On equating $(\partial^2 J/\partial n_2 \partial y)$ to $(\partial^2 J/\partial y \partial n_2)$, we obtain the desired identity 6.

6. $$-\left(\frac{\partial \alpha}{\partial n_2}\right)_y = \left(\frac{\partial (G_2)_\alpha}{\partial y}\right)_{n_2}.$$ Q.E.D.

To recover Eq. (4.13b) from line 5, we note that G_2 is the limit of $(G_2)_y$ as $y \to 0$. Hence

7. $$\left(\frac{\partial G_2}{\partial y}\right)_{n_2} = \lim_{y \to 0}\left(\frac{\partial (G_2)_y}{\partial y}\right)_{n_2} = \lim_{y \to 0}\left[y\left(\frac{\partial^2 \alpha}{\partial n_2 \partial y}\right)\right].$$

Since $y \to 0$, line 7 reduces to Eq. (4.13b) if (i) $(\partial^2 \alpha/\partial n_2 \partial y)$ is finite as $y \to 0$ or (ii) $(\partial^2 \alpha/\partial n_2 \partial y)$ becomes infinite but its product with y goes to zero. We will state without proof that case (i) obtains in all physically realized cases, since n_2 must be finite. Moreover, case (ii) obtains in the limit as y goes to zero and n_2 goes to zero. Equation (4.13b) is therefore general. Since n_2 is constant in line 7, and since T, P, and $\{n \neq n_2\}$ are constant in all lines, the error-tolerant system is a closed system at constant T and P. Note also that there are no restrictions on the nature of the reaction.

In order to prove strain tolerance as in Eq. (4.13a), we note that at constant T, P, $\{n\}$ we obtain the following:

8. $$(\partial G_2/\partial \alpha)_{y=0} = [(\partial G_2/\partial y)(\partial y/\partial \alpha)]_{y=0}.$$

$\partial G_2/\partial y = 0$, according to line 7, and $\partial y/\partial \alpha$ is finite. (See, for example, Fig. 4.2.) Hence $(\partial G_2/\partial \alpha)_{y=0} = 0$. Q.E.D.

General Corollary. To expand the analytical proof from a single reaction to a set of reactions, the first step is to classify the reactions on the basis of half-life. On the given observational time scale, some reactions go to equilibrium while others are "stuck" or mired off equilibrium. The Tolerance Theorem applies at equilibrium, so we may ignore the reactions that are stuck off equilibrium.

Let there be r independent reactions that *do* reach equilibrium, and let us consider the tolerance relationship for the kth reaction. The active variables will be n_2 and either α_k or y_k. The inactive variables will be T, P, $\{n \neq n_2\}$, and the molar stresses $y_1, y_2, \ldots, y_{k-1}, y_{k+1}, \ldots, y_r$ of the $r - 1$ independent reactions that *did* reach equilibrium. These molar stresses will be zero. To derive the two forms of the Tolerance Theorem, namely $\partial G_2/\partial y_k = 0$, and $\partial G_2/\partial \alpha_k = 0$, we enlarge the Memo in line 1 of the preceding proof by adding that $y_1, y_2, \ldots, y_{k-1}, y_{k+1}, \ldots, y_r = 0$. Then we proceed through lines 2 to 8 as before. This procedure will work for any of the r independent reactions.

Q.E.D.

Maximum or Minimum?

As shown in Fig. 4.3, the Tolerance Theorem derives from the fact that $\partial G/\partial \alpha$ vanishes at equilibrium at *any* formal composition. The result that $\partial G_i/\partial \alpha = 0$ then follows purely by mathematics and should be taken at mathematical face value. That is, when $\alpha = \alpha_{eq}$, the partial free energy G_i may be a minimum, a maximum, or simply a point on a horizontal line.

To find out, we need the *curvature* $(\partial^2 G_i/\partial \alpha^2)$ of the plot of G_i versus α at the equilibrium point α_{eq}. Since $(\partial^2 G_i/\partial \alpha^2) = [\partial(\partial^2 G/\partial \alpha^2)/\partial n_i]$, we shall plot $(\partial^2 G/\partial \alpha^2)$ versus n_i, at constant T, P, and $\{n \neq n_i\}$. If $(\partial^2 G/\partial \alpha^2)$ increases with n_i, then G_i is a minimum. If it decreases, then G_i is a maximum. And if the slope is zero, then G_i versus α is a horizontal line.

Figure 4.4 shows such plots for two reactions: (i) the isomerization $U \rightleftarrows V$ in composition tree (T-4.1) and (ii) the reaction $X + Y \rightleftarrows P + Q$ in composition tree (T-4.3). In both cases, the amount of solvent is constant at 1 kg, so that the formula-weight number, n_i, is equal to the formality m_i.

For the isomerization $U = V$, the derivation of $(\partial^2 G/\partial \alpha^2)$ has been given in Eq. (4.5). We shall let $K = 1.5$. The resulting plot of $(\partial^2 G/\partial \alpha^2)$ versus m_2 appears on the right-hand side of Fig. 4.4. Since the slope is positive, G_2 is a minimum.

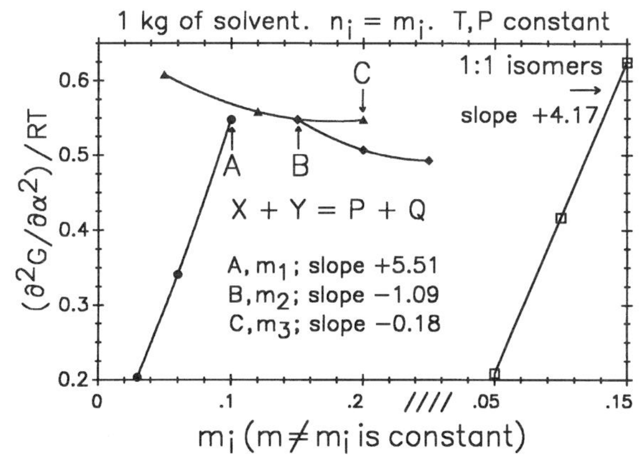

Figure 4.4. Graphical characterization of the extremum in Eq. (4.13a). A positive slope indicates a minimum. See text.

Composition Tree (T-4.3): X + Y = P + Q

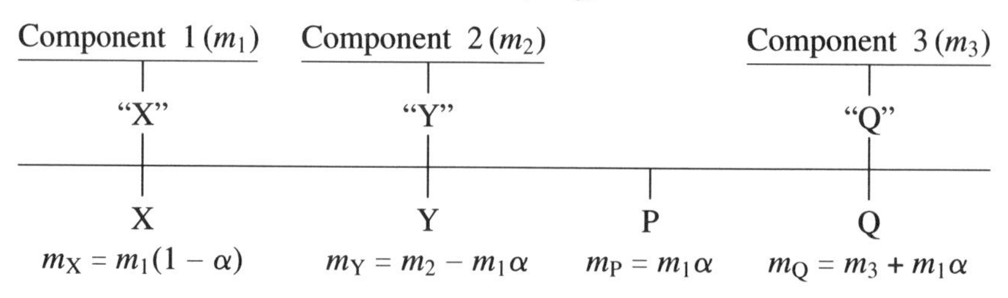

In (T-4.3), the reaction X + Y = P + Q involves three components, whose respective formalities are m_1, m_2, and m_3, with m_1 being the limiting amount. The expression for G as a function of G_X, G_Y, G_P, G_Q, and the molar concentrations resembles Eq. (4.2), except for being longer, and the subsequent derivation of $(\partial^2 G/\partial\alpha^2)$ follows the precepts of Eqs. (4.3)–(4.5). In calculating α_{eq} we assume that $K = 1.5$. In curve A, $m_2 = 0.15$, $m_3 = 0.2$, and m_1 is variable. The positive slope shows that G_1 is a minimum at α_{eq}. In curve B, $m_1 = 0.1$, $m_3 = 0.2$, and m_2 is variable. The negative slope shows that G_2 is a maximum. In curve C, $m_1 = 0.1$, $m_2 = 0.15$, and m_3 is variable. The negative slope shows that G_3 is a maximum. If there were also an unreactive component 4, α_{eq} would be independent of m_4 and the slope of $\partial^2 G/\partial\alpha^2$ versus m_4 would be zero. The preceding examples show that at equilibrium, G_i for any ith component may be a maximum, a minimum, or a point on a level line, depending on the formal composition and the stoichiometry of the reaction.

When Euler's equation was introduced in Chapter 2, we noted that partial properties are indispensable in chemical thermodynamics because of their mathematical (as opposed to physical) analogy to properties of formal components. The present results emphasize the absence of a close physical analogy. When a reaction (such as 2HOAc $\rightarrow$ $(HOAc)_2$) goes to equilibrium in a formal component (such as 1 formula weight of glacial acetic acid), $\partial G/\partial\alpha$ goes to zero, just like $\partial G_i/\partial\alpha$ in homogeneous solution, but the extremum of $\partial G/\partial\alpha$ is governed by the Second Law and must be a minimum.

Theorem for Standard Partial Free Energies

Noting once again that the Tolerance Theorem applies to partial free energies of components rather than of molecular species, we shall now extend it to *standard* partial free energies.

By definition, the formal concentrations m_1, $m_2, \ldots$ of the components in a closed system are constant. But when a chemical reaction goes to equilibrium in a closed system, the *molar* concentrations of the molecular species go to their equilibrium values and the progress variable α goes to α_{eq}. As a result, the microscopic composition changes and the change in α in principle produces a medium effect.

As depicted in Fig. 3.4 in Chapter 3, there are two conventions for handling a medium effect on the partial free energy of a component. In one convention we write

$$G_i = G_i^0(\text{medium}) + RT \ln m_i,$$

that is, the standard partial free energy G_i^0 is a variable, being a function of the medium. In the other convention we write

$$G_i = G_i^0 + RT \ln m_i + RT \ln \gamma_i.$$

Here, G_i^0 is constant and the variable term $RT \ln \gamma_i$ accounts for the medium effect.

In the following derivation we shall adopt the convention in which G_i^0 is a function of the medium. For a closed system in which a reaction is going to equilibrium at constant T and P, $G_i^0(\text{medium})$ then is a function of α:

$$G_i = G_i^0(\alpha) + RT \ln m_i.$$

Because the system is closed, the formal composition, including m_i, is constant. Hence $\partial \ln m_i/\partial\alpha = \partial \ln m_i/\partial y = 0$. The Tolerance Theorem (4.14) for G_i^0 then follows directly from that [Eq. (4.13)] for G_i:

Tolerance Theorem for G_i^0

For a closed system (constant formal composition) at equilibrium ($y = 0$) at constant T and P

$$\left(\frac{\partial G_i^0(\alpha)}{\partial\alpha}\right) = \left(\frac{\partial G_i}{\partial\alpha}\right) - RT\left(\frac{\partial \ln m_i}{\partial\alpha}\right)_{m_i} = 0 \qquad (4.14a)$$

Similarly,

$$\left(\frac{\partial G_i^0(\alpha)}{\partial y}\right) = \left(\frac{\partial G_i}{\partial y}\right) = 0 \qquad (4.14b)$$

Although Eqs. 4.14 are general, we shall apply them mostly to solutes in dilute solutions. In that case, these equations imply that *standard* partial free energies of dilute solute components are not sensitive to minor (or even moderate) deviations from chemical equilibrium in the solution.

5

LINEAR FREE-ENERGY RELATIONSHIPS, TRANSITION-STATE THEORY, AND ERROR TOLERANCE

The topics of this chapter—linear free-energy relationships and transition-state theory—are "extrathermodynamic" (outside the domain of traditional thermodynamics), yet they are plainly relevant to the reaction thermodynamics of molecular species. Linear free-energy relationships are manifestations of the Principle of Analogy, which, ranging from the sweeping generality of the Periodic Law to specific parallels within classes of reactions, is the chemist's chief guide for predicting reactivity. The transition-state theory builds a bridge between thermodynamic and kinetic properties of molecular species, and extends the use of linear free-energy relationships from equilibrium constants to rate constants. And thermodynamic error tolerance allows the linear free-energy relationships to be more accurate and more broadly applicable than they otherwise would be.

Figure 5.1 shows experimental ln K versus ln K equilibrium-constant relationships for the addition of a series of nucleophiles, such as methylamine, to several aldehyde substrates [Sander and Jencks, 1968]:

$$CH_3NH_2 + RC(H){=}O \rightleftharpoons RC(H)(OH)(NHCH_3)$$

The nucleophiles comprise quite a variety: amines, alcohols, amino acids, peroxides, and thiols. In the Fig. 5.1a the variation of ln K with the nucleophile is compared for formaldehyde (HCHO) and pyridine-4-carboxaldehyde

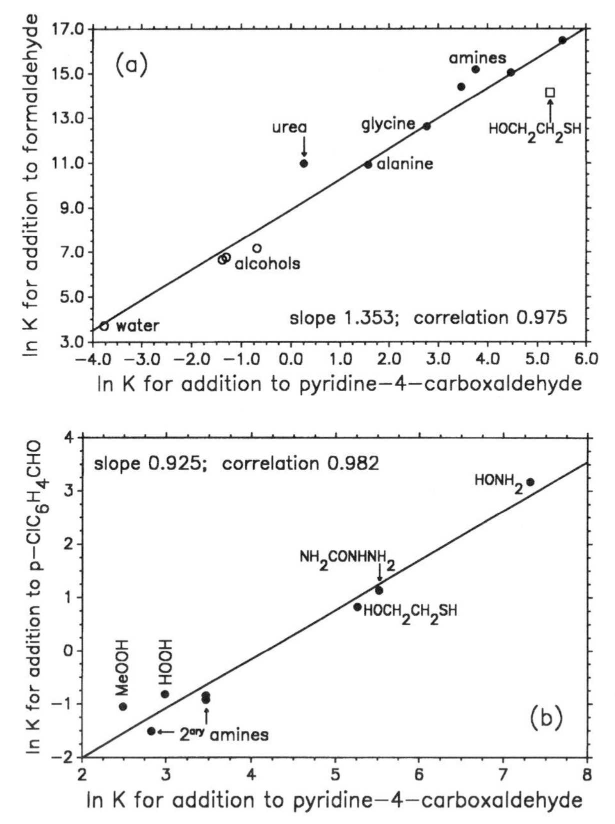

Figure 5.1. Equilibrium-constant relationships in water at 25°C, for addition of nucleophiles to aldehydes [Sander and Jencks, 1968].

(4-PyCHO, Py = $—C_5H_4N$). Figure 5.1b shows a similar comparison for *p*-chlorobenzaldehyde (*p*-ClC_6H_4CHO). In both figures the data show linear correlations with respectable correlation coefficients. Since $RT \ln K = -\Delta G^0$, relationships such as these are called *linear free-energy relationships*. Note, however, that the linearity is approximate and statistical, rather than strict and by mathematical law.

REACTION SERIES AND FREE-ENERGY PERTURBATIONS

In a *reaction series* the nominal reaction is constant, and a fixed substrate reacts with a series of functionally related reactants or in a series of solvents. In a *free-energy comparison*, standard free energies of reaction are compared for two reaction series. The comparison may involve the same nominal reaction or different nominal reactions, the same substrate and solvent or different substrates and solvents, the same set of reactants, or corresponding structural changes in different reactants. If the plotted points show a linear correlation with a respectable correlation coefficient, a linear free-energy relationship is said to exist.

Experience shows that linearity is most likely when the structural changes—the substituents introduced into the reactants—occur far enough from the reaction zone so that direct steric encroachments upon the reaction zone are avoided. The effects of the substituents may then be regarded as physical perturbations, such as electric fields that polarize the reaction zone or as perturbations of the molecular wave functions that modify the electron density in the reaction zone.

Accordingly, the notation used to describe free-energy relationships uses two kinds of mathematical operators. The familiar reaction operator Δ (as in ΔG^0) denotes a change in the reaction zone that converts reactants to products, while the perturbation operator δ denotes a change due to an added substituent (δ_R) or a change of solvent (δ_M). For example, $\delta_R(\Delta G^0)$ or $\delta_R \Delta G^0$ (without the parentheses) denotes a substituent effect on ΔG^0, and $\Delta(\delta_M G^0)$ or $\Delta\delta_M G^0$ denotes the effect of a reaction on $\delta_M G^0$.

The reaction operator (Δ) and the perturbation operators (δ_R, δ_M) commute. For example, $\delta_R(\Delta G) = \Delta(\delta_R G)$, and $\delta_M(\delta_R G) = \delta_R(\delta_M G)$. This property follows from simple arithmetic since the Δ and δ operations are subtractions.

PROOF. It will be sufficient to show that $\delta_R(\Delta G) = \Delta(\delta_R G)$. Let $A \rightleftarrows B$ denote the original reaction, and let $A' \rightleftarrows B'$ be the reaction in the presence of a substituent. The required operations then are as follows:

$$\Delta G = G_B - G_A, \qquad \Delta G' = G'_B - G'_A, \qquad \delta_R G_A = G'_A - G_A$$
$$\delta_R G_B = G'_B - G_B, \qquad \delta_R(\Delta G) = \Delta G' - \Delta G$$

Therefore

$$\begin{aligned}\delta_R(\Delta G) &= (G'_B - G'_A) - (G_B - G_A) = (G'_B - G_B) - (G'_A - G_A) \\ &= \delta_R G_B - \delta_R G_A = \Delta(\delta_R G)\end{aligned} \qquad \text{Q.E.D.}$$

The plots shown in Fig. 5.2 use the $\delta\Delta$ operators. Figure 5.2a compares the effects of *m*- and *p*-substituents on ΔG^0 for benzoic acid dissociation [McDaniel and Brown, 1958] with the effects of the same substituents on ΔG^0 for formanilide formation [Davis, 1912]:

$$m\text{- or } p\text{-}XC_6H_4COOH + HOH \rightarrow m\text{- or } p\text{-}XC_6H_4CO_2^- + H_3O^+$$

compared with

$$m\text{- or } p\text{-}XC_6H_4NH_2 + HCOOH \rightarrow m\text{- or } p\text{-}XC_6H_4\text{—N(H)—C(H)=O} + HOH$$

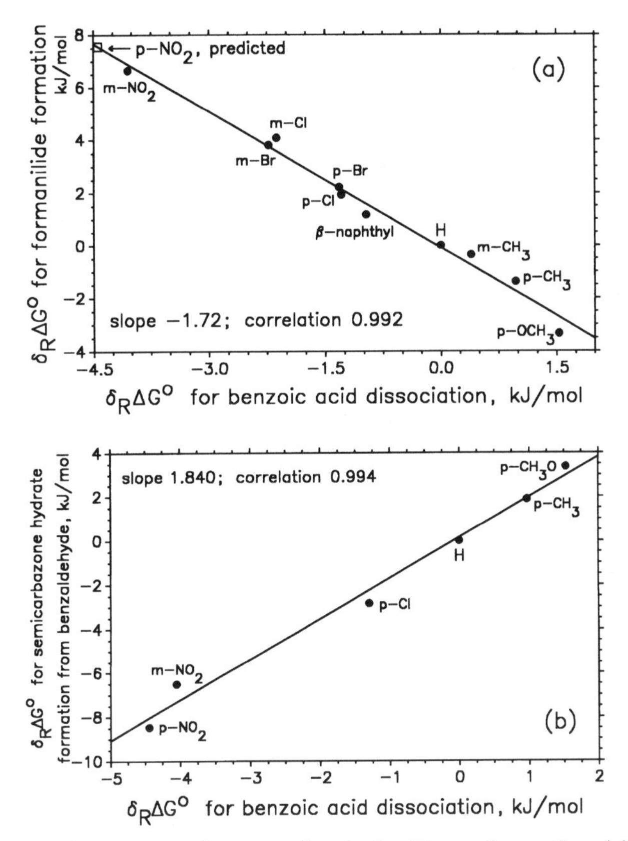

Figure 5.2. Linear free-energy plots according to the Hammett equation. (a) Formanilide formation in 67% pyridine–water at 100°C [Davis, 1912]. (b) Benzaldehyde semicarbazone hydrate formation in 25 vol% ethanol–water at 25°C [Anderson and Jencks, 1960].

Figure 5.2b compares benzoic acid dissociation with benzaldehyde semicarbazone hydrate formation [Anderson and Jencks, 1960]:

$$m\text{- or } p\text{-}XC_6H_4CHO + H_2NNHCONH_2 \rightarrow$$

$$m\text{- or } p\text{-}XC_6H_4\underset{}{\overset{\displaystyle OH \atop |}{C}H}\text{—}NHNHCONH_2$$

Both plots show good linear trends with correlation coefficients above 0.99. Note that the slope in Fig. 5.2a is negative, showing that the perturbations of ΔG^0 by *m*- and *p*-substituents in acid dissociation and formanilide formation are of opposite sign. Figure 5.2a shows also how the plot might be used for predictive purposes. Owing to limited solubility, the equilibrium for the formation of *p*-nitroformanilide could not be measured as part of the series. However, a short extrapolation to the abscissa for *p*-nitrobenzoic acid yields the prediction that $\delta_R \Delta G^0$ is 7.56 kJ/mol.

LINEAR FREE-ENERGY PLOTS INVOLVING RATE CONSTANTS

A strikingly important feature of linear free-energy relationships is that they apply to rate constants as well as to equilibrium constants. For example, Fig. 5.3 compares *m*- and *p*-substituent effects on the acid dissociation constant K of benzoic acid [McDaniel and Brown, 1958] with the effects of the same substituents on the second-order rate constants k for (a) semicarbazone hydrate formation from benzaldehyde and (b) hydrogen-ion-catalyzed dehydration of benzaldehyde semicarbazone hydrate [Anderson and Jencks, 1960].

Just as relationships involving equilibrium constants may be used to predict further equilibrium constants, relationships involving rate constants may be used to predict further rate constants. For example, the acid dissociation constant of

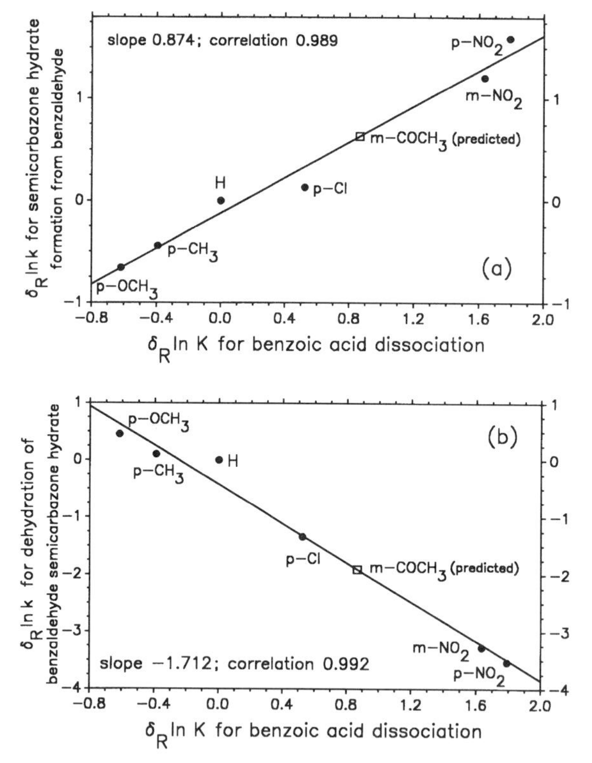

Figure 5.3. Hammett linear free-energy plots involving rate constants. (a) Kinetically second-order semicarbazone hydrate formation from benzaldehyde [Anderson and Jencks, 1960]. (b) Hydrogen-ion-catalyzed dehydration of benzaldehyde semicarbazone hydrate. Kinetic measurements in 25 vol% ethanol–water at 25°C [Anderson and Jencks, 1960].

m-CH_3CO–benzoic acid is known. When this information is entered on the plots of Fig. 5.3, the effect of an *m*-CH_3CO substituent on the rate constants can be predicted by linear interpolation. Direct measurement of these rate constants might be complicated since the *m*-CH_3CO substituent introduces a potentially interfering carbonyl group. But quite apart from complications: Being able to predict a needed reactivity semiquantitatively, from known reactivities by simple mathematics, can be a true blessing, and linear free-energy methods often fill that bill [Hammett, 1940, 1970; Leffler and Grunwald, 1989; Taft, 1956]. In fact, chemists and engineers have developed schemes for predicting equilibrium constants, rate constants, solubilities, distribution ratios, enzymatic activities, pesticidal and fungicidal activities, toxicity levels, and even sweetness and adhesion. Formally, these *structure–energy* and *structure–activity* correlations utilize linear free-energy approaches, but, owing to the complexity of the phenomena being described, the abscissas tend to be linear combinations of standard free-energy changes for model processes, together with physical *descriptors* such as molar volumes or spectroscopic transition energies [Abboud et al., 1981; Abraham, 1993; Shorter, 1973; Hansch and Leo, 1979; Austel, 1983; Fujita and Iwamura, 1983; Lipnick, 1989].

Returning to orthodox linear free-energy relationships such as those in Figs. 5.1–5.3, their chief application is in the field of reaction mechanisms. When the logarithmic plots of the rate constants for two reaction series are closely linear, the Principle of Analogy predicts a high probability that the underlying reaction mechanisms belong to the same category. This makes it possible to classify reactions into mechanistic categories and to characterize their transition states by category rather than individually, thus gaining coherent perspectives. It is clear, therefore, that linear free-energy relationships are an integral part of the study of reaction mechanisms [Hammett, 1940, 1970; Leffler and Grunwald, 1989; Taft, 1956]. In the rest of this section we shall give brief descriptions of three relationships whose importance is historical as well as intrinsic: the Hammett plot, which applies to substituents that are well separated from the reaction zone; the Brønsted catalysis law, which correlates the rate constants in general acid–base catalysis with the acid–base dissociation constants of the catalysts; and the correlation of solvolysis rates, which organizes the solvent effects in polar reactions.

The Hammett Plot

As shown in Figs. 5.2 and 5.3, Hammett plots compare the effects of *m*- and *p*-phenyl substituents in aromatic side-chain reactions. It had long been known that *m*- and *p*-substituents modify side-chain reactivity in a regular, well-behaved manner. Hammett [1935] reviewed the data and found linear free-energy relationships. He further introduced a consistent method of plotting so that the relationships might elucidate features of the reaction mechanism [Hammett, 1940, 1970]. As illustrated in Figs. 5.2 and 5.3, the y-axis of the Hammett plot gives the effect of *m*- and *p*-substituents in a reaction of choice, while the x-axis gives the effect of the same substituents in the acid dissociation of benzoic acid in water at 25°C. The analysis of the slopes of such plots, and of the deviations for specific substituents, has produced a wealth of inferences about the modes of interaction between the substituents and the well-separated reaction zone—that is, about the electric charges, charge densities, and degree of delocalization and anisotropy of the polarizabilities [Streitwieser, 1963; Dewar et al., 1971; Swain et al., 1983]. *Ortho*-substituents, which in most cases intrude sterically on the reaction zone, produce reactivities that almost always deviate from the linear relationships established by the *m*- and *p*-substituents. Judicious choice of reactivities other than benzoic acid dissociation, for plotting along the x-axis, has expanded the scope of the linear relationships to fit *m*- and *p*-effects in literally hundreds of benzenoid reactions, including electrophilic substitution on the benzene ring [McGary et al., 1955; Okamoto and Brown, 1957, 1958; Tsuno et al., 1959].

Brønsted Catalysis Law

The rate law for an acid- or base-catalyzed reaction often contains not only the familiar term for catalysis by hydrogen ion or hydroxide ion, but also additional terms proportional to the concentrations of weaker proton acids or bases that might be present in the solution. For example, the acid-catalyzed dehydration (5.1a) of 1,1-ethanediol (acetaldehyde hydrate) in acetic acid buffers follows the rate law (5.1b):

$$\mathrm{CH_3CH(OH)_2} \xrightarrow[\text{catalyst}]{\text{acid}} \mathrm{CH_3CHO + HOH} \tag{5.1a}$$

$$\text{rate} = (k_{\mathrm{H^+}}[\mathrm{H^+}] + k_{\mathrm{HOAc}}[\mathrm{HOAc}])[\mathrm{CH_3CH(OH)_2}] \quad \text{in acetic acid buffers} \tag{5.1b}$$

Acid-catalyzed reactions in which each acid in the solution contributes a specific term [as in Eq. (5.1b)] are said to be general-acid-catalyzed. The complementary base catalysis in which each base in the solution contributes a specific kinetic term is said to be general-base-catalyzed. Experience has shown that the catalytic rate constants k_{A} [e.g., k_{HOAc} in Eq. (5.1b)] and k_{B} vary according to the *Brønsted catalysis law*, formulated by Brønsted and Pedersen [1924] in their historic study of the general-base-catalyzed dehydration of nitramide. Qualitatively, k_{A} tends to increase with the acid dissociation constant K_{A} of the acid catalyst, and k_{B} tends to increase with the base dissociation constant K_{B} of the base catalyst. In most cases these relationships are described with good correlation coefficients by the linear equations (5.2).

$$\ln k_{\mathrm{A}} = \beta_{\mathrm{A}} \ln K_{\mathrm{A}} + \ln g_{\mathrm{A}} \tag{5.2a}$$

$$\ln k_{\mathrm{B}} = \beta_{\mathrm{B}} \ln K_{\mathrm{B}} + \ln g_{\mathrm{B}} \tag{5.2b}$$

The slopes (β_{A} or β_{B}) and intercepts (g_{A} or g_{B}) are specific parameters of the given reaction series. The magnitudes of the slopes are usually in the range 0 to 1, although slopes greater than 1 are possible [Bordwell and Boyle, 1972].

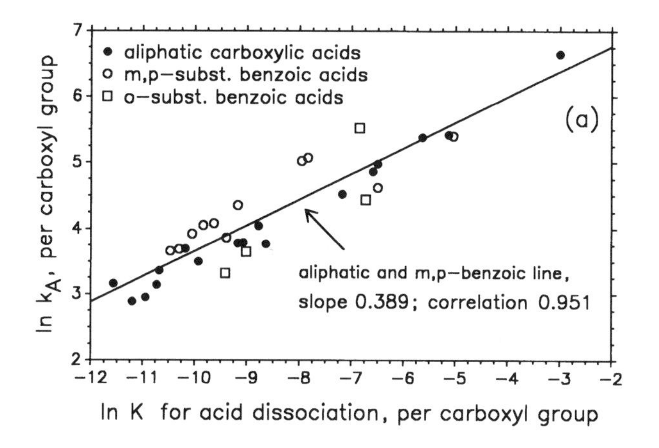

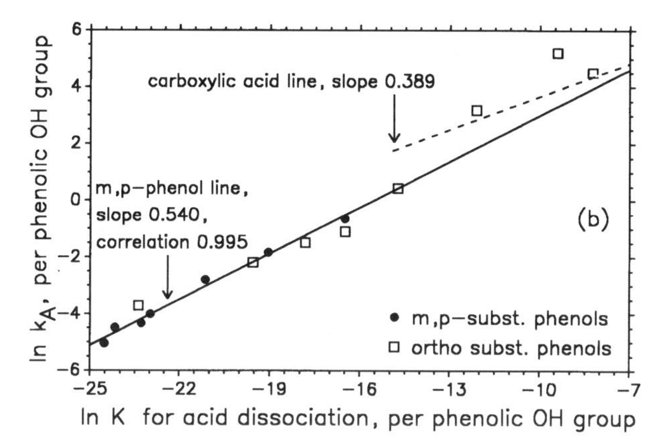

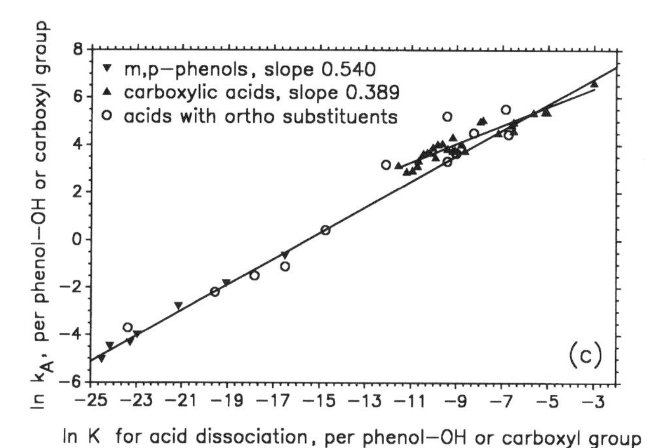

Figure 5.4. Brønsted catalysis law applied to data for the general-acid-catalyzed dehydration of 1.1-ethanediol in 92.5% acetone–water at 25°C. (a) Carboxylic acid catalysts [Bell and Higginson, 1949]. (b) Phenol catalysts [Bell and Higginson, 1949]. (c) Composite of (a) and (b).

An exceptionally detailed experimental study of the Brønsted catalysis law is summarized in Fig. 5.4. The catalyzed reaction is the dehydration of 1,1-ethanediol [Eq. (5.1a)] in 92.5% acetone–water at 25°C [Bell and Higginson, 1949]. Figure 5.4a shows results for carboxylic acid catalysts, Fig. 5.4b shows data for phenols, and Fig. 5.4c gives the overall picture. The rate constants vary by six orders of magnitude; and in view of this large variation, the fit to a straight line is quite respectable.

In contrast to the Hammett equation, which deals with *perturbations* of reaction rates by remote substituents, the Brønsted catalysis law deals with straight reactivity—that is, with the effectiveness of the catalyst. As the catalytic driving force measured by ln K_A increases, Eq. (5.2a) predicts a linear increase in ln k_A, with a slope β. On the other hand, intuition buttressed by first-order theory anticipates "diminishing returns": The increase in ln k_A per unit increase in catalytic driving force should decrease at high reactivity, because the potential for further gains in reactivity becomes more limited. Accordingly, the plot of ln k_A versus ln K_A should be concave down. In principle, the data in Fig. 5.4 can probe this issue because the carboxylic acids are better catalysts than the *m*- and *p*-phenols. In fact, the least-squares slope for the carboxylic acid points (0.389) is smaller than that (0.540) for the *m*- and *p*-phenol points, consistent with curvature in the expected direction, but the scatter of the data about a smooth curve is too great to endow this fact with statistical significance.

Correlation of Solvolysis Rates

Our third example will describe a linear free-energy relationship for solvent effects that, historically, played a role in assigning reactions to mechanistic categories.

Organic halides and strong-acid esters, such as *t*-butyl chloride or ethyl *p*-toluenesulfonate (EtOTs), undergo substitution reactions in hydroxylic solvents in which the solvent molecules act as nucleophiles. For example: In methanol, EtOTs + MeOH $\rightarrow$ EtOMe + H^+ + OTs^-. In glacial acetic acid, *t*-BuCl + HOAc $\rightarrow$ *t*-BuOAc + HCl. This kind of reaction is called *solvolysis*, since the solvent's O–H bonds come apart.

There are two principal mechanistic categories for nucleophilic substitution, called S_N1 and S_N2 [Ingold, 1953]. In S_N2 the nucleophile acts directly in the rate-determining step, whereas in S_N1 it does not. The rate-determining step in S_N1 is essentially an ionization to produce an ion pair. For EtOTs and *t*-BuCl these categories are shown schematically in Eq. (5.3), in which Nu denotes a nucleophile:

$$\begin{aligned} S_N2\text{:}\quad & \text{Nu} + \text{Et—OTs} \rightarrow \text{Nu}\cdot\cdot\text{Et}\cdot\cdot\text{OTs (transition state)} \rightarrow \\ & \qquad \text{Nu—Et}^+ + \text{OTs}^- \end{aligned} \tag{5.3a}$$

$$S_N1: \quad t\text{-Bu—Cl} \rightarrow t\text{-Bu}^{+\delta}\cdot\cdot\text{Cl}^{-\delta} \text{ (transition state)} \rightarrow t\text{-Bu}^{+}\cdots\text{Cl}^{-}$$

$$\text{Nu} + t\text{-Bu}^{+}\cdots\text{Cl}^{-} \xrightarrow{\text{fast}} \text{Nu—}t\text{-Bu}^{+} + \text{Cl}^{-} \tag{5.3b}$$

The two mechanisms can be distinguished because, other things being equal, the rate by S_N2 increases with the strength and/or concentration of the nucleophile, while the rate by S_N1 is independent of that. This criterion works well when Nu is a dilute solute, and under such conditions EtOTs reacts by S_N2 and *t*-BuCl reacts by S_N1. In solvolysis, on the other hand, Nu is a molecule of the solvent, and neither the concentration nor the nature of Nu can be changed without simultaneously changing the solvent medium; and when one changes the solvent medium, one changes a host of properties whose resultant represents the solvent polarity. Because of the generation of ionic charges in solvolysis, the rate constants are sensitive to solvent polarity, and this must be taken into account before inferences can be made about the reaction mechanism. Yet the mechanism is of interest, because the solvent is already in place around the reaction zone and need not surmount a diffusion barrier before acting as a nucleophile.

Grunwald and Winstein [1948] approached the problem via linear free-energy relationships. First they found that substrates that, by other criteria, react by the S_N1 mechanism give nearly linear ln k versus ln k plots for solvolysis rate constants in a series of hydroxylic solvents. An example is shown in Fig. 5.5a for α-phenylethyl chloride versus *t*-butyl chloride [Fainberg and Winstein, 1957]. The plot spans six orders of magnitude in rate constant and is respectably linear, even though it covers two distinct levels of solvent nucleophilicity: (1) a lesser level based on the carboxyl group in acetic acid and formic acid and (2) a higher level based on the OH group of water, methanol, and ethanol. The scatter of the points about the least-squares line is clearly *not* correlated with the nucleophilicity of the solvent.

By contrast, Fig. 5.5b shows a plot ln k for ethyl *p*-toluenesulfonate, an S_N2 substrate, versus ln k for *t*-butyl chloride, an S_N1 substrate [Winstein et al., 1951]. At the nucleophilic level of the OH group in water, methanol, and ethanol, the data fall on a smooth line. But at the lesser nucleophilic level of the carboxyl group in acetic acid, the rate constant for ethyl *p*-toluenesulfonate falls decisively below that line, showing clearly that ethyl *p*-toluenesulfonate is more sensitive than *t*-butyl chloride to the decrease in nucleophilicity. Figure 5.5b does *not* imply that *t*-butyl chloride is completely insensitive—that inference must be made inductively from plots such as Fig. 5.5a.

Physical Problems That Must Be Faced

In spite of the empirical success of the linear free-energy approach, basic objections might be raised on physical grounds. For one thing, linear free-energy relationships treat equilibrium constants and rate constants on an equal foot-

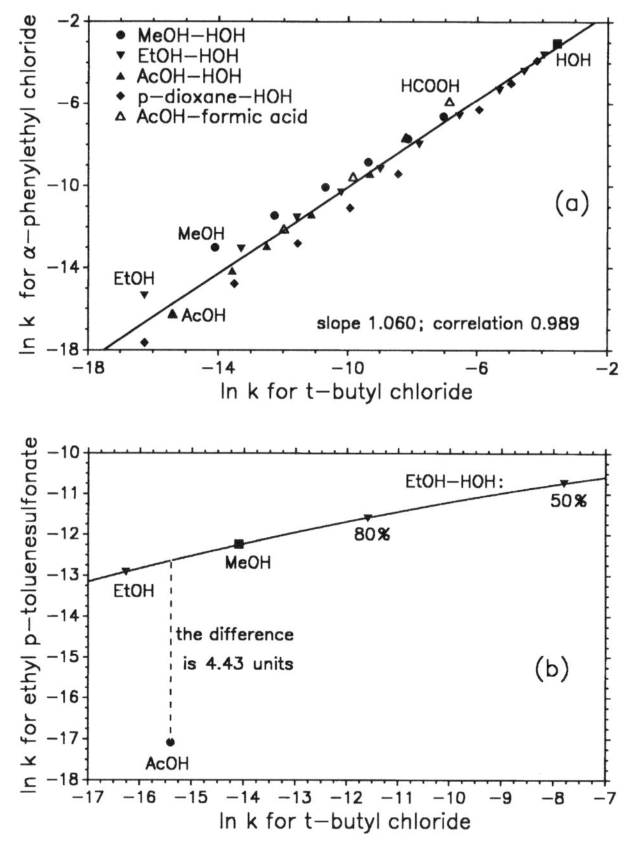

Figure 5.5. Correlations of solvolysis rates. (a) α-Phenylethyl chloride versus *t*-butyl chloride [Fainberg and Winstein, 1957]. (b) Ethyl *p*-toluenesulfonate versus *t*-butyl chloride [Winstein et al., 1951]. Judging by the smooth line, the deviant point for AcOH hints that the lesser nucleophilicity of acetic acid affects the rate constant for ethyl *p*-toluenesulfonate by a factor of 1/80.

ing. But objections on those grounds are largely neutralized by the success of transition-state theory, which supports the notion and virtually integrates chemical kinetics into equilibrium thermodynamics. This will be taken up in the next section.

For another thing, the linear free-energy relationships are too simple. They employ scalar variables (ln K and ln k), while physical descriptions require vector and tensor variables to allow for the three-dimensional nature of interacting charge distributions. Even though the linear free-energy fits are not exact and the processes being compared are chosen deliberately on the basis of chemical analogy, some dependence on specific molecular geometries is bound to remain, so that the quality of fit is probably helped by error tolerance. The specific form of error tolerance is *configurational error tolerance*. It derives from the fact that equilibrium configurations are identified with potential minima. We will consider this matter in a later section.

TRANSITION-STATE THEORY

The transition-state theory is a modification of Arrhenius' theory of reaction rates [Arrhenius, 1889], modified by adding assumptions that permit the calculation of rate constants by methods of statistical thermodynamics [Eyring, 1935; Glasstone et al., 1941; Laidler and Laidler, 1987; Steinfeld et al., 1989]. The basic notion is that the reacting molecules must surmount a potential-energy barrier, as sketched in Fig. 5.6 for the symbolic bimolecular reaction: $A + B \rightarrow P + Q$. The potential energies of the separated reactants and products correspond to the asymptotic stable minima; that of the transition state is the unstable maximum.

There are two points of view regarding the nature of the transition state. To many theoreticians the transition state is a geometric surface that demarcates the boundary between reactants and products. To experimentalists who seek to probe molecules in the process of passing over the barrier, it is a distinct molecular species with a very short but accessible lifetime, probably in the subpicosecond range. According to the definitions given in Chapter 1, the transition state is a molecular species, so we are committed to the second point of view.

As a molecular species, the transition state is identified with the population in a narrow line segment around the maximum in Fig. 5.6. The precise length δ of that segment may be left indefinite because δ cancels out in the derivation

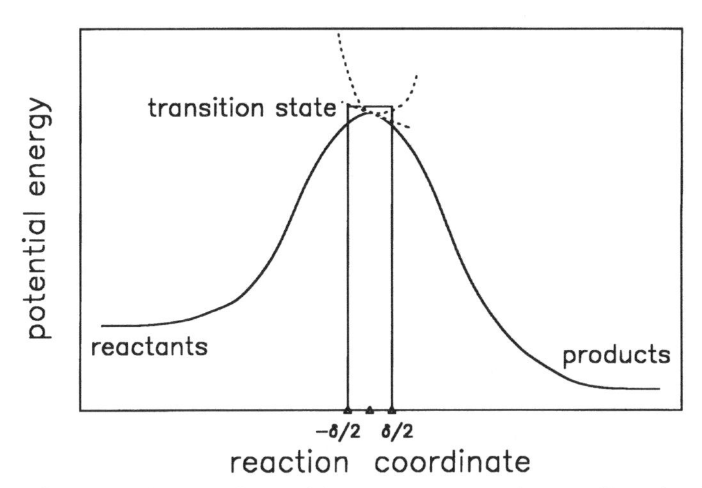

Figure 5.6. Schematic diagram of potential energy versus reaction coordinate for a bimolecular reaction-step, $A + B \rightarrow P + Q$.

of the rate constant. But if the lifetime is indeed in the subpicosecond range, δ must be well below the distance traveled by a molecule in 10^{-12} s, almost certainly less than 1 Å.

For simplicity, in Fig. 5.6 the reaction coordinate is projected onto a straight line from reactants to products, while in actual configuration space the reaction coordinate is a curve. Figure 5.7 shows an unusually simple example, the potential-energy contour diagram for a hypothetical triatomic reaction, X + Y–Z → X–Y + Z. The angle X·Y·Z is fixed at 180°, so there remain only two configuration variables, the X ··· Y distance r_1 and the Y ··· Z distance r_2.

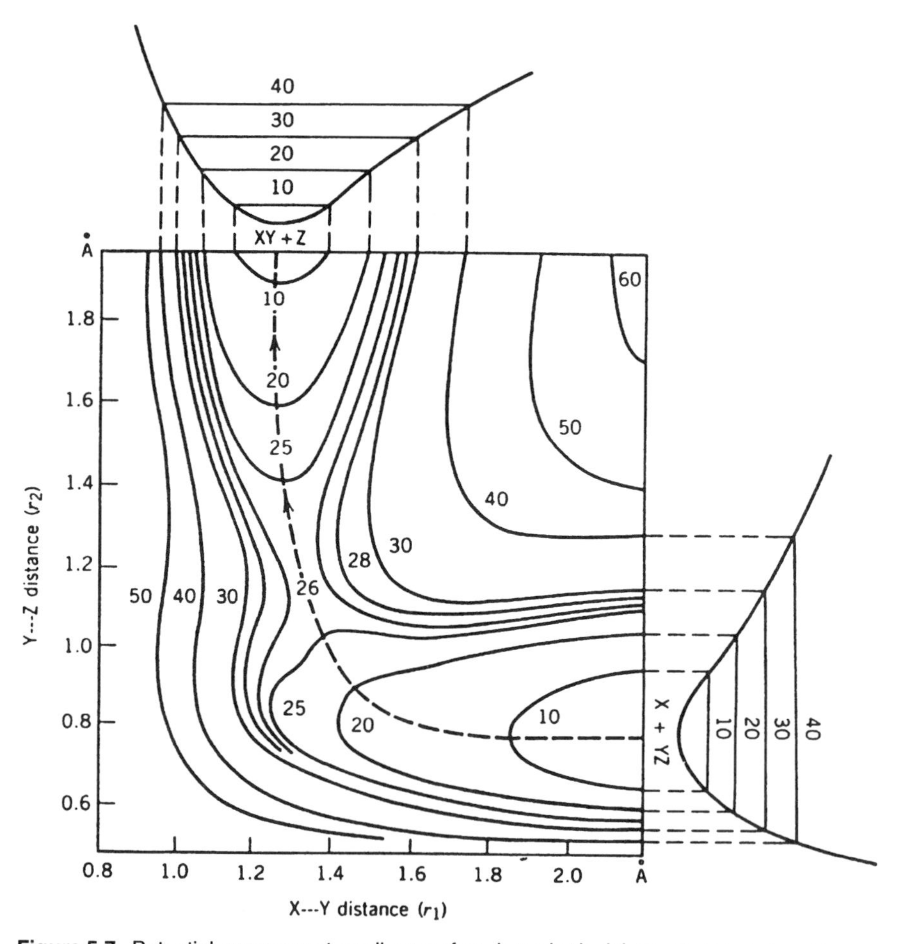

Figure 5.7. Potential-energy contour diagram for a hypothetical three-atom reaction. Energy contours are in kilocalories per mole. The angle X·Y·Z is constant at 180°. The dashed line is the reaction coordinate. The potential functions at the ends of the valleys are, respectively, for X–Y and Y–Z bond stretching. (Reproduced from Glasstone et al. [1941], with consent of Professor K. J. Laidler.)

The reaction coordinate, which is shown as a dashed line, is the locus of points of minimum potential energy for motion along the other coordinates. In Fig. 5.6, the presence of such modes is suggested by the dashed curve at the transition state.

In the full quantum-mechanical description of reaction paths, the reacting molecules can also tunnel *through* the barrier, not necessarily under the reaction coordinate [Bell, 1980; Marcus and Coltrin, 1977; Skodje et al., 1981]. But if the mass number of the particle being transferred is greater than 4, the fraction of reaction by tunneling at 300 K is less than 10% and can for our purposes be neglected.

Approximate Amount of Tunneling. The fraction of reaction by tunneling can be estimated by a method due to Christov [1972]. Let

$$T_c = [24.3 \times 10^{-8}/(\delta r)_{1/2}] \cdot (E_{\text{barrier}}/\mathcal{L})^{1/2}$$

where $(\delta r)_{1/2}$ is the half-width of the barrier (half the width at half the height, in centimeters), $\mathcal{L}$ is the mass number of the particle being transferred, and $E_{\text{barrier}} \approx E_{\text{act}}$ (in kilojoules per mole); and let $x = \pi T_c/2T$. Then, in first approximation, the fraction of reaction by tunneling equals $[1-(\sin x)/x]$. For example, when $E_{\text{barrier}} = 50\text{kJ/mol}$, $\mathcal{L} = 4$, and $(\delta r)_{1/2} = 8 \times 10^{-9}\text{cm}$, $T_c = 107$ K. At 300 K, the fraction by tunneling then is 0.05.

Straight Passage

Transition-state theory is based on the concept of *straight passage*, which may be stated as follows:

> Molecules do not move back and forth across the top of the potential barrier while residing in the transition state. Reactants, having crossed the barrier in the direction of products, go straight on to products. Products, having crossed the barrier in the direction of reactants, go straight on to reactants. The population of the transition state may thus be divided into separate forward-moving and backward-moving subspecies.

This innocuous concept has weighty corollaries. First, because molecules exit from the transition state after a single crossing of the barrier, the mean residence time τ in the transition state is very short, in the subpicosecond range, which bears on the attainment of thermal equilibrium.

Consider a transient species whose molecules, at the time of entering the species, have an energy distribution that deviates from thermal equilibrium. Let T_1 denote the relaxation time for relaxation of the energy distribution to thermal equilibrium, and τ the mean residence time. If $T_1 > \tau$, the energy distribution cannot reach equilibrium. Such a species can exist at thermal equilibrium *only if the molecules enter with a thermal energy distribution.* The relationship $T_1 > \tau$ applies to transition states.

At the same time, the Second Law requires that when thermal equilibrium exists at a temperature T, this temperature must be uniform throughout. Statistical thermodynamics then defines "uniform throughout" to mean that the molecular energy distributions of all molecular species in the system are equilibrium distributions, characteristic of the temperature T and independent of the molecular residence times τ. At constant T, the molecular energy distribution must therefore be an equilibrium distribution, both for the forward-moving and for the background-moving subspecies of the transition state. This does not follow from straight passage but simply because the temperature is constant. Straight passage causes the inequality $T_1 > \tau$, which implies that the required equilibrium distribution cannot be reached while the molecules reside in the transition state. Because of straight passage the molecules must therefore *enter* with an equilibrium distribution.

When the reaction proceeds at a constant and uniform temperature, straight passage causes segmentation of the equilibrium. Because the forward-moving transition-state species is formed *only* from the reactants, and the backward-moving transition-state species only from the products, the former must exist in equilibrium with the reactants at the given temperature, and the latter must exist in equilibrium with the products at the given temperature, *but the reactants and products need not exist in equilibrium with each other.*

We find, therefore, that straight passage divides the transition-state population into two noninteracting subspecies: (1) a forward-moving subspecies that is in equilibrium with the reactants and (2) a backward-moving subspecies that is in equilibrium with the products. We shall find that the rate constants of chemical kinetics thereby become endowed with equilibrium properties.

Rate Equation

In addition to the key assumption of straight passage, transition-state theory also makes a subsidiary assumption:

> Motion along the reaction coordinate at the transition state in a thermal reaction is equivalent to a normal mode of translation, with a Boltzmann energy distribution.

The actual derivation of the transition-state rate equation is based on statistical thermodynamics and will be given in outline only. Explicit derivations will be found in textbooks on chemical kinetics such as those cited earlier [e.g., Steinfeld et al., 1989]. In the following notation, algebraic subscripts (+, −) will indicate macroscopic variables for the forward and reverse reaction, while arrow subscripts ($\rightarrow, \leftarrow$) indicate microscopic variables. The superscript ‡ indicates properties of the transition state. K is the equilibrium constant, k_+ is the forward rate constant, c is the concentration, r is the forward reaction rate, τ is the mean residence time of a molecule in the transition state, v is the mean velocity along the reaction coordinate, and δ is the coordinate length assigned to the transition state (see Fig. 5.6); δ will cancel out in the end. Accordingly, $\tau_{\rightarrow} = \delta/v_{\rightarrow}$. Moreover, owing to the assumption of straight passage, r_+ equals the turnover rate $c_{\rightarrow}/\tau_{\rightarrow}$ of the forward-moving transition-state subspecies.

For definiteness we will consider a bimolecular reaction: A + B $\rightarrow$ products. The forward rate then is given by $r_+ = c_{\rightarrow}/\tau_{\rightarrow} = k_+ c_A c_B$, and the rate constant is given by Eq. (5.4a):

$$k_+ = \frac{c_{\rightarrow}}{c_A c_B} \cdot \frac{1}{\tau_{\rightarrow}} \tag{5.4a}$$

In the traditional formulation of transition-state theory, one specifies that the reaction is proceeding at dynamic equilibrium, so that $r_+ = r_-$. It then follows that $c_{\rightarrow} = c_{\leftarrow}$, $\tau_{\rightarrow} = \tau_{\leftarrow} = \delta/v_{\rightarrow}$, and the *total* transition-state concentration c_{act} equals $2c_{\rightarrow}$.

Let $K_{\text{act}} = c_{\text{act}}/c_A c_B = 2c_{\rightarrow}/c_A c_B$, the equilibrium constant for the conversion of A+B to the overall transition-state that includes both subspecies. Substitution in Eq. (5.4a) then yields Eq. (5.4b):

$$k_+ = \frac{K_{\text{act}}}{2\tau_{\rightarrow}} = \frac{K_{\text{act}} v_{\rightarrow}}{2\delta} \tag{5.4b}$$

Transition-state theory then makes an ingenious substitution that eliminates unknown microscopic variables and causes k_+ at constant T to become simply proportional to a quasi-equilibrium constant. According to statistical thermodynamics,

$$K_{\text{act}} = \frac{z_{\text{act}}}{z_A z_B} e^{-(\Delta\varepsilon_{0,\text{act}}/kT)} \tag{5.5a}$$

where the z's are molecular partition functions and $\Delta\varepsilon_{0,\text{act}}$ is the height of the energy barrier from the ground level of the reactants to the ground level of the transition state—the 0,0-energy of activation. Since translation (trl) along the reaction coordinate is an independent mode of motion, it contributes an independent factor $z^{\ddagger}_{\text{trl}}$ to z_{act}; that is, $z_{\text{act}} = z^{\ddagger}_{\text{trl}} \cdot z^{\ddagger}$, where $z^{\ddagger}$ comprises *all modes except translation along the reaction coordinate*. The result is Eq. (5.5b), whose substitution in Eq. (5.4b) yields Eq. (5.5c):

$$K_{\text{act}} = K^{\ddagger} \cdot z^{\ddagger}_{\text{trl}}$$

$$K^{\ddagger} = \frac{z^{\ddagger}}{z_A z_B} e^{-(\Delta\varepsilon_{0,\text{act}}/kT)} \tag{5.5b}$$

$$k_+ = K^{\ddagger} \frac{z^{\ddagger}_{\mathrm{trl}} v_{\rightarrow}}{2\delta} \tag{5.5c}$$

Using the model of a statistical-equilibrium ensemble of particles, in translational motion in both directions along a single coordinate, moving between two points separated by a distance δ in a flat potential, the factor $(z^{\ddagger}_{\mathrm{trl}} v_{\rightarrow}/2\delta)$ can be evaluated by orthodox statistical thermodynamics. The result is $(z^{\ddagger}_{\mathrm{trl}} v_{\rightarrow}/2\delta) = (kT/h)$, where k denotes Boltzmann's constant and h is Planck's constant. Substitution of this result in Eq. (5.5c) yields the desired *transition-state rate equations* (5.6). In cgs units, $k/h = 2.0836 \times 10^{10}(\mathrm{s}^{-1}\mathrm{K}^{-1})$.

$$k_+ = \frac{kT}{h} K^{\ddagger} \tag{5.6a}$$

$$k_+ = \frac{kT}{h} \frac{z^{\ddagger}}{z_A z_B} e^{-(\Delta\varepsilon_{0,\,\mathrm{act}}/kT)} \tag{5.6b}$$

Scope and Validity

Although Eqs. (5.6) were derived for a bimolecular reaction step, they can be generalized to any kinetic order simply by replacing $z_A z_B$ with $\Pi(z_J)^j$, where j denotes the kinetic order with respect to any molecular species J, and the multiplication extends over all molecular species in the rate law.

Equation (5.6a) predicts that the rate constant k_+ is proportional to a parameter $K^{\ddagger}$ that, except for omitting the mechanically unstable mode of translation along the reaction coordinate, is tantamount to an equilibrium constant. Furthermore, because the prefactor kT/h is nonspecific, $K^{\ddagger}$ gathers up everything in the rate constant that is reaction-specific, including of course substituent effects and medium effects. The transition-state rate equation therefore rationalizes the fact that linear free-energy relationships of substituent effects and medium effects work as well for rate constants as for equilibrium constants.

One would of course like to have direct tests of validity. In principle, such tests are possible because the transition-state theory is an *absolute* theory. The molecular partition functions $(z_A, z_B, z^{\ddagger})$ and the 0,0-energy of activation $(\Delta\varepsilon_{0,\,\mathrm{act}})$ in Eq. (5.6b) can be derived from the potential-energy surface for the reacting molecular system—in principle by pure quantum mechanics [Eyring, 1935; Evans and Polanyi, 1935]. Unfortunately, this is easier said than done. At the moment, the only purely quantum-mechanical potential surface with the requisite accuracy is the H_3 surface [Valentini and Phillips, 1989; Siegbahn and Liu, 1978; Liu, 1984; Truhlar and Horowitz, 1978; Garrett et al., 1986], whose properties determine the dynamics of the reaction, $H\cdot + H_2 \rightarrow H_2 + H\cdot$, and of its isotopic variants. This may not be the easiest reaction to study kinetically, but consistent rate constants by independent methods are now available for $H\cdot + D_2 \rightarrow HD + D\cdot$ and for $D\cdot + H_2 \rightarrow DH + H\cdot$, in the gas phase, over the extraordinarily wide temperature range 200–2000 K [Mitchell and

LeRoy, 1973; Westenberg and de Haas, 1967; Michael, 1990; Michael and Fisher, 1990; Michael et al., 1990].

The two parts of Fig. 5.8 compare the measured rate constants for $D \cdot + H_2 \rightarrow DH + H \cdot$ with those based on absolute theoretical prediction. The predictions are based on specialized versions of Eq. (5.6b) that employ variational transition-state theory and make full allowance for quantum-mechanical tunneling [Garrett et al., 1986]. The "variational" calculation places the dividing surface between reactants and products—the transition state—so as to minimize the forward-moving flux of molecules through the transition state and thus optimizes agreement with the straight-passage assumption of transition-state theory. The experimental rate constants span more than seven orders of magnitude and agree substantially with the theoretical curve.

The fit for the alternate exchange reaction $H \cdot + D_2 \rightarrow HD + D \cdot$, is equally

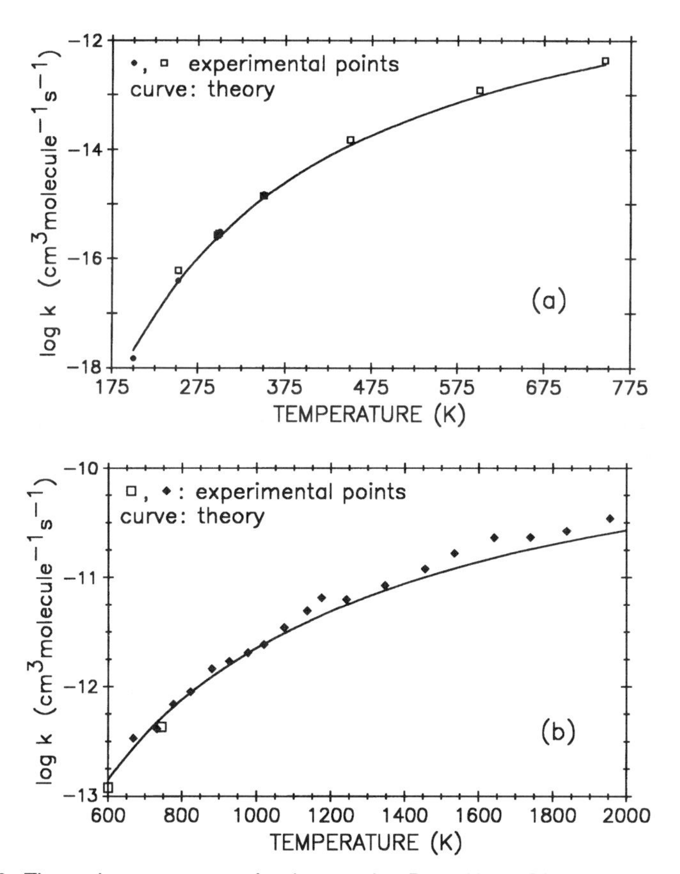

Figure 5.8. Thermal rate constants for the reaction $D \cdot + H_2 \rightarrow DH + H \cdot$, compared with absolute theoretical prediction. (a) Circles, Mitchell and LeRoy [1973]; squares, Westenberg and de Hass [1967]; theory, Garrett et al. [1986]. (b) Squares, Westenberg and de Haas [1967]; diamonds, Michael and Fisher [1990]; theory, Michael et al. [1990].

good. At the lower temperatures, both exchange reactions proceed largely by quantum-mechanical tunneling. But above about 500 K, reaction by passage over the potential barrier according to the transition-state model becomes dominant, so that the scientists who made the high-temperature measurements were able to state, with evident pleasure: "The present comparisons firmly demonstrate that (the proposition of the Eyring–Polanyi theory of reaction rates [1935]) is completely true within the random error of the experimental data" [Michael et al., 1990].

There have been questions about the transition-state theory ever since it was first proposed in 1935, and the beautiful example shown in Fig. 5.8 may not allay them entirely because (as the Talmud says) "an example is not proof." We shall deal briefly with two interrelated questions that have often been raised: the existence of equilibrium for the forward-moving transition state when reactants and products are not at equilibrium, and the validity of straight passage.

The forward-moving transition-state subspecies may be off equilibrium owing to an error either in the concentration $c_{\rightarrow}$ or in the molecular energy distribution at the given $c_{\rightarrow}$. The Tolerance Theorem of Chapter 4 signifies that even a substantial error in the molecular energy distribution would cause only a small error in the prediction of k_+ [see, for example, Eq. (4.10)] and thus would not be a serious problem. An error in $c_{\rightarrow}$, on the other hand, would be hard to estimate and would propagate directly into k_+.

The concept of straight passage is in the domain of molecular dynamics, and deviations from straight passage are related to forces derived from the potential energy surface for the reaction [Levine and Bernstein, 1987; Jordan, 1979]. To illustrate this dependence, Fig. 5.9 shows two stylized, simplified potential-energy contour diagrams for the region around the transition state (TS). The reaction coordinate advances along the vertical axis (y), at right angles to the dotted contours, and passes through a maximum at TS; while in the horizontal direction (x), the potential energy increases with distance from the reaction coordinate in the manner shown by the solid contours. The landscape at TS resembles that of a mountain pass, but the details are constrained by rules governing normal modes of motion. In particular, the probability of straight passage depends on the anharmonicity of the potential surface.

The two contour diagrams shown in Fig. 5.9 differ only in the anharmonicity of the potential surface. Figure 5.9a depicts a strictly harmonic (quadratic) potential function—a hyperbolic paraboloid centered on TS—whose salient feature is a progressive constriction of the approach path to TS. An approaching molecule can reach TS only if it moves fairly accurately in the direction of TS, with the required degree of accuracy depending on its kinetic energy. If the molecule moves otherwise, it collides with a potential wall and turns back. But once the molecule is past TS, it can no longer turn back, because the forces generated by the harmonic potential propel it toward the products. When the product side is reached, the mode of motion along the reaction coordinate has a high probability of becoming deactivated, either by intramolecular energy transfer to other modes or by collisional deactivation. This mode of motion is there-

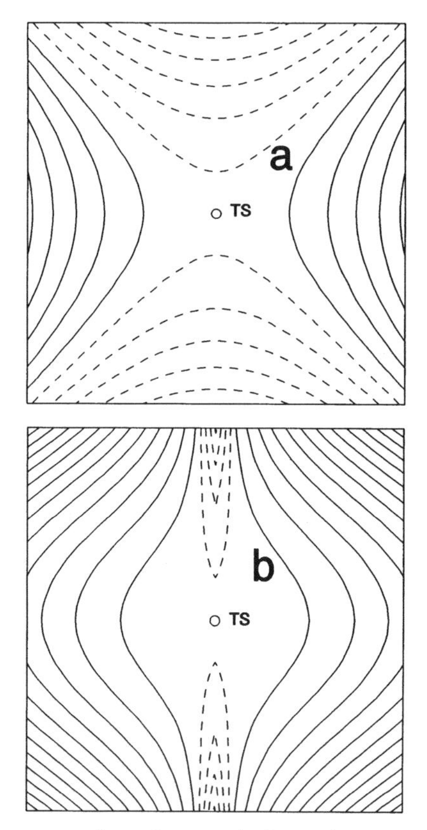

Figure 5.9. Potential energy contour diagrams in the region around the transition state for (a) the harmonic potential $V = V_{TS} - y^2 + x^2$ and (b) an unusually anharmonic potential, $V = V_{TS} - y^2 + x^2 + 1.2y^2\sqrt{|x|}$. The reaction coordinate y passes vertically through the transition state TS; x and y vary from -10 to $+10$. The dashed contours indicate elevations below the transition state; the solid contours indicate those above it. The contour interval is 20 units of V in (a) and 30 units in (b). In both diagrams the transition state is a saddle point. These diagrams were computer-generated by I. Lengyel.

fore straight passage. In summation: A harmonic potential function restricts the approach to the transition state, but molecules getting past react in straight passage.

On the other hand, the potential function with unrealistically high anharmonicity shown in Fig. 5.9b forces a different molecular dynamics. The constriction of the approach path has disappeared, and the landscape around TS has changed from a mountain pass to a mountain park. A molecule entering this park at a sufficient angle to the reaction coordinate is temporarily trapped. It bounces from wall to wall inside the park until a suitable rebound directs it into an exit channel. The particle thus has a significant chance of returning to reactants.

Potential functions for real molecules are anharmonic in the region around the transition state, because the curvature of the reaction coordinate (e.g., Fig. 5.7) imposes it. But ordinarily the anharmonicity is mild, so that Fig. 5.9a is a better model than Fig. 5.9b. Indeed, judging by the success of linear free-energy relationships involving rate constants and by the agreement of experimental rate constants with absolute theory (e.g., Fig. 5.8), straight passage is a viable concept.

THERMODYNAMIC PROPERTIES OF TRANSITION STATES

Equation (5.6) is readily translated into the language of thermodynamics. The key equation is (5.7), which is analogous to (3.11c). The thermodynamic system is thereby treated as a chemical system. $\Delta G^\ddagger$ is the *standard free energy of activation*, or simply *free energy of activation*. Substitution in (5.6a) then yields the Eyring equation, (5.8).

$$-RT\ln K^\ddagger = \Delta G^\ddagger \tag{5.7}$$

$$k_+ = \frac{kT}{h} e^{-\Delta G^\ddagger/RT} \tag{5.8}$$

Dissection of $\Delta G^\ddagger$ into an enthalpic and entropic term [Eq. (5.9)] allows Eq. (5.8) to be written in terms of a standard enthalpy ($\Delta H^\ddagger$) and entropy ($\Delta S^\ddagger$) of activation, Eq. (5.10).

$$\Delta G^\ddagger = \Delta H^\ddagger - T\Delta S^\ddagger \tag{5.9}$$

$$k_+ = \frac{kT}{h} e^{-\Delta H^\ddagger/RT} e^{\Delta S^\ddagger/R} \tag{5.10}$$

On taking the logarithm of Eq. (5.8), solving for $\Delta G^\ddagger$, and differentiating with respect to T, we express $\Delta S^\ddagger$ and $\Delta H^\ddagger$ as functions [Eqs. (5.11) and (5.12)] of the temperature derivative of the rate constant, which can be measured. Expressions for the heat capacity of activation [$\Delta C_P^\ddagger$, Eq. (5.13)] and volume of activation [$\Delta V^\ddagger$, Eq. (5.14)] then follow from basic definitions for chemical systems.

$$\Delta S^\ddagger = -(\partial \Delta G^\ddagger/\partial T)_P = RT(\partial \ln k_+/\partial T)_P - (\Delta G^\ddagger/T) - R \tag{5.11}$$

$$\Delta H^\ddagger = \Delta G^\ddagger + T\Delta S^\ddagger = RT^2(\partial \ln k_+/\partial T)_P - RT \tag{5.12}$$

$$\Delta C_P^\ddagger = (\partial \Delta H^\ddagger/\partial T)_P \tag{5.13}$$

$$\Delta V^\ddagger = (\partial \Delta G^\ddagger/\partial P)_T = -RT(\partial \ln k_+/\partial P)_T \tag{5.14}$$

Parameters of the Arrhenius Equation

In the Arrhenius theory of reaction rates, values of the rate constant k_+ depend on two parameters: the activation energy E_{act} and the preexponential factor A, Eqs. (5.15) and (5.16):

$$k_+ = Ae^{-E_{act}/RT} \tag{5.15}$$

$$E_{act} = RT^2(\partial \ln k_+/\partial T)_P \tag{5.16}$$

Algebraic manipulation then shows that these parameters are related to $\Delta H^{\ddagger}$ and $\Delta S^{\ddagger}$ by the conversion equations (5.17) and (5.18):

$$E_{act} = \Delta H^{\ddagger} + RT \tag{5.17}$$

$$\ln A = \ln(kT/h) + 1 + \Delta S^{\ddagger}/R \tag{5.18a}$$

$$= 24.760 + \ln T + \Delta S^{\ddagger}/R \tag{5.18b}$$

ERROR TOLERANCE IN LINEAR FREE-ENERGY RELATIONSHIPS

We will now return to the physical problem that linear free-energy relationships are logically too simple, because they allow only for coarse features of molecular geometry and neglect fine details. For example, the Hammett equation distinguishes between meta and para substituents but neglects torsional angles about single bonds, even though there is good reason to believe that these angles are not the same in different reaction series. For example, in the Hammett plot of $\ln K_A$ for the dissociation of phenylacetic acid (19a) versus that of benzoic acid (19b), an m-CH_2Cl group interacts electrostatically with the COOH group by dipole–dipole forces, and with the COO^- group both by charge–dipole and dipole–dipole forces.

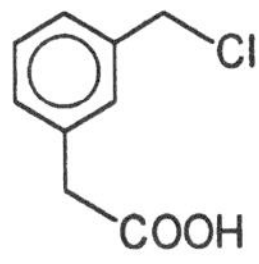

Cl, COOH (19a) Cl, COOH (19b) (5.19)

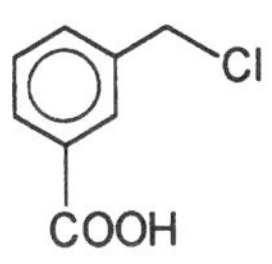

The interacting groups respond to these forces so as to optimize the interaction energy: They become polarized and make geometric adjustments, including some rotation with respect to the plane of the benzene ring, some *cis–trans* shifting of the proton in the COOH group, and in (19a) some rotation about the C_{Ar}–CH_2 bond axis. Experience with geometry reoptimization in quantum-mechanical perturbation calculations shows that the magnitudes of the geometric adjustments, and the resulting changes in physical vectors, are specific. Yet in spite of neglecting these variables, the Hammett plot for phenylacetic acids versus benzoic acids is linear with a slope of 0.489 and a correlation coefficient of 0.981 [Jaffe, 1953]. Apparently there exists a mode of error tolerance that dampens the effect of changes in molecular geometries on the standard partial free energies.

The concern over matters physical extends to the numbers of independent variables. Charge–charge interactions depend only on the distance between the charges, charge–dipole interactions depend on the distance and one angle, and dipole–dipole interactions depend on the distance and two angles. More general physical approaches involve still more geometric variables, as shown by the prevalence of vector and tensor variables in the literature of chemical physics. By contrast, a linear free-energy relationship has only two parameters available for adjustment, namely, the slope and the intercept; and when the fit is good, one congratulates oneself on having bypassed the formidable complexity of the physical models.

Of course, the physical equations must be so complicated because nature endows the molecules with so many degrees of freedom. If freedoms were eliminated by forcing geometrical constraints upon the reactants, products, and transition states, the constraints would distort the natural equilibrium configurations, but could be designed so as to realign the interacting charge distributions until the free-energy relationships were exactly linear. A precedent for this stratagem appears in Greek mythology, where the landlord Procrustes imposed good "fits" on his overnight visitors: If the bed didn't fit, he stretched or pared the shape of the visitor—fatally if necessary, rather than let him have a better-fitting bed. Fortunately, molecular shapes are more elastic, and we may mentally adjust them to fit the "bed" of a linear free-energy relationship.

The advantage of the Procrustean view of linear free-energy relationships is that the distorted reactants, products, and transition states by hypothesis give an exactly straight line. This allows us to equate the deviations from exact free-energy linearity with the work of producing the required distortions, since the actual molecules react in their equilibrium configurations. We shall show that such distortions are subject to thermodynamic error tolerance and that, when chemical analogies exist, the deviations can be quite small.

Procrustean Approach

Consider the reaction series:

$$\begin{array}{ll} \mathrm{A} + \mathrm{X}_1 \rightarrow \mathrm{P}_1, & \mathrm{A} + \mathrm{Y}_1 \rightarrow \mathrm{Q}_1 \\ \mathrm{A} + \mathrm{X}_2 \rightarrow \mathrm{P}_2, & \mathrm{A} + \mathrm{Y}_2 \rightarrow \mathrm{Q}_2 \\ \cdots & \cdots \\ \mathrm{A} + \mathrm{X}_n \rightarrow \mathrm{P}_n, & \mathrm{A} + \mathrm{Y}_n \rightarrow \mathrm{Q}_n \end{array}$$

Let i be the general index, and let $\{\Delta G_{\mathrm{A,X}_i}, \Delta G_{\mathrm{A,Y}_i}\}$ and $\{\Delta W_{\mathrm{A,X}_i}, \Delta W_{\mathrm{A,Y}_i}\}$

denote the standard free energy changes and free energies of distortion for the ith pair of reactions. The Procrustean approach then takes the form of Eq. (5.20).

$$[\Delta G_{\mathrm{A,Y}_i} + \Delta W_{\mathrm{A,Y}_i}] = a + b[\Delta G_{\mathrm{A,X}_i} + \Delta W_{\mathrm{A,X}_i}]$$

$$\therefore \quad \Delta G_{\mathrm{A,Y}_i} = \underbrace{a + b\Delta G_{\mathrm{A,X}_i}}_{\text{Experimental points}} + \underbrace{[b\Delta W_{\mathrm{A,X}_i} - \Delta W_{\mathrm{A,Y}_i}]}_{\text{Distortional resultants}} \tag{5.20}$$

Equation (5.20) implies that a least-squares fit of the experimental points to a straight line automatically adjusts the molecular distortions so that the distortional resultants are minimized. Introduction of explicit physical or geometrical constraints is not required.

Configurational Error Tolerance

To show that departures of molecular configurations from equilibrium proceed with error tolerance, we express the potential energy E of a molecule as a quadratic function (5.21) of the modewise displacements $q_1, \ldots, q_j, \ldots, q_m$ from equilibrium, where m is the number of independent vibrational modes ($3n - 6$ for nonlinear stable molecules and $3n - 7$ for nonlinear transition states); $k_1, \ldots, k_m$ are the corresponding force constants.

$$E = E_0 + \tfrac{1}{2}k_1 q_1^2 + \cdots + \tfrac{1}{2}k_j q_j^2 + \cdots + \tfrac{1}{2}k_m q_m^2 \tag{5.21}$$

Higher-order terms may be neglected because the following derivation will consider only small displacements. E is at a minimum at the equilibrium configuration so that $(\partial E/\partial q_j)_{\mathrm{eq}} = 0$ for any j, which implies error tolerance in E for any normal-mode displacement from equilibrium. To make a displacement reversible, the displacing force f_j is only infinitesimally greater than the holding force $(\partial E/\partial q_j) = k_j q_j$, and we shall in fact write $f_j = k_j q_j$. The work w_j done on the molecule by distorting the jth mode, from an initial equilibrium configuration ($q_j[i] = 0$) to a final configuration ($q_j[f] = q_j$), is therefore expressed as

$$w_j = \int_0^{q_j} f_j \, dq_j = \tfrac{1}{2}k_j q_j^2 \tag{5.22}$$

So far so good! But we are not interested in w_j for a single molecule, but in W_j for one mole of a statistical ensemble of such molecules. Let $\langle q_j \rangle$ denote the ensemble average of q_j in a given macroscopic state. In the initial state we have $\langle q_j \rangle_i = 0$, since the distribution function for q_j in the natural unperturbed state is symmetrical about $q_j = 0$ as required by the quadratic form of Eq. (5.21). Forces then are applied to distort the molecules until in the final state,

$\langle q_j \rangle_f = \tilde{q}_j$:

$$W_j = L \int_0^{\tilde{q}_j} \langle f_j \rangle \, d\langle q_j \rangle = \tfrac{1}{2} L k_j \tilde{q}_j^2 \tag{5.23a}$$

The mean force $\langle f_j \rangle$ during displacement from equilibrium is $k_j \langle q_j \rangle$. Hence W_j per mole, as defined in Eq. (5.23a), is subject to error tolerance according to Eq. (5.23b):

$$(\partial W_j / \partial \langle q_j \rangle) = L k_j \tilde{q}_j = 0 \quad \text{when} \quad \tilde{q}_j = 0 \quad \text{(i.e., at configurational equilibrium)} \tag{5.23b}$$

To derive error tolerance for the partial free energy, the distorting forces are applied reversibly at constant T and P. The work of distortion then is free-energy work. That is, the free energy $G(\langle q \rangle)$ per mole of the distorted ensemble is the sum of the standard free energy G^0 of the natural molecular species, plus the distortion work $\sum_j W_j$ per mole summed over all vibrational modes, Eq. (5.24). Error tolerance to distortion of any jth mode then follows at once, Eq. (5.25).

$$G(\langle q \rangle) = G^0 + \sum_j W_j = G^0 + \tfrac{1}{2} L \cdot \sum_j k_j \tilde{q}_j^2 \tag{5.24}$$

Error tolerance to distortion of any jth mode at constant T and P:

$$(\partial G(\langle q \rangle) / \partial q_j) = (L \cdot k_j \tilde{q}_j) = 0 \quad \text{when} \quad \tilde{q}_j = 0 \tag{5.25}$$

Since the derivative of W_j equals $k_j \tilde{q}_j$ [Eq. (5.23b)], the effectiveness of the error tolerance is proportional to k_j, the vibrational force constant. A high k_j implies that even a small distortion may perturb the free energy noticeably, while a low k_j implies that even a moderate distortion may be innocuous. As a general rule, force constants vary in the following sequence: covalent bond stretching > bending of angles between covalent bonds > mutual rocking or bending of connected parts of a molecule > torsional motions about flexible bonds. Thus, if a linear free-energy relationship fits well, the substituent and medium effects on bond distances are probably similar, while shifts in the angles between interacting dipoles can probably be offset, at small cost in distortion energy, by torsional motions about flexible bonds. The low sensitivity of linear free-energy relationships to shifts in the relative orientations of the reaction zones and substituents in two reaction series can therefore be rationalized.

A further consideration is the number of molecular species over which the required distortions can be spread. Consider, for example, the reaction $A+X_i \rightarrow$

P_i, and suppose that an angle θ between a reference direction and a dipole in the ith substituent must be changed. For definiteness, let θ_P denote the normal undistorted angle in P_i and let θ_X denote that in X_i, and require that $(\theta_P - \theta_X)$ be distorted by an amount of $+q_\theta$. This could be done by changing θ_P to $(\theta_P + q_\theta)$, or by changing θ_X to $(\theta_X - q_\theta)$, or by changing both angles by lesser amounts—for instance, θ_P to $(\theta_P + q_\theta/2)$ and θ_X to $(\theta_X - q_\theta/2)$. Assuming equal force constants k_θ and a quadratic potential as in Eq. (5.21), the respective distortion energies will be $k_\theta q_\theta^2$, $k_\theta(-q_\theta)^2$, and $k_\theta q_\theta^2/2$ (i.e., $k_\theta[q_\theta/2]^2 + k_\theta[-q_\theta/2]^2$). Clearly the distortion energy is less when the distortion is distributed over two molecules.

In Eq. (5.20) the distortional resultant, $[b\Delta W_{A,X_i} - \Delta W_{A,Y_i}]$, derives from two reactions: $A + X_i \rightarrow P_i$ and $A + Y_i \rightarrow Q_i$. A distortion can therefore be spread not only over X_i and P_i but also over Y_i and Q_i—over four species rather than two, with a further reduction in the distortional resultant.

The full benefit of configurational error tolerance is obtained in free-energy relationships only if the molecular species in both reaction series are configurationally at statistical equilibrium. This condition is satisfied by the reactants, transition states, and products in ordinary thermal reactions but *not* by esoteric "descriptors," such as electronic spectral transition energies or nuclear magnetic resonance chemical shifts or electric dipole moments of the reactants, that are sometimes adopted as surrogates. In short, the traditional method of comparing the free-energy changes in related series of reactions is expected to work best.

6

ENTHALPY–ENTROPY COMPENSATION AND DORMANT INTERACTIONS

In common with linear free-energy relationships, enthalpy–entropy compensation is outside the domain of traditional chemical thermodynamics but within the scope of the thermodynamics of molecular species. There is an important difference, however. Methods of measuring free energies of reaction and activation are relatively accurate, while those for enthalpies of reaction and activation are more error-prone. As a result, comparative free-energy plots for reaction series, such as those in Chapter 5, are normally accepted at face value, while comparative enthalpy or enthalpy–entropy plots for reaction series are more likely to be probed for inaccuracies than for inherent information. One might say that there is a tendency to discount enthalpy–entropy data, a tendency that is not likely to change unless, or until, observed enthalpy–entropy patterns can be authenticated by thermodynamic prediction. Traditional chemical thermodynamics cannot help, because enthalpy–entropy relationships are outside of its domain.

In this chapter we apply the thermodynamics of molecular species to juxtapose theoretical prediction with experimental data. The treatment is descriptive and general, preparing for explicit developments (and proofs) to be given in later chapters.

We will deal particularly with enthalpy–entropy compensation. The model of full enthalpy–entropy compensation may be stated as follows. Let $\mathcal{D}H$ denote either ΔH^0 for a reaction, or δH_i for the perturbation of a component, or $\delta\Delta H^0$ in a reaction series; and let $\mathcal{D}S$ and $\mathcal{D}G$ denote the conjoint changes in entropy and free energy. $\mathcal{D}H$ may have any finite value, positive or negative. Since $\mathcal{D}G = \mathcal{D}H - T\,\mathcal{D}S$, the following holds: When $\mathcal{D}H = T\,\mathcal{D}S$, we obtain $\mathcal{D}G = 0$ and we have full enthalpy–entropy compensation.

In fact, full enthalpy–entropy compensation is a mirage, and we shall cite thermodynamic proof. However, for processes in liquid solutions there is undoubtedly a *propensity* toward enthalpy–entropy compensation—a propensity that can be rationalized thermodynamically.

A THEORETICAL OVERVIEW

The partial properties of formal components in a homogeneous phase, as defined by the Euler equation (2.7), are exact mathematical analogs of the properties of pure components—but the analogy is mathematical, not physical. For example, the partial enthalpy, $H_2 = \partial H/\partial n_2$, of a solute component 2 in a liquid solvent gathers up all contributions to ∂H, the change in enthalpy when ∂n_2 formula weights are added. $\partial H/\partial n_2$ consists of a basic term for the solute itself, plus add-on terms for the solute-induced enthalpy changes in the original medium, per mole of solute. There might be shifts of chemical equilibria or physical changes in, for example, the packing of solvent molecules, associated with the addition of the component. The explicit expression for the partial property of a formal component in a condensed phase is therefore a *sum*:

Partial property of a formal component = a basic term + coupled add-ons

The word "add-on" is used deliberately, because the mathematics of partial differentiation causes each mechanism of coupled change to contribute a specific additive term to the partial property.

If we let the partial property be either the standard free energy, enthalpy, or entropy, we obtain

$$G_2^0 = G_2^0(\text{basic}) + \sum G_2(\text{coupled add-on}) \tag{6.1a}$$

$$H_2^0 = H_2^0(\text{basic}) + \sum H_2(\text{coupled add-on}) \tag{6.1b}$$

$$S_2^0 = S_2^0(\text{basic}) + \sum S_2(\text{coupled add-on}) \tag{6.1c}$$

We have seen [Eqs. (4.13) and (4.14)] that at constant T and P, the Gibbs free energy is unique because of error tolerance with respect to chemical changes. Many of the phenomena coupled to the addition of a component can be classified as chemical—that is, component-induced shifts in chemical equilibria, as measured by changes in progress variables. Such add-ons will be called *chemically coupled*. Equations (6.1) then take the more detailed form (6.2):

$$G_2^0 = G_2^0(\text{basic}) + \sum G_2(\text{chem. coupled}) + \sum G_2(\text{other coupled}) \tag{6.2a}$$

$$H_2^0 = H_2^0(\text{basic}) + \sum H_2(\text{chem. coupled}) + \sum H_2(\text{other coupled}) \tag{6.2b}$$

$$S_2^0 = S_2^0(\text{basic}) + \sum S_2(\text{chem. coupled}) + \sum S_2(\text{other coupled}) \tag{6.2c}$$

Because of error tolerance, the chemically coupled add-ons are subject to full enthalpy–entropy compensation: For each mode of chemical coupling, G_2 (chem. coupled) = 0. Proofs will be given in Chapter 7. At this point we shall state only that the theorem, G_2 (chem. coupled) = 0, is of the same genre as the Tolerance Theorem [Eqs. (4.13) and (4.14)] and is logically related to it. Given that G_2 (chem. coupled) = 0, it follows that H_2 (chem. coupled) = TS_2 (chem. coupled).

When nonchemical add-ons [called *other coupled* in Eq. (6.2)] are significant, there is a good chance that some of them will also conform to the model of full enthalpy–entropy compensation. The Tolerance Theorem is very general. Although, in the derivation of Eq. (4.13), α was the progress variable for a chemical reaction, the derivation works equally well for any physical variable with analogous properties. Thus α may be any distinct physical displacement variable whose mean value in a molecular ensemble can be measured and displaced controllably from equilibrium, either in a real experiment or in a physically valid thought experiment.

The Basic Term

On the other hand, full enthalpy–entropy compensation does *not* embrace the basic term in Eq. (6.2) and thus *cannot* extend to the overall $\{G_2, H_2, S_2\}$ of the formal component.

PROOF. It will be sufficient to consider $\{G_2, H_2, S_2\}$ for a pure component in the gas phase at low pressure, because medium effects and affiliated add-ons then are absent. In the absence of add-ons, $\{G_2, H_2, S_2\}$ characterizes the basic term.

A detailed quantum-statistical proof that negates full enthalpy–entropy compensation for a component in a dilute gas phase was given by Rhodes [1991]. However, instead of repeating his argument, we will sketch a brief proof adapted from original work by Lumry [1995]. This proof focuses on thermal excitation and shows that full compensation of H and TS does not obtain.

Let T_1 denote the temperature at which thermal excitation (EXC) is considered, and examine the contribution to TS. In the following integrals, the component remains hypothetically a gas. From calculus, $d(TS) = S\ dT + T\ dS$. Thus

$$(TS_{\text{EXC}})_{T_1} = \int_0^{T_1} d(TS) = \int_0^{T_1} S\ dT + \int_0^{T_1} T\ dS$$

$$= \int_0^{T_1} S\ dT + \int_0^{T_1} T(C_P/T)\, dT = \underbrace{\int_0^{T_1} S\ dT}_{\text{positive}} + \underbrace{[H_{T_1} - H_0]}_{\text{positive}}$$

Thus $(TS_{\text{EXC}})_{T_1} = (H_{\text{EXC}})_{T_1} + \int_0^{T_1} S\ dT$.

In words, the thermal-excitation part of TS fails to compensate the thermal-excitation part of H by a positive amount equal to $\int S\, dT$. Q.E.D.

It follows as a corollary that in a chemical reaction in a gas phase at low pressure,

$$\Delta(TS_{\mathrm{EXC}}) = \Delta(H_{\mathrm{EXC}}) + \Delta \int S\, dT$$

The relative magnitude of the two terms on the right-hand side varies with the nature of the reaction, and there may well be reactions for which $\Delta \int S\, dT$ is relatively small. Lumry [1995], Benzinger [1971], and Benzinger and Hammer [1981] have offered this concept as a mechanism for close approach to full compensation, believing that the concept is especially useful when the reactant molecules are large and flexible. Their thermodynamic formulations are reasonable, but the resulting equations do not clearly support their belief.

For a broader outlook, let us examine the parameters involved in the quantum-statistical prediction of ΔG^0 for gas-phase reactions. The structure-dependent parameters can be divided into two sets. One set consists of parameters that largely determine ΔH^0, while the other set consists of those that largely determine ΔS^0. The prediction of ΔH^0 focuses on differences in the potential energies of equilibrium configurations and on zero-point energies; that is, on manifestations of bonded interactions and on modes of motion with widely spaced quantum states. By contrast, the prediction of ΔS^0 focuses on modes of motion with closely spaced quantum states, including torsional motions about single bonds and flexible conformational motions, all of which are sensitive to nonbonded interactions. Because bonded and nonbonded interaction mechanisms are fundamentally distinct, it follows that the effects of structure on enthalpies and entropies are not well correlated.

Theoretical Summary

The standard free energy, enthalpy, and entropy of components are sums of a basic term and coupled add-ons, according to Eqs. (6.2). The basic term shows only partial enthalpy–entropy compensation, the chemical add-ons show full compensation, and other add-ons show either partial or full compensation. Thus, for a reaction series, it is never strictly true that $\delta\Delta G^0$ is precisely zero throughout the series, but the departures from full compensation can vary, depending on the magnitude of the fully compensated add-ons, relative to that of the partially compensated add-ons and the basic term.

In later chapters we will analyze two chemical mechanisms that produce fully compensated add-ons: the molar-shift mechanism and the solvation mechanism. In the molar-shift mechanism, an added solute causes shifts in existing chemical equilibria. In the solvation mechanism, solvent molecules transfer from a purely solvent environment to sites adjacent to the solute molecules (*solvent reorganization*). Both mechanisms apply to liquid solutions. These add-ons are most likely to be significant for solutes in hydrogen-bonding solvents, especially in water and in mixed aqueous solvents.

AN EXPERIMENTAL OVERVIEW

In this section we present a sampling of enthalpy–entropy plots, all of which are based on data that should be taken at face value. We will first examine cases that show a strong propensity toward enthalpy–entropy compensation. Then we will look at reaction series in which there is a recognizable propensity but also significant scatter, and we will give guides for interpreting the enthalpy–entropy plots. Most of the reactions take place in liquid solution, but there are also a few surface reactions.

Strong Propensities Toward Compensation

In the examples of this section, ΔG^0 (or $\Delta G^\ddagger$) is relatively constant throughout a reaction series, while ΔH^0 (or $\Delta H^\ddagger$) varies substantially.

Figure 6.1 shows the solvent effect on the proton transfer equilibrium (6.3) in ethanol–water mixtures [Millero et al., 1969; Grunwald and Berkowitz, 1951].

$$ClCH_2COOH + CH_3COO^- \rightleftarrows ClCH_2COO^- + CH_3COOH \tag{6.3}$$

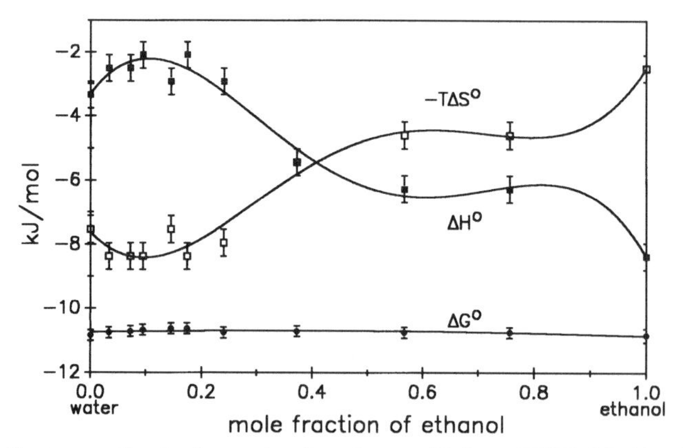

Figure 6.1. Compensation in the proton-transfer equilibrium, $ClCH_2CO_2H + CH_3CO_2^- \rightleftarrows ClCH_2CO_2^- + CH_3CO_2H$, in ethanol–water mixtures at 298 K. ΔG^0 was measured potentiometrically; ΔH^0 was measured calorimetrically. $T\Delta S^0$ was obtained by difference. The error bars represent standard deviations as reported by Millero et al. [1969] and by Grunwald and Berkowitz [1951].

ΔH^0 and $T\Delta S^0$ each varies by more than 4 kJ/mol, while ΔG^0 varies by less than 0.2 kJ/mol. The compensation is especially convincing because the individual curves for ΔH^0 and $T\Delta S^0$ are complex, showing statistically significant maxima and minima. It is probable that the interactions that are compensated in ΔG^0 include solute-induced shifts of one or more hydrogen-bonded linkage equilibria in the ethanol–water solvent.

Figure 6.2 shows data for the binding of a series of zwitterionic amino acids with Cryptand-222, a bicyclic host whose host–guest complex with an amino acid is depicted in (6.4). The solvent is methanol [Danil de Namor et al., 1991a].

(6.4)

In this case, ΔG^0 and ΔH^0 are plotted versus ΔS^0. ΔG^0 is nearly constant, with a range of less than 4 kJ/mol, while the range of ΔH^0 is 37 kJ/mol, more than nine times greater. The plot of ΔH^0 versus ΔS^0 accordingly is close to a straight line with slope nearly equal to the experimental temperature of 298 K. Although

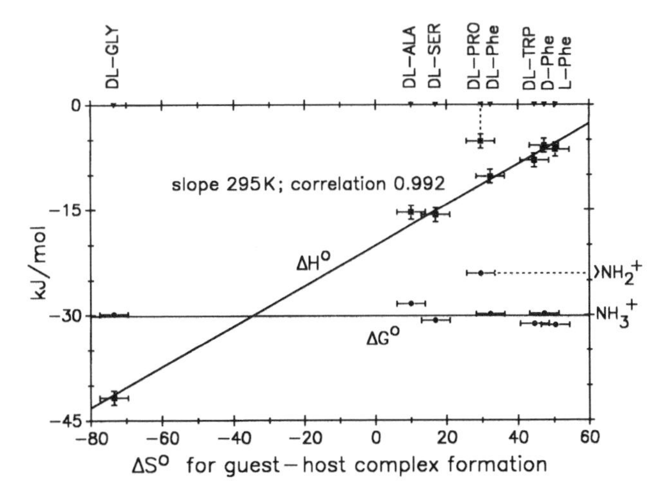

Figure 6.2. Equilibrium results for guest–host binding of amino acids to Cryptand-2,2,2 at 298 K in methanol [Danil de Namor et al., 1991a]. The amino acids exist almost entirely in the zwitterion form.

the ΔH^0 values for D,L-phenylalanine are puzzling, the near-constancy of ΔG^0 guarantees the existence of near-compensation.

Figure 6.3 shows a kinetic example of the compensation effect—the radical decomposition of phenylazotriphenylmethane in a series of regular, non-hydrogen-bonding solvents [Alder and Leffler, 1954].

$$\mathrm{Ph{-}N{=}N{-}C(Ph)_3 \rightarrow Ph\cdot + N{\equiv}N + \cdot C(Ph)_3} \qquad (6.5)$$

Once again, there is a strong propensity toward compensation. The free energy of activation, $\Delta G^{\ddagger}$, varies by less than 1 kJ/mol, while $\Delta H^{\ddagger}$ and $T\Delta S^{\ddagger}$ vary by nearly 20 kJ/mol. The figure shows that $\Delta H^{\ddagger}$ and $T\Delta S^{\ddagger}$ are *not* simple functions of the dielectric constant (arguing against an electrostatic mechanism). On the other hand, the aromatic solvents form a nearly constant group that differs from malonic ester and cyclohexane.

A subtle variation on the theme of nearly complete compensation appears in Fig. 6.4, which shows kinetic medium effects in the neutral hydrolysis of 1-benzoyl-3-phenyl-1,2,4-triazole in water-rich *t*-butyl alcohol–water mixtures [Haak, 1986]. The abscissa is the formula-weight fraction N_2 of alcohol in the solvent.

```
PhCO —N—N=C —Ph + HOH → PhCOOH + HN—N=C —Ph   (6.6)
      |    |                      |    |
    (C H) =N                    (CH)  =N
```

The presence of compensation is evident from the huge difference in the complexity of the relationships. $\Delta H^{\ddagger}$ and $T\Delta S^{\ddagger}$ are complicated, shapely functions

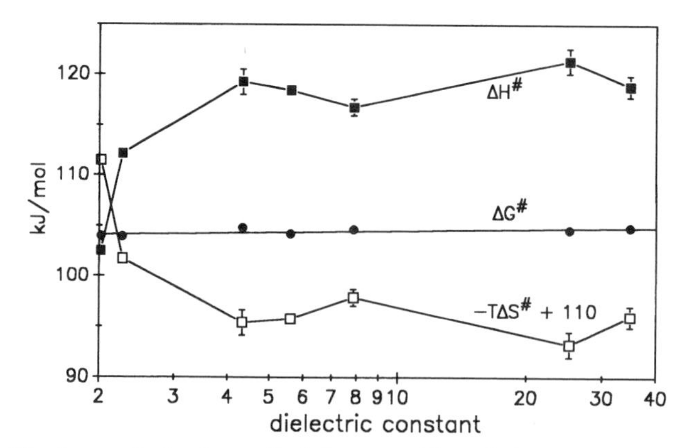

Figure 6.3. Kinetic results for the radical decomposition of phenylazotriphenylmethane at 323 K. The solvents are (left to right) cyclohexane, benzene, anisole, chlorobenzene, malonic ester, benzonitrile, and nitrobenzene. The error bars represent standard deviations [Alder and Leffler, 1954].

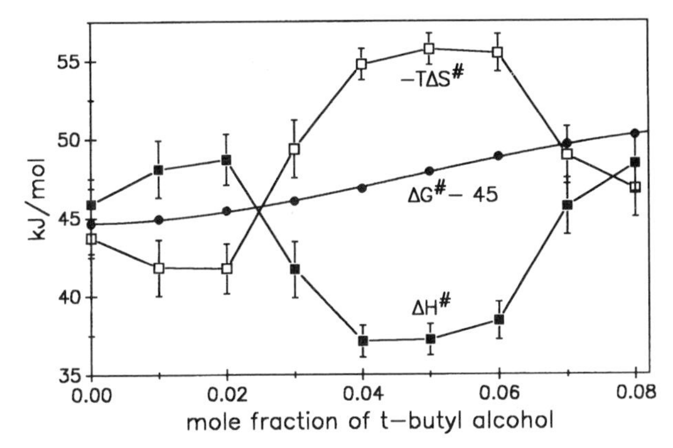

Figure 6.4. Compensation in the rate of neutral hydrolysis of 1-benzoyl-3-phenyl-1,2,4-triazole in *t*-butyl-alcohol–water mixtures at 298 K [Haak, 1986]. The error bars represent *two* standard deviations. On the scale of this figure, the data for $\Delta G^{\ddagger}$ are error-free.

of N_2, with compensating maxima and minima, while $\Delta G^{\ddagger}$ is simply monotonic. In terms of actual numbers, the medium effect on $\Delta G^{\ddagger}$ is only slightly smaller than that on $\Delta H^{\ddagger}$ and $T\Delta S^{\ddagger}$.

Surface reactions on solid catalysts also show propensities toward compensation, although unambiguous examples are rare [Galwey, 1977]. Figures 6.5 and 6.6 show the results of two independent kinetic studies in which the propensities are so striking that no plausible experimental error can upset them.

Figure 6.5 shows compensation in the atomic dissociation of carbon monoxide chemisorbed on a series of molybdenum(110) surfaces [Erickson and

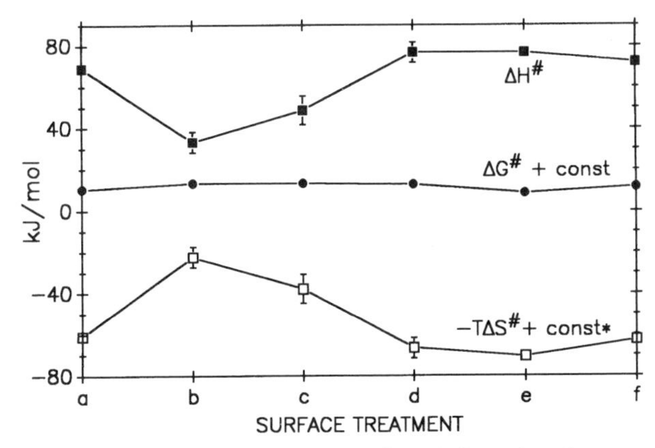

Figure 6.5. Activation thermodynamics for the dissociation of carbon monoxide chemisorbed on modified Mo(110) surfaces at 300 K and 25% completion. Data points, left to right: Original unmodified surface, C/Mo = 1.0, O/Mo = 0.4, (C + O)/Mo = 0.1, S/Mo = 0.1, K/Mo = 0.1 [Erickson and Estrup, 1986].

Estrup, 1986]. Before exposure to carbon monoxide, the Mo(110) surfaces were outgassed, purified, and then modified by treatment with controlled amounts of C, O, S, or K. The nature of the treatment has a marked effect on $\Delta H^{\ddagger}$ and $T\Delta S^{\ddagger}$, which varies by 43 and 48 kJ/mol, respectively, while $\Delta G^{\ddagger}$ varies by less than 5 kJ/mol. Experimental errors in this study range from 1 to 7 kJ/mol for $\Delta H^{\ddagger}$ and $T\Delta S^{\ddagger}$ and are less than 0.5 kJ/mol for $\Delta G^{\ddagger}$.

Figure 6.6 shows the Arrhenius parameters for the rate of desorption of CO from tungsten(110) surfaces covered with palladium. The fraction $(\theta/\theta_{\max})$ of palladium ranges from 0.08 to 0.75 [Zhao and Gomer, 1990]. The abscissa shows log A, the logarithm of the Arrhenius prefactor, whose variation is essentially that of $\Delta S^{\ddagger}$ [cf. Eq. (5.18)]. The ordinate shows E_{act}, which equals $\Delta H^{\ddagger} + RT$. The linear relationship, indicative of strong compensation, spans a 77-kJ/mol range of E_{act}. Judging by the typical stability of metal–CO bonds in molecular complexes of W, Mo, and Pd [Collman et al., 1987], the adsorption of CO is chemisorption, probably with delocalized electron orbitals formed from the metal atoms in the annealed surfaces. The work function—that is, the energy required to dislodge an electron from these surfaces—increases monotonically with E_{act} for the desorption of carbon monoxide and varies by 48 kJ/mol in the experiments shown in the figure [Zhao and Gomer, 1990].

Enthalpy–Entropy Vector Diagrams

Let δG, δH, δS denote either ΔG^0, ΔH^0, ΔS^0 of reaction, or $\Delta G^{\ddagger}$, $\Delta H^{\ddagger}$, $\Delta S^{\ddagger}$ of activation. According to Eq. (6.1), δG, δH, δS then consists of a basic part with partial compensation, as well as coupled add-ons—some with partial and others with full compensation. Let δ_i denote the component of δ with full compensa-

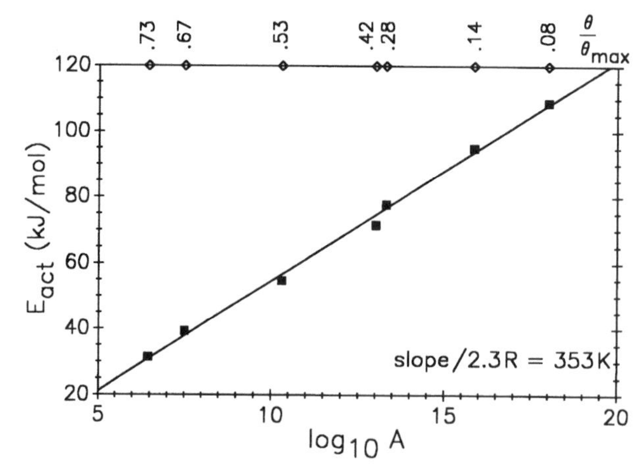

Figure 6.6. Arrhenius parameters for the desorption of CO from a tungsten (1,1,0) surface partly covered with palladium. The coverage by palladium (as fraction $\theta/\theta_{\max}$ of a monolayer) is shown on the upper abscissa [Zhao and Gomer, 1990].

tion, and let δ_{ii} gather up all parts with partial compensation. This dissection generates Eq. (6.7), where T [in Eq. (6.7d)] is the experimental temperature.

$$\delta G = \delta_{ii} G \tag{6.7a}$$

$$\delta H = \delta_i H + \delta_{ii} H \tag{6.7b}$$

$$\delta S = \delta_i S + \delta_{ii} S \tag{6.7c}$$

$$\delta_i H / \delta_i S = T \tag{6.7d}$$

To bring out the inherent complexity, let us think of δS and δH as components of a vector **V**. The vector **V** then becomes the sum of a vector $\mathbf{V}_i$ with components $(\delta_i S, \delta_i H)$ and a vector $\mathbf{V}_{ii}$ with components $(\delta_{ii} S, \delta_{ii} H)$, as indicated in Fig. 6.7a. This analysis is theoretical: Experiment yields only the vector sum.

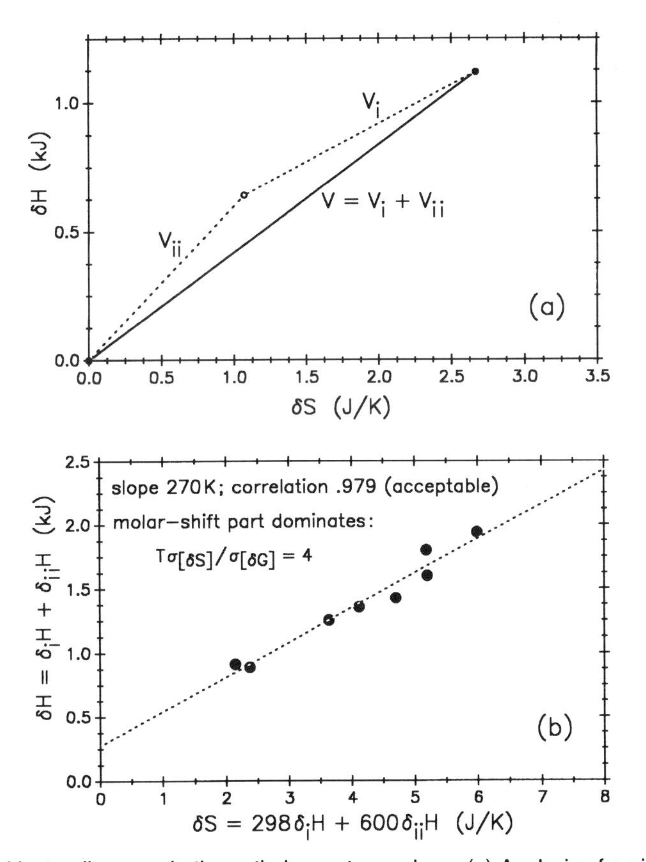

Figure 6.7. Vector diagrams in the enthalpy–entropy plane. (a) Analysis of a single vector. (b) Enthalpy–entropy diagram for a reaction series when the molar-shift terms are dominant. (c) On next page: Enthalpy–entropy diagram for a reaction series in which the isomolar and molar-shift parts are of comparable importance.

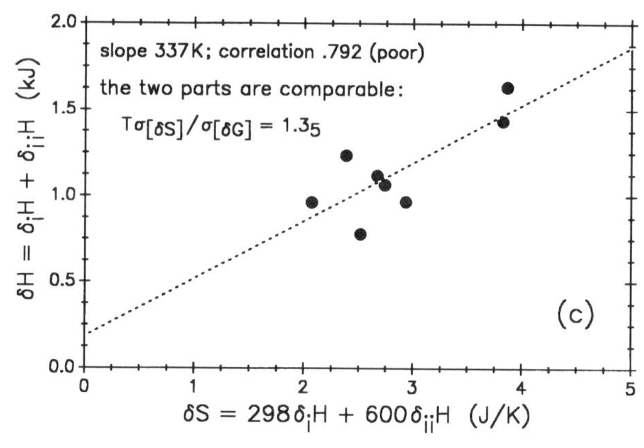

Figure 6.7. *(Continued)*

Figures 6.7b and 6.7c show the results of such vector additions for an entire reaction series. The analysis here assumes the following: (a) There is no correlation between $\delta_i H$ and $\delta_{ii} H$; (b) $\delta_i H/\delta_i S = 298$—that is, $T = 298$K; (c) $\delta_{ii} H = 600\delta_{ii} S$, mimicking Leffler's equation (6.8) (below) with $\beta = 600$ K. The two figures use the same base values for $\delta_i H$ and $\delta_{ii} H$, but with different scaling. In Fig. 6.7b the fully compensated part dominates: The standard deviation σ of the plotted values of δG is only one-fourth that of the plotted values of δS. This plot is acceptably linear, with slope near 298 K.

In Fig. 6.7c neither part dominates. The standard deviation σ of the plotted values of δG is three-fourths that of the plotted values of δS. This plot is essentially a scatter diagram. If one nevertheless fits a straight line, the fitted slope is near 298 K, but with a large statistical error.

Modest or Weak Propensities Toward Compensation

Given the simulations of Fig. 6.7, we may now examine propensities whose significance is blurred by scatter. Figure 6.8 shows data for the binding of a series of cations by gramicidin A [Hinton et al., 1988]. This linear polypeptide was dispersed in lipid lysophosphocholine, thereby causing the gramicidin units to assemble into structures containing channels that specifically bind cations from aqueous solution. In Fig. 6.8a the cations are arranged so that the strength of binding, as measured by $-\Delta G^0$, increases from left to right. ΔH^0 and $T\Delta S^0$ always have the same sign but differ in magnitude, thus showing a propensity toward compensation. Figure 6.8b examines this propensity more closely. Note that ΔH^0 and ΔG^0 vary monotonically, so that ΔG^0 remains a roughly constant fraction of ΔH^0. The fitted line, with slope 0.44, provides an index of the degree of compensation.

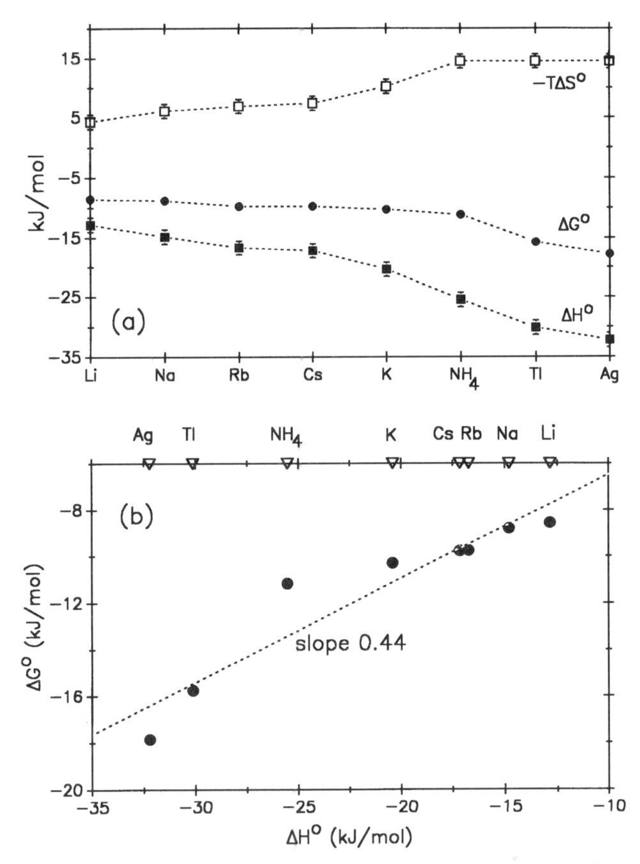

Figure 6.8. Binding of aqueous cations to gramicidin A dispersed in lysophosphocholine at 313 K. (a) The cations are arranged in order of increasing strength of binding [Hinton et al., 1988]. (b) Plot of ΔG^0 versus ΔH^0 [Hinton et al., 1988].

Figure 6.9 shows data for the acid dissociation of substituted benzoic acids in water which, it will be recalled, is the reference reaction series for Hammett linear free-energy plots. The enthalpy changes (ΔH^0) were measured both by direct calorimetry [Matsui et al., 1974] and by measuring the equilibrium constant as a function of temperature and applying the equation, $\Delta H^0 = RT^2(\partial \ln K/\partial T)_P$. Because the calculation of a derivative entails loss of accuracy, the equilibrium constants used to generate $\partial \ln K/\partial T$ were quite precise [Bolton et al., 1972]. The results in Fig. 6.9a were obtained by calorimetry, and the error bars are realistic [Matsui et al., 1974]. A propensity for compensation exists when most of the substituent effects, denoted in the figure by $\delta\Delta H^0$ and $-T\delta\Delta S^0$, lie on opposite sides of the horizontal line through the origin, as is in fact the case.

On the other hand, the propensity for compensation is so slight that only data

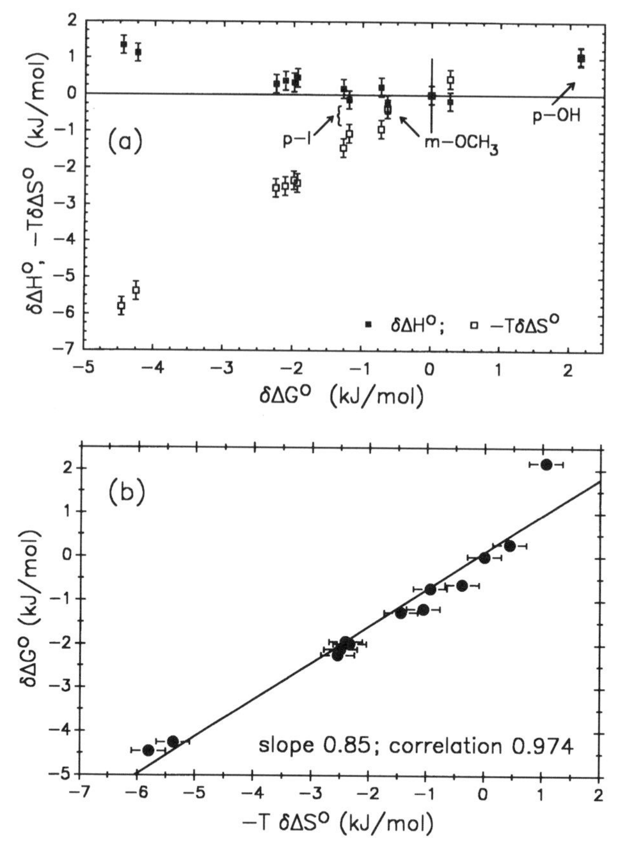

Figure 6.9. Substituent effects in the acid dissociation of benzoic acid in water at 298 K. The substituents are *m*-F, *m*-Cl, *m*-Br, *m*-I, *m*-OH, *m*-OCH_3, *m*-CH_3, *m*-NO_2, and *p*-Br, *p*-I, *p*-OH, and *p*-NO_2. (a) $\delta\Delta H^0$ and $-T\ \delta\Delta S^0$ versus $\delta\Delta G^0$ [Bolton et al., 1972; Matsui et al., 1974]. (b) Correlation of $-T\ \delta\Delta S^0$ with $\delta\Delta G^0$ [Bolton et al., 1972; Matsui et al., 1974].

of high accuracy can reveal it. Figure 6.9b shows this fact by way of a plot of $\delta\Delta G^0$ versus $-T\ \delta\Delta S^0$. The slope, 0.85, is close enough to unity to indicate that $\delta\Delta G^0$ is nearly entropy-controlled.

Figure 6.10 shows partial compensation in a series of adsorption equilibria [Garrone et al., 1989]. ΔH_{ads} was measured calorimetrically. The substrates are arranged so that the equilibrium ratio for adsorption increases from left to right. The functions ΔH_{ads}, ΔG_{ads}, and $T\Delta S_{\mathrm{ads}}$ vary (a) monotonically and (b) with a linear correlation between ΔH_{ads} and ΔS_{ads}. The range of ΔH_{ads} is 35 kJ/mol.

Some Enthalpy–Entropy Patterns

Given linear free-energy relationships, Hammett [1940] predicted that there might be reaction series in which $\delta\Delta H^0$ is proportional to $\delta\Delta S^0$. Leffler [1955]

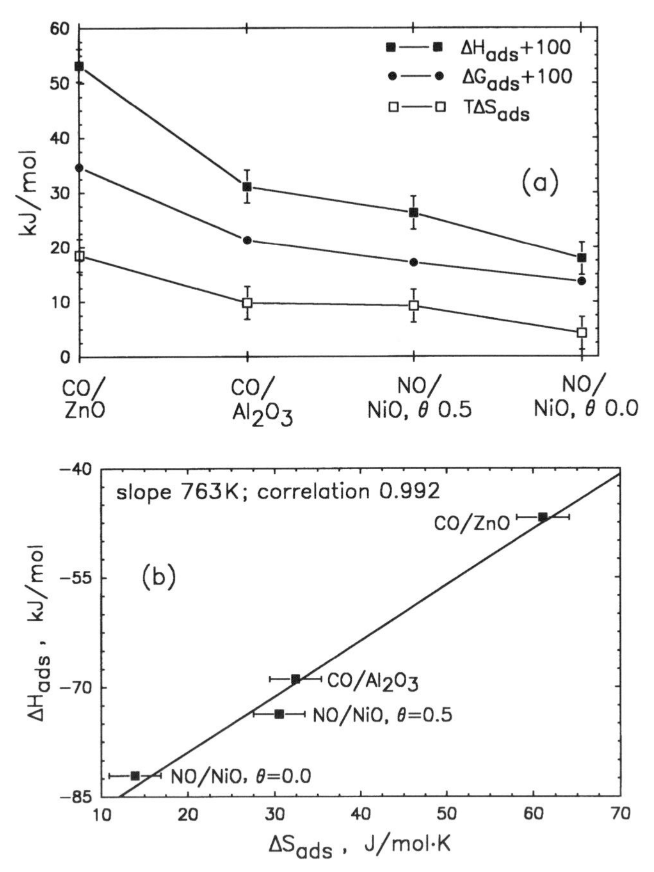

Figure 6.10. (a) Thermodynamic equilibrium data for the adsorption of CO and NO on a series of crystallographically defined surfaces [Garrone et al., 1989]. (b) Same data, plot of ΔH_{ads} versus ΔS_{ads} [Garrone et al., 1989].

reviewed the literature to test this hypothesis and in 1955 was able to cite 50 reaction series for which a linear relationship, the Leffler equation (6.8), is a good empirical approximation. In (6.8), $\delta\Delta G$, $\delta\Delta H$, $\delta\Delta S$ denotes either $\delta\Delta G^0$, $\delta\Delta H^0$, $\delta\Delta S^0$ of reaction or $\delta\Delta G^{\ddagger}$, $\delta\Delta H^{\ddagger}$, $\delta\Delta S^{\ddagger}$ of activation; the parameter β is characteristic of the reaction series.

$$\delta\Delta H = \beta\, \delta\Delta S \tag{6.8a}$$

$$\delta\Delta G = [(\beta - T)/\beta]\, \delta\Delta H \tag{6.8b}$$

One of Leffler's correlation plots is reproduced in Fig. 6.11. Other, more recent examples are shown in Figs. 6.6 and 6.8–6.10.

When observing a linear correlation, one has to decide whether the correlation is an accident emerging from physical complexity (as in the model shown

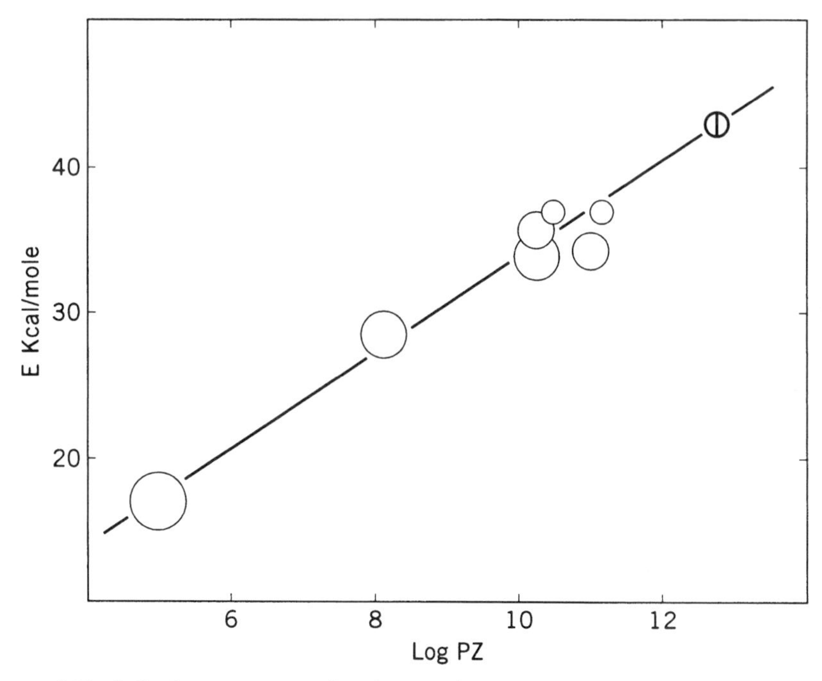

Figure 6.11. Activation parameters for *cis–trans* isomerization of stilbene and substituted stilbenes in the liquid phase (without solvent), of stilbene in the gas phase, and of *p*-amino-*p*′-nitrostilbene in xylene. The slope $(\delta E/\delta \log PZ) = 2.3R\beta$, yielding $\beta = 720$ K. The correlation is 0.985. The mean experimental temperature T is 470 K. (Reproduced from Leffler [1955], with permission of the American Chemical Society.)

in Fig. 6.7b) or whether it truly represents physical simplicity. When the linear correlation is of broad scope, the case for accepting it at face value becomes very strong. Leffler [1955] did just that. Noting that β has the dimensions of a temperature and assuming that β is independent of T, he called β the *isokinetic temperature*—"kinetic" because the evidence consisted mostly of activation parameters, and "iso" because $\delta\Delta G$ is zero when $T = \beta$ [Eq. (6.8b)]. This interpretation leads to the inference that the *sign* of $\delta\Delta G^0$ or $\delta\Delta G^\ddagger$ inverts when T passes through β. That is, substituent effects and/or medium effects should change sign when $T = \beta$!

From a physical point of view this inference cannot be right—especially for substituent effects. Many substituent effects are dominated by potential energy terms that are virtually independent of the temperature. For example, owing to the relatively high electronegativity of the Cl atom, the C–Cl bond is electron-removing relative to the C–H bond, and this relationship does not invert at any likely reaction temperature. The validity of the inference was therefore tested independently, by high-powered analysis of the statistical fit of Leffler's equation (6.8) [Exner, 1964]. It turns out that contrary to appearances, the linear fits need not, and probably do not, indicate a single mode of physical linearity. Thus there is no necessary conflict between the empirical equation (6.8) and the more intricate thermodynamic equations (6.1) and (6.2).

Leffler's β-parameters range from 200 to 800 K overall, but many values are near the common experimental temperature of 300 K. This is especially true for physical processes of biopolymers in water, where $\delta\Delta H$ tends to be so close to $T\ \delta\Delta S$ that one might as well omit treating the slope as an adjustable parameter and proceed directly to Eq. (6.9) [Lumry and Rajender, 1970; Lumry and Gregory, 1986].

$$\delta\Delta H = T\ \delta\Delta S \tag{6.9}$$

In short, whether using a fitted slope or a slope equal to the experimental temperature, the incremental enthalpy–entropy changes in these reaction series border on full compensation. A dramatic example is given in Fig. 6.2; further examples will appear in Chapter 11.

DORMANT INTERACTIONS

When we deal with changes in partial free energies at constant T and P, the fully compensated coupled add-ons in Eqs. (6.1) and (6.2) are dormant, and we deal only with the partly compensated basic part and the partly compensated add-ons. We say "dormant" because, even though the contributions to the partial free energies vanish, the coupling mechanisms themselves do *not* vanish and make contributions to all other add-on properties. In particular, values of H_2 (add-on) and S_2 (add-on) are nonzero, except in unusual special cases. The phenomenology of partial enthalpies and entropies is therefore more complicated than that of partial free energies, because the fully compensated interactions now enter. This is crucial in theoretical modeling. Models for changes in partial free energies at constant T and P may omit—*must* omit—such fully compensated interactions as solvent reorganization and chemically coupled shifts. Models for all other partial properties, including those for changes in partial enthalpies and entropies, must include them.

Scientists are taught as part of their training that theoretical models must be internally consistent. Here we encounter the paradox that thermodynamic consistency requires a change in the number of consequential interaction mechanisms. The uninitiated may find that this takes some getting used to.

Effect of Observational Time Scale

The general formulations of enthalpy–entropy relationships [Eqs. (6.1) and (6.2)] depend on the observational time scale because the processes responsible for the coupled add-ons must be at equilibrium.

When the basic measurement is that of an equilibrium constant, the observational time scale τ_{obs} must be long compared to the time scale τ_K on which the measured reaction goes to equilibrium. The coupled add-ons during such a measurement then reflect those processes whose relaxation times τ_{add-on} are short enough for add-on equilibrium to be reached during τ_{obs}. Supposing that these conditions are met, the relative magnitudes of τ_K and τ_{add-on} are irrelevant.

In kinetic measurements the constraint on τ_{add-on} is somewhat different, because now τ_{obs} must be *short* compared to τ_{kin}, the kinetic half-time of the monitored reaction. As that reaction proceeds, the add-ons must vary in equilibrium with the instantaneous molar composition—they must relax in step. This condition is met when τ_{add-on} is less than τ_{kin}. The observational time scale τ_{obs}, which is also less than τ_{kin}, is now irrelevant. Because τ_{add-on} must be less than τ_{kin}, the number of add-ons that are significant depends in principle on the speed of the monitored reaction.

7

EFFECT OF MOLAR SHIFTS ON STATE FUNCTIONS

Molar shifts are changes in a progress variable associated with a chemical reaction. When the reaction exists at equilibrium, the shifts occur in response to changes in the variables of state. For changes of T or P in a chemical system, the shifts follow the Principle of Le Châtelier. For changes in the formula-weight numbers of the reactant or product components, they follow the Law of Mass Action. For changes in concentration of a dilute solute in a solvent that is a mixture of molecular species, they include solute-induced medium effects. And always, the shifts include physical continuum effects, such as those caused by a change of dielectric constant or of ionic strength.

Molar shifts are important in the thermodynamics of molecular species because they generate a broad spectrum of chemically coupled add-ons, of the sort introduced in Eq. (6.2). While the add-ons themselves are highly specific, the thermodynamic coupling mechanisms are general, and the mathematical derivations are exact. In this chapter we shall derive general equations that apply to any molar shift in any homogeneous phase, regardless of the specific composition model. The results *do* depend, additively, on the number of independent progress variables.

THE TWO REPRESENTATIONS

There are two natural variables for representing molar shifts: (1) the progress variable α for the reaction and (2) the molar stress $y = \partial G/\partial \alpha$. Each has inherent advantages. The molar stress y is the natural variable for specifying chemical equilibrium, where $y = 0$. The progress variable α is the natural variable for specifying molar (not formal) composition and for treating molar shifts. In this section we will examine the transformation $y \leftrightarrow \alpha$, and define some useful nomenclature.

For definiteness, consider composition tree (T-7.1A), which is formally equivalent to (T-2.1) in Chapter 2. We will specify that the relaxation time τ_{rlx} for equilibration of the molecular species C and T is short compared to the observational time scale τ_{obs}, so that C and T exist at equilibrium. The inverse case, $\tau_{\text{rlx}} \gg \tau_{\text{obs}}$, will be considered later.

Composition Tree (T-7.1A)

($\tau_{\text{rlx}} < \tau_{\text{obs}}$ so that C and T exist at equilibrium)

Component 1 (n_1)	Component 2 (n_2)	
$\text{A}\,(n_\text{A} = n_1)$	$\text{C}\,(n_\text{C} = [1-\alpha]n_2)$ $\rightleftarrows$	$\text{T}\,(n_\text{T} = \alpha n_2)$

The partial free energy G_2 of component 2 at equilibrium is traditionally defined by $G_2 = (\partial G/\partial n_2)_{n_1}$, at constant T and P, with the understanding that $y = 0$; that is, C and T remain in equilibrium as ∂n_2 formula weights are added:

$$G_2 = \left(\frac{\partial G}{\partial n_2}\right)_{T,P,n_1,y} \quad \text{with} \quad y = 0$$

The independent variables are (T, P, n_1, n_2, y); n_2 is the active independent variable, and (T, P, n_1, y) are inactive.

On the level of molecular species the corresponding variables are $(T, P, n_\text{A}, n_\text{C}, n_\text{T})$ or, alternatively, (T, P, n_1, n_2, α). The latter set is of interest here because the transformation from (T, P, n_1, n_2, α) to (T, P, n_1, n_2, y) requires the transformation of only one variable. Moreover, since $y = \partial G/\partial\alpha$, the transformation $\alpha \leftrightarrow y$ is a Legendre transformation of the same type as the transformation $P \leftrightarrow V$ (where $V = \partial G/\partial P$), which was considered in Chapter 2.

For example, let us transform $(\partial G/\partial T)_\alpha$ to $(\partial G/\partial T)_y$, using the formal procedure described in Chapter 2:

(P, n_1, n_2) = constant through Eq. (7.2)

(a) $G(T, \alpha)$: $\quad dG = \left(\frac{\partial G}{\partial T}\right)_\alpha dT + \left(\frac{\partial G}{\partial \alpha}\right)_T d\alpha$

(b) $\alpha(T, y)$: $\quad d\alpha = \left(\frac{\partial \alpha}{\partial T}\right)_y dT + \left(\frac{\partial \alpha}{\partial y}\right)_T dy$

(c) Substitute (b) in (a) to obtain $G(T, y)$:

$$dG = \left(\frac{\partial G}{\partial T}\right)_\alpha dT + \left(\frac{\partial G}{\partial \alpha}\right)_T \left(\frac{\partial \alpha}{\partial T}\right)_y dT + \left(\frac{\partial G}{\partial \alpha}\right)_T \left(\frac{\partial \alpha}{\partial y}\right)_T dy$$

$(\partial G/\partial T)_y$ is the coefficient of dT:

$$\left(\frac{\partial G}{\partial T}\right)_y = \left(\frac{\partial G}{\partial T}\right)_\alpha + \left(\frac{\partial G}{\partial \alpha}\right)_T \left(\frac{\partial \alpha}{\partial T}\right)_y \tag{7.1a}$$

Let $(S)_y = -(\partial G/\partial T)_y$, and $(S)_\alpha = -(\partial G/\partial T)_\alpha$. Equation (7.1a) then takes the form (7.1b):

$$\underset{\text{Constant stress}}{(S)_y} = \underset{\text{Isomolar}}{(S)_\alpha} \underset{\text{Molar-shift term}}{- \left(\frac{\partial G}{\partial \alpha}\right)_T \left(\frac{\partial \alpha}{\partial T}\right)_y} \tag{7.1b}$$

At equilibrium, $\partial G/\partial \alpha = y = 0$, and $-(\partial G/\partial T)_{y=0}$ is the entropy S of traditional equilibrium thermodynamics. Equation (7.1b) then becomes Eq. (7.2):

$$\underset{\text{Entropy}}{S} = \underset{\text{Isomolar entropy}}{(S)_\alpha} \tag{7.2}$$

Equations (7.1b) and (7.2) indicate the nomenclature we shall use for thermodynamic partial derivatives. A derivative at constant y denotes the given property at constant stress. A derivative at $y = 0$ denotes the same property with maintenance of equilibrium; it will be named by its traditional name. For instance, $-(\partial G/\partial T)_{y=0}$ is simply the entropy S, without an adjective, as mandated by a century of tradition.

A derivative at constant α denotes the given property at constant molar (not formal) composition. It will be called *isomolar*; α may be constant at any chosen value, including the equilibrium value α_{eq} for the given conditions.

A term proportional to a molar shift $\partial\alpha$, such as the term proportional to $(\partial\alpha/\partial T)_y$ in Eq. (7.1b), will be called a molar-shift term. When the composition is off-equilibrium ($y \neq 0$), molar-shift terms vanish only rarely. But at equilibrium, molar-shift terms vanish whenever the infinitesimal change $\partial\alpha$ (or its finite small scale-up $\delta\alpha$) is negligible. For instance, $\partial\alpha/\partial T$ or $\delta\alpha/\delta T$ might be negligible because the standard enthalpy change ΔH^0 for the process is close to zero. At equilibrium ($y = 0$), molar shift terms vanish also when the general expression for the term is proportional to $\partial G/\partial\alpha = y$, as in (7.1b) $\rightarrow$ (7.2). As a result, in Eq. (7.2) the traditional entropy S is equal to the isomolar entropy $(S)_\alpha$. It should be noted, however, that equality at the equilibrium point does not imply identity. In general, $(S)_y$ and $(S)_\alpha$ are separate functions, which just happen to coincide when $y = 0$. The preceding nomenclature is summarized in Table 7.1.

Molar-shift terms arise by mathematical necessity whenever the representation is transformed from the y-basis to the α-basis. We shall find that molar-

Table 7.1. Summary of nomenclature in various representations

Derivative	Symbol	Name
	(1) y-Set	
$-(\partial G/\partial T)_y$	$(S)_y$	No formal name
$(\partial^2 G/\partial P\, \partial n_2)_y$	$(V_2)_y$	No formal name
$-(\partial G/\partial T)_{y=0}$	S	Entropy[a]
$(\partial^2 G/\partial P\, \partial n_2)_{y=0}$	V_2	Partial volume[b]
	(2) α-Set	
$-(\partial G/\partial T)_\alpha$	$(S)_\alpha$	Isomolar entropy[a]
$(\partial^2 G/\partial P\, \partial n_2)_\alpha$	$(V_2)_\alpha$	Isomolar partial volume[b]

[a]For the homogeneous phase.
[b]For component 2.

shift terms vanish at equilibrium for all first derivatives of the free-energy G; but that, for second and higher derivatives of G, the molar-shift terms in general do not vanish. In terms of Eq. (6.2), the isomolar term may be identified with the basic term, and the molar-shift term may be identified with a chemically coupled add-on. This chapter will contain sufficient evidence to support these identifications.

The composition tree (T-7.1A) assumes that $\tau_{rlx} < \tau_{obs}$, so that C and T exist at equilibrium. When in fact $\tau_{rlx} \gg \tau_{obs}$, C and T during an observation are mired off equilibrium and thereby acquire the status of components. The composition tree then changes to (T-7.1B). Because $\tau_{rlx} \gg \tau_{obs}$, the chemical relaxation required for molar shifts does not have time to take place, and molar-shift terms—terms proportional to $\partial\alpha$—become negligible.

Composition Tree (T-7.1B)

($\tau_{rlx} >> \tau_{obs}$ so that C and T are components)

Component 1 (n_1)	Component 2 (n_2)	Component 3 (n_3)
A ($n_A = n_1$)	C ($n_C = n_2$)	T ($n_T = n_3$)

DERIVATIONS FOR A SINGLE PROGRESS VARIABLE

The systems treated in this section consist of a single phase with a single chemical reaction. The number of components is irrelevant, but, for definiteness, we shall assume two components. More important, the nature and stoichiometry of the reaction is also irrelevant. For example, the composition tree might be

(T-7.1A) where the reaction is localized in component 2, or (T-7.2) where the reaction embraces both components—but not (T-7.1B) where the short observational time scale precludes reaction. Mathematically, what matters is that the free energy G is a function of (T, P, n_1, n_2) and one additional variable, which may be either the progress variable α or the molar stress y. Given these conditions, we shall transform partial properties from the α-set to the y-set by straight mathematics.

Composition Tree (T-7.2)

Component 1 (n_1)		Component 2 (n_2)
"A"		"U"
$A\,(n_1 - n_2\alpha)$ +	$U\,(n_2[1-\alpha])$ $\rightleftarrows$	$A\cdot U\,(n_2\alpha)$

Partial Free Energy

Starting with the free energy as a function $G(T, P, n_1, n_2, \alpha)$, we shall seek a relationship between $(G_2)_y$ and $(G_2)_\alpha$. By definition, $(G_2)_y = (\partial G/\partial n_2)_y$ and $(G_2)_\alpha = (\partial G/\partial n_2)_\alpha$. The traditional G_2 equals $(\partial G/\partial n_2)_{y=0}$. All derivatives are taken at constant (T, P, n_1). The operative variables thus are (n_2, y) and (n_2, α). The desired relationship is obtained by a simple transformation from $G(n_2, y)$ to $G(n_2, \alpha)$, followed by substitution of appropriate notation. The mathematics in Eqs. (7.3) and (7.4) is analogous to that in Eqs. (7.1) and (7.2).

(T, P, n_1) = constant; transform from (n_2, α) to (n_2, y)

$$\left(\frac{\partial G}{\partial n_2}\right)_y = \left(\frac{\partial G}{\partial n_2}\right)_\alpha + \left(\frac{\partial G}{\partial \alpha}\right)_{n_2}\left(\frac{\partial \alpha}{\partial n_2}\right)_y \qquad (7.3a)$$

$$(G_2)_y = \underbrace{(G_2)_\alpha}_{\text{Isomolar}} + \underbrace{y\,(\partial \alpha/\partial n_2)_y}_{\text{Molar-shift term}} \qquad (7.3b)$$

At equilibrium,

$$(G_2)_{y=0} \equiv G_2 = (G_2)_\alpha \qquad (7.4)$$

As in Eqs. (7.1) and (7.2), the property in the y-set is equal to the sum of the property in the α-set plus a molar-shift term. The latter vanishes at equilibrium.

Because the initial transformation involves mathematics rather than physics, an analogous result, Eqs. (7.5), is obtained when G is differentiated with respect to any state variable v, which may be T, P, or a formula-weight number.

$$\left(\frac{\partial G}{\partial v}\right)_y = \left(\frac{\partial G}{\partial v}\right)_\alpha + \left(\frac{\partial G}{\partial \alpha}\right)_v \left(\frac{\partial \alpha}{\partial v}\right)_y \tag{7.5a}$$

At equilibrium,

$$(\partial G/\partial v)_{y=0} \equiv (\partial G/\partial v) = (\partial G/\partial v)_\alpha \tag{7.5b}$$

The disappearance of the molar-shift term at equilibrium is characteristic of first derivatives of the free energy. Second and higher derivatives retain molar-shift terms at equilibrium, as we shall see.

Partial Entropy

Continuing with an analysis of partial properties, we will give the full mathematics for the partial entropy and then write equations for other second derivatives of G by analogy. By definition, $(S_2)_\alpha = -(\partial^2 G/\partial n_2\, \partial T)_{P,n_1,\alpha}$ and $(S_2)_y = -(\partial^2 G/\partial n_2\, \partial T)_{P,n_1,y}$. We will begin with Eq. (7.3) and differentiate both sides with respect to T at constants n_2 and y. P and n_1 remain constant.

Equation (7.3) is rewritten below, with each partial derivative assigned a number.

(P, n_1) = constant in Eqs. (7.3) and (7.6)–(7.10)

$$\begin{array}{ccccccc} \left(\frac{\partial G}{\partial n_2}\right)_{y,T} & = & \left(\frac{\partial G}{\partial n_2}\right)_{\alpha,T} & + & \left(\frac{\partial G}{\partial \alpha}\right)_{n_2,T} & & \left(\frac{\partial \alpha}{\partial n_2}\right)_{y,T} \\ (1) & = & (2) & + & (3) & \times & (4) \end{array} \tag{7.3}$$

(1) and (4) are functions of (T, n_2, y), and partial differentiation with respect to T at constant y is straightforward. But (2) and (3) are functions of (T, n_2, α), and partial differentiation with respect to T at constant y requires a change of variables. To make this change, let ϕ be any function $\phi(T, n_2, \alpha)$; the required change then takes the form of Eq. (7.6):

$$(\partial \phi/\partial T)_{y,n_2} = (\partial \phi/\partial T)_{\alpha,n_2} + (\partial \phi/\partial \alpha)_{n_2,T}(\partial \alpha/\partial T)_{y,n_2} \tag{7.6}$$

Terms (2) and (3) therefore yield Eqs. (7.7):

$$\left(\frac{\partial (2)}{\partial T}\right)_{n_2,y} = \left(\frac{\partial^2 G}{\partial n_2 \partial T}\right)_\alpha + \left(\frac{\partial^2 G}{\partial n_2 \partial \alpha}\right)_T \left(\frac{\partial \alpha}{\partial T}\right)_{n_2,y} \tag{7.7a}$$

$$\left(\frac{\partial(3)}{\partial T}\right)_{n_2,y} = \left(\frac{\partial^2 G}{\partial\alpha\partial T}\right)_{n_2} + \left(\frac{\partial^2 G}{\partial\alpha^2}\right)_{T,n_2}\left(\frac{\partial\alpha}{\partial T}\right)_{n_2,y} \tag{7.7b}$$

These equations are simplified by noting that $(\partial G/\partial\alpha)_{n_2,T} \equiv y(\alpha, n_2, T)$. Applying the chain rule of partial differentiation,

$$\text{at constant } T, \quad (\partial y/\partial n_2)_{\alpha,T} = -(\partial y/\partial\alpha)_{n_2,T}(\partial\alpha/\partial n_2)_{y,T}$$
$$\text{at constant } n_2, \quad (\partial y/\partial T)_{\alpha,n_2} = -(\partial y/\partial\alpha)_{n_2,T}(\partial\alpha/\partial T)_{y,n_2}$$

Moreover, $(\partial(3)/\partial T)_{n_2,y} = (\partial y/\partial T)_{n_2,y} = 0$. Reintroducing $(\partial G/\partial\alpha)_{n_2,T}$ for y and applying these relations, we obtain Eqs. (7.8). (The minus sign is correct.)

$$\left(\frac{\partial(2)}{\partial T}\right)_{n_2,y} = \left(\frac{\partial^2 G}{\partial n_2\partial T}\right)_{\alpha} - \left(\frac{\partial^2 G}{\partial\alpha^2}\right)_{n_2,T}\left(\frac{\partial\alpha}{\partial T}\right)_{n_2,y}\left(\frac{\partial\alpha}{\partial n_2}\right)_{T,y} \tag{7.8a}$$

$$\left(\frac{\partial(3)}{\partial T}\right)_{n_2,y} = 0 \tag{7.8b}$$

Accordingly, partial differentiation of Eq. (7.3) with respect to T yields Eq. (7.9):

$$\left(\frac{\partial^2 G}{\partial n_2\partial T}\right)_{y} = \left(\frac{\partial^2 G}{\partial n_2\partial T}\right)_{\alpha} - \left(\frac{\partial^2 G}{\partial\alpha^2}\right)_{n_2,T}\left(\frac{\partial\alpha}{\partial T}\right)_{n_2,y}\left(\frac{\partial\alpha}{\partial n_2}\right)_{T,y} + \left(\frac{\partial G}{\partial\alpha}\right)_{n_2,T}\left(\frac{\partial^2\alpha}{\partial n_2\partial T}\right)_{y} \tag{7.9}$$

To obtain an expression for $(S_2)_{y=0}$ (i.e., for S_2) we let y [$= \partial G/\partial\alpha$] be zero and introduce $(S_2)_\alpha = -(\partial^2 G/\partial n_2\,\partial T)_\alpha$. The result is (7.10).

$$S_2 = \underbrace{(S_2)_\alpha}_{\text{Isomolar}} + \underbrace{\left(\frac{\partial^2 G}{\partial\alpha^2}\right)_{T,n_2}\left(\frac{\partial\alpha}{\partial n_2}\right)_{T,y=0}\left(\frac{\partial\alpha}{\partial T}\right)_{n_2,y=0}}_{\text{Molar-shift term}} \tag{7.10}$$

The molar-shift term does not vanish because (a) $\partial^2 G/\partial\alpha^2 > 0$ (G is at a minimum), and (b) the equilibrium value of α, at least in principle, is a function of T and m_2. In practice, the molar-shift term is often negligible compared to experimental errors in S_2. But there are important examples in which it is quite significant.

Partial Enthalpy. Propensity Toward Compensation

By definition, $H = G + TS$. Hence in the $y = 0$ representation, $H_2 = G_2 + TS_2$ [Eq. (7.11a)]. In the α-representation, the expression $H = G + TS$ takes the form $(H)_\alpha = (G)_\alpha + T(S)_\alpha$. (Recall that $-(\partial G/\partial T)_\alpha = (S)_\alpha$ [Eq. (7.1b)].) Partial differentiation at constant α then yields Eq. (7.11b).

$$y = 0 \text{ representation:} \qquad H_2 = G_2 + TS_2 \tag{7.11a}$$

$$\alpha\text{-representation:} \qquad (H_2)_\alpha = (G_2)_\alpha + T(S_2)_\alpha \tag{7.11b}$$

For simplicity of notation, we will denote the molar-shift term in Eq. (7.10) by $(S_2)_{\text{shift}}$ and add Eq. (7.4). The result is Eq. (7.12), where $(H_2)_{\text{shift}} = T(S_2)_{\text{shift}}$. At equilibrium,

$$\begin{aligned}
&\text{(a)} & TS_2 &= T(S_2)_\alpha + T(S_2)_{\text{shift}} \quad \text{from (7.10)} \\
&\text{(b)} & G_2 &= (G_2)_\alpha & (7.4) \\
&\text{(a)} + \text{(b)} & G_2 + TS_2 &= (G_2)_\alpha + T(S_2)_\alpha + T(S_2)_{\text{shift}} \\
&\therefore & H_2 &= (H_2)_\alpha + T(S_2)_{\text{shift}} \equiv (H_2)_\alpha + (H_2)_{\text{shift}} & (7.12)
\end{aligned}$$

On writing the molar-shift term in full we obtain Eq. (7.13):

$$H_2 = \underbrace{(H_2)_\alpha}_{\text{Isomolar}} + \underbrace{T\left(\frac{\partial^2 G}{\partial \alpha^2}\right)_{T,n_2}\left(\frac{\partial \alpha}{\partial n_2}\right)_{T,y=0}\left(\frac{\partial \alpha}{\partial T}\right)_{n_2,y=0}}_{\text{Molar-shift term}} \tag{7.13}$$

Note that the molar-shift term for H_2 is exactly T times the molar-shift term for S_2:

$$(G_2)_{\text{shift}} = 0; \qquad \text{hence } (H_2)_{\text{shift}} = T(S_2)_{\text{shift}} \tag{7.14}$$

On comparing Eq. (7.14) with Eq. (6.2), it becomes logical to identify the molar-shift terms of Eq. (7.14) with the chemically coupled add-ons of Eq. (6.2). To be observable, $(H_2)_{\text{shift}}$ must be substantial enough to make its presence felt, when added to $(H_2)_\alpha$. Relative magnitudes will be taken up in Chapter 8.

Partial Volume. Interaction Coefficient

The mathematics that was used to derive Eq. (7.9) is general. Thus, for any second derivative $\partial^2 G/\partial v \partial w$, Eq. (7.15) gives the relationship at equilibrium between the $y = 0$ and the α-representations; v and w are variables in the set

$(T, P, \{n\})$; v may be identical to w (as in $\partial^2 G/\partial P^2$). At equilibrium,

$$\left(\frac{\partial^2 G}{\partial v\,\partial w}\right)_{y=0} = \underbrace{\left(\frac{\partial^2 G}{\partial v\,\partial w}\right)_{\alpha}}_{\text{Isomolar}} - \underbrace{\left(\frac{\partial^2 G}{\partial \alpha^2}\right)_{v,w}\left(\frac{\partial \alpha}{\partial v}\right)_{w,y=0}\left(\frac{\partial \alpha}{\partial w}\right)_{v,y=0}}_{\text{Molar-shift term}} \tag{7.15}$$

It follows that any second derivative of the Gibbs free energy, measured with maintenance of equilibrium ($y = 0$), is a sum consisting of the corresponding isomolar second derivative and a molar-shift term.

For example, when $v = n_2$ and $w = P$, the left-hand side of Eq. (7.15) represents $(\partial^2 G/\partial n_2\,\partial P)_{y=0}$, which is identical to the partial volume V_2 at equilibrium. Similarly, $(\partial^2 G/\partial n_2\,\partial P)_\alpha = (V_2)_\alpha$ in our notation. Substitution in Eq. (7.15) then yields Eqs. (7.16). Note that the negative sign is now part of $(V_2)_{\text{shift}}$.

(T, n_1) = constant

$$V_2 = (V_2)_\alpha + \left(\frac{-\partial^2 G}{\partial \alpha^2}\right)_{n_2,P}\left(\frac{\partial \alpha}{\partial n_2}\right)_{P,y=0}\left(\frac{\partial \alpha}{\partial P}\right)_{n_2,y=0} \tag{7.16a}$$

$$= (V_2)_\alpha + (V_2)_{\text{shift}} \tag{7.16b}$$

For another example, consider a three-component system in which 1 is the solvent and 2 and 3 are nondilute solutes. The solutes interact with an interaction coefficient $\partial G_2/\partial m_3$, which is defined by $(\partial G_2/\partial m_3)_{y=0} = (n_3/m_3)\cdot(\partial^2 G/\partial n_3\,\partial n_2)_{y=0}$. Thus Eq. (7.15) applies, with $v = n_2$ and $w = n_3$, and the coefficient $\partial G_2/\partial m_3$ in principle has a molar-shift term. Qualitatively, $\partial G_2/\partial m_3$ measures the medium effect on G_2 due to component 3. However, $n_3/m_3 = n_2/m_2$; hence $\partial G_2/\partial m_3 = \partial G_3/\partial m_2$. There is symmetry with respect to interchange of subscript labels for the solutes, and the medium effect of 3 on G_2 is identical to that of 2 on G_3.

Partial Heat Capacity

In the $y = 0$ representation, $C_{P,2} = -T(\partial^3 G/\partial n_2\,\partial T^2)_{y=0}$; in the α-representation, $(C_{P,2})_\alpha = -T(\partial^3 G/\partial n_2\,\partial T^2)_\alpha$. Both third derivatives are at constant (P, n_1). To derive the relationship between the two, we start with Eq. (7.9), differentiate with respect to T, and then multiply the result by $-T$.

Term-by-term differentiation of Eq. (7.9), with the aid of Eqs. (7.6) and (7.7b), yields the rather formidable result (7.17) for $y = 0$. On introducing the definitions for $C_{P,2}$ and $(C_{P,2})_\alpha$ and representing the remaining terms by numbers, we obtain Eq. (7.18).

***P* and n_1 are constant; $y = 0$**

$$\left(\frac{\partial^3 G}{\partial n_2 \partial T^2}\right)_y = \left(\frac{\partial^3 G}{\partial n_2\, \partial T^2}\right)_\alpha + \underbrace{\left(\frac{\partial^3 G}{\partial n_2\, \partial T\, \partial \alpha}\right)\left(\frac{\partial \alpha}{\partial T}\right)_{y, n_2}}_{(11)}$$

$$- \underbrace{\left(\frac{\partial^3 G}{\partial \alpha^2 \partial T}\right)_{n_2}\left(\frac{\partial \alpha}{\partial T}\right)_{y, n_2}\left(\frac{\partial \alpha}{\partial n_2}\right)_{y, T}}_{(12)}$$

$$- \underbrace{\left(\frac{\partial^3 G}{\partial \alpha^3}\right)_{n_2, T}\left(\frac{\partial \alpha}{\partial T}\right)^2_{n_2, y}\left(\frac{\partial \alpha}{\partial n_2}\right)_{T, y}}_{(13)}$$

$$- \underbrace{\left(\frac{\partial^2 G}{\partial \alpha^2}\right)_{n_2, T}\left(\frac{\partial^2 \alpha}{\partial T^2}\right)_{n_2, y}\left(\frac{\partial \alpha}{\partial n_2}\right)_{T, y}}_{(14)}$$

$$- \underbrace{\left(\frac{\partial^2 G}{\partial \alpha^2}\right)_{n_2, T}\left(\frac{\partial^2 \alpha}{\partial n_2\, \partial T}\right)_{y}\left(\frac{\partial \alpha}{\partial T}\right)_{n_2, y}}_{(15)} \tag{7.17}$$

$$\underset{\text{Isomolar} +}{C_{P,2} = (C_{P,2})_\alpha} + \underset{\text{Molar-shift term}}{(-T) \cdot [(11) - (12) - (13) - (14) - (15)]} \tag{7.18}$$

These general expressions for $C_{P,2}$ are probably too abstract for their physical meanings to be obvious, but their specific solutions will be clear enough. For example, we shall find that in dilute solutions, the molar-shift term includes an obvious contribution from the endothermic heat of reaction as α_{eq} shifts per degree rise in temperature. It also includes less obvious contributions from other interactions.

PARTIAL PROPERTIES FOR MULTIPLE PROGRESS VARIABLES

In the case of two or more independent reactions, the total molar-shift term is simply a sum of molar-shift terms for the individual reactions, each of the form already derived. This theorem will now be proved. The mathematics is basically the same as before, but of course it involves more variables.

The homogeneous system at constant T and P consists of f formal components whose molecular species engage in k independent reactions of unspecified nature. The progress variables are $\alpha_1, \ldots, \alpha_g, \ldots, \alpha_k$, and the corresponding molar stresses are $y_1, \ldots, y_g, \ldots, y_k$. The symbols $\{n\}$, $\{\alpha\}$, and $\{y\}$ denote full sets of variables; while $\{n \neq n_i\}$ and $\{\alpha \neq \alpha_g, \alpha_h\}$ denote partial sets.

$[\{n\} = \text{constant}]$ denotes a closed system, and $[\{y\} = 0]$ denotes full equilibrium. Partial properties will be derived for component 2.

Uncoupling Theorem

The key to additivity resides in an uncoupling theorem that eliminates cross terms in second (and higher) derivatives. This will be stated first, since the proof is instructive.

Uncoupling Theorem. As any ith reaction goes to equilibrium at constant T and P, the molar stress $y_i = (\partial G/\partial \alpha_i)_{\{n\}, \{\alpha \neq \alpha_i\}}$ vanishes even if the inactive set $\{\alpha \neq \alpha_i\}$ includes α's that are not at equilibrium. As a corollary, when the ith reaction is at equilibrium, a change in any other progress variable α_h has no effect on $\partial G/\partial \alpha_i$:

$$\left(\frac{\partial^2 G}{\partial \alpha_i \, \partial \alpha_h} \right)_{\{\alpha \neq \alpha_i, \alpha_h\}; \alpha_i = \alpha_{i,\text{eq}}} = 0 \quad \text{for any } h \neq i \qquad (7.19)$$

Either reaction i or h or both must be at equilibrium.

PROOF. The derivative $(\partial G/\partial \alpha_i)_{\{n\}, \{\alpha \neq \alpha_i\}}$ specifies a closed system with one active (α_i) and $k - 1$ mathematically inactive progress variables. In the corresponding physical problem, the active reaction relaxes to equilibrium while the inactive reactions stay put—some because they exist at equilibrium, and the others because they are mired or "frozen" off-equilibrium. Basically, as explained in Chapter 4, a reaction whose progress has been frozen—by whatever mechanism—is a nonreaction. Thus, when the set $\{\alpha \neq \alpha_i\}$ is constant, the chemistry effectively simplifies to that of the ith

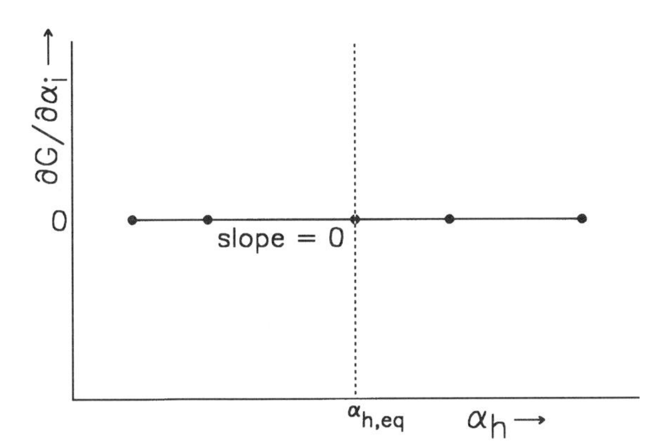

Figure 7.1. Graphical proof of Eq. (7.19).

reaction. And, as for any single reaction coming to equilibrium, the Gibbs free energy reaches a minimum. The fact that a minimum is reached depends in no way on the values at which the other α's are frozen.

Given this premise, Eq. (7.19) can be derived graphically from the plot of $\partial G/\partial \alpha_i$ versus α_h, as in Fig. 7.1. Along the horizontal line the ith reaction exists at equilibrium, so that $(\partial G/\partial \alpha_i)$ is constant at zero. The "frozen" value, α_h, for the other reaction has no effect, as shown by the solid circles. Accordingly, $\partial^2 G/\partial \alpha_h\, \partial \alpha_i = 0$. While the ith reaction must be at equilibrium, reaction h may be, but need not be, at equilibrium. Q.E.D.

Partial Free Energy

The development parallels that for a single progress variable. First Eq. (7.20) transforms variables from $(T, P, \{n\}, \{\alpha\})$ to $(T, P, \{n\}, \{y\})$; then Eq. (7.21) gives the solution at full equilibrium, where $\partial G/\partial \alpha_g = 0$ for all k reactions.

$(T, P, \{n \neq n_2\})$ = constant in Eqs. (7.20) and (7.21)

$$\left(\frac{\partial G}{\partial n_2}\right)_{\{y\}} = \underbrace{\left(\frac{\partial G}{\partial n_2}\right)_{\{\alpha\}}}_{\text{Isomolar}} + \underbrace{\sum_{g=1}^{k} \left(\frac{\partial G}{\partial \alpha_g}\right)_{n_2, \{\alpha \neq \alpha_g\}} \left(\frac{\partial \alpha_g}{\partial n_2}\right)_{\{y\}}}_{\text{Molar-shift term}} \tag{7.20}$$

$$\left(\frac{\partial G}{\partial n_2}\right)_{\{y\}=0} = \underbrace{\left(\frac{\partial G}{\partial n_2}\right)_{\{\alpha\}=\{\alpha_{\text{eq}}\}}}_{\text{Isomolar}} \qquad \therefore \quad G_2 = (G_2)_\alpha \tag{7.21}$$

It follows from Eq. (7.21) that partial free energies measured at operational equilibrium are always isomolar, no matter how complex the chemistry.

Regardless of the time-scale on which equilibrium is established, the system probably includes some reactions that are "frozen" or mired off equilibrium. By definition, such "frozen" reactions do not shift, and hence do not contribute molar-shift terms. They need not be considered here.

Partial Entropy

To derive an expression for S_2, Eq. (7.20) is differentiated with respect to T at constant P and $\{n \neq n_2\}$ with the help of Eq. (7.22), which is analogous to Eq. (7.6). As before, ϕ is a function of n_2, T, and α. In the present differentiation, ϕ is either $(\partial G/\partial n_2)_{\{\alpha\}}$ or $(\partial G/\partial d_g)_{n_2, \{\alpha \neq \alpha_g\}}$. Moreover, an entire set of terms vanishes in the differentiation owing to Eq. (7.23). The result is Eq. (7.24).

$$\left(\frac{\partial \phi}{\partial T}\right)_{n_2,\{y\}} = \left(\frac{\partial \phi}{\partial T}\right)_{n_2,\{\alpha\}} + \sum_{h=1}^{k}\left(\frac{\partial \alpha_h}{\partial T}\right)_{n_2,\{y\}}\left(\frac{\partial \phi}{\partial \alpha_h}\right)_{T,n_2,\{\alpha \neq \alpha_h\}} \tag{7.22}$$

$$\left(\frac{\partial y_g}{\partial T}\right)_{n_2,\{y\}} = 0 \quad \text{for } g = 1 \text{ to } k \tag{7.23}$$

P, $\{n \neq n_2\}$= constant

$$\left(\frac{\partial^2 G}{\partial n_2\, \partial T}\right)_{\{y\}} = \left(\frac{\partial^2 G}{\partial n_2\, \partial T}\right)_{\{\alpha\}} + \sum_{h=1}^{k}\left(\frac{\partial \alpha_h}{\partial T}\right)_{\{y\},n_2}\left(\frac{\partial^2 G}{\partial n_2\, \partial \alpha_h}\right)_{T,\{\alpha \neq \alpha_h\}}$$
$$+ \sum_{g=1}^{k} y_g \left(\frac{\partial^2 \alpha_g}{\partial n_2\, \partial T}\right)_{\{y\}} \tag{7.24}$$

When Eq. (7.24) is evaluated at operational equilibrium, the y_g's are zero and the final summation vanishes. The term on the left-hand side becomes $-S_2$, the first term on the right-hand side becomes $-(S_2)_\alpha$, and the middle summation represents the molar-shift term. Each term in the middle summation will now be transformed into the same form as that of the molar-shift term in Eq. (7.10). P and $\{n \neq n_2\}$ continue to be constant.

The gist of the transformation is that the uncoupling theorem (7.19) simplifies $(\partial^2 G/\partial n_2\, \partial \alpha_h)_{T,\{\alpha \neq \alpha_h\}}$ by effectively liberating $(\partial G/\partial \alpha_h)$ from dependence on $\{\alpha \neq \alpha_h\}$. This follows because $(\partial^2 G/\partial \alpha_h\, \partial \alpha_i) = 0$ implies that $(\partial G/\partial \alpha_h)$ is independent of α_i, when α_h is at equilibrium, as shown in Fig. 7.1. The theorem is valid for all possible combinations of i and h in which $i \neq h$.

Thus $(\partial^2 G/\partial n_2\, \partial \alpha_h)_{T,\{\alpha \neq \alpha_h\}}$ in Eq. (7.24) becomes simply $(\partial^2 G/\partial n_2\, \partial \alpha_h)_T$, which in turn equals $(\partial y_h/\partial n_2)_{T,\alpha_h}$. Furthermore, because y_h is a function simply of α_h (rather than of the full set $\{\alpha\}$), α_h is a function simply of y_h, and $(\partial \alpha_h/\partial T)_{n_2,\{y\}}$ becomes simply $(\partial \alpha_h/\partial T)_{n_2,y_h}$. The result is Eq. (7.25).

(P, $\{n \neq n_2\}$) = constant in Eqs. (7.25) and (7.26)

$$\left(\frac{\partial^2 G}{\partial n_2\, \partial \alpha_h}\right)_{T,\{\alpha \neq \alpha_h\}} = \left(\frac{\partial y_h}{\partial n_2}\right)_{T,\alpha_h} = -\left(\frac{\partial y_h}{\partial \alpha_h}\right)_{T,n_2}\left(\frac{\partial \alpha_h}{\partial n_2}\right)_{T,y_h} \tag{7.25}$$

Identification of S_2 and $(S_2)_\alpha$ and introduction of Eq. (7.25) converts Eq. (7.24) into the simple form (7.26):

$$S_2 = \underset{\text{Isomolar +}}{(S_2)_\alpha} + \underbrace{\sum_{h=1}^{k} \left(\frac{\partial^2 G}{\partial \alpha_h^2} \right)_{T,n_2} \left(\frac{\partial \alpha_h}{\partial n_2} \right)_{T,y_h=0} \left(\frac{\partial \alpha_h}{\partial T} \right)_{n_2,y_h=0}}_{\text{Sum of independent molar-shift terms}} \tag{7.26}$$

Note that each of the molar-shift terms in the summation depends on one, and only one, progress variable, and that each term is analogous to that for the single reaction treated in Eq. (7.10).

Other Second Derivatives. Partial Enthalpy

By a straightforward extension of the preceding proof, all second derivatives of the Gibbs free energy in the case of multiple progress variables lead to additive relationships of the same form as Eq. (7.26). For example, the partial enthalpy H_2 is expressed by Eq. (7.27):

$(P, \{n \neq n_2\})$ = constant

$$H_2 = \underset{\text{Isomolar}}{(H_2)_\alpha} + T \underbrace{\sum_{h=1}^{k} \left(\frac{\partial^2 G}{\partial \alpha_h^2} \right)_{T,n_2} \left(\frac{\partial \alpha_h}{\partial n_2} \right)_{T,y_h=0} \left(\frac{\partial \alpha_h}{\partial T} \right)_{n_2,y_h=0}}_{\text{Sum of independent molar-shift terms}} \tag{7.27}$$

INTERCONNECTIONS

When the Euler equation is used to represent an extensive property, the resulting partial properties are interconnected, as shown for example by the Gibbs–Duhem equation (2.8d). In this section we will describe a representative medley of such interconnections.

Permutation of Subscripts

Relationships among physical variables must be independent of the arbitrary choices we make in labeling the variables. The labels for physically different but logically equivalent variables may therefore be permuted.

In the preceding derivations, our decision to differentiate with respect to n_2 and call the results "typical" implies equal logical status for all components. We may therefore permute the number subscripts. For example, by interchanging the 1,2 subscripts in Eqs. (7.16) we obtain a valid expression for V_1:

(n_2, T) = constant

$$V_1 = (V_1)_\alpha + \left(\frac{-\partial^2 G}{\partial \alpha^2} \right)_{n_1,P} \left(\frac{\partial \alpha}{\partial n_1} \right)_{P,y=0} \left(\frac{\partial \alpha}{\partial P} \right)_{n_1,y=0} \tag{7.28}$$
$$= (V_1)_\alpha + (V_1)_{\text{shift}}$$

Similarly, all reactions that exist at equilibrium on the given time scale enjoy equal logical status. We may therefore permute the subscript labels for the progress variables. For example, Fig. 7.1 remains valid if the i and h subscripts are interchanged, a fact that was exploited in the derivation of Eq. (7.25).

When permuting subscripts to generate relationships, it is best to do so before rather than after introduction of a specific model, because the model may create changes in logical status. For example, in the dilute solution model the status of the solvent is different from that of the solutes, because direct solute–solute interactions are modeled as statistically insignificant compared to solute–solvent and solvent–solvent interactions. In the same vein, the interionic attraction theory creates differences in status between nonelectrolytes and ionic strength-producing electrolytes.

Separateness of Isomolar/Molar-Shift Parts

Let X be any extensive linear function of a derivative of the Gibbs free energy; for example, X might be the heat capacity $C_P = -T(\partial^2 G/\partial T^2)_{P,\{n\}}$. Owing to the given definition, X is a sum of an isomolar and molar-shift part (7.29a), or a weighted sum of Euler partial derivatives (7.29b). The latter, likewise, are sums of isomolar and molar-shift parts (7.29c) and (7.29d).

$$X = (X)_\alpha + (X)_{\text{shift}} \tag{7.29a}$$

$$X = n_1 X_1 + n_2 X_2 \tag{7.29b}$$

$$\begin{aligned} X_1 &= \partial X/\partial n_1 = (X_1)_\alpha + (X_1)_{\text{shift}} \\ X_2 &= \partial X/\partial n_2 = (X_2)_\alpha + (X_2)_{\text{shift}} \end{aligned} \tag{7.29c}$$

$$X = (X)_\alpha + (X)_{\text{shift}} = n_1[(X_1)_\alpha + (X_1)_{\text{shift}}] + n_2[(X_2)_\alpha + (X_2)_{\text{shift}}] \tag{7.29d}$$

Because the isomolar and molar-shift parts are separate functions, the corresponding terms in Eq. (7.29d) are independently equal and may be separated to yield the interconnections of Eq. (7.30).

$$(X)_\alpha = n_1(X_1)_\alpha + n_2(X_2)_\alpha \tag{7.30a}$$

$$(X)_{\text{shift}} = n_1(X_1)_{\text{shift}} + n_2(X_2)_{\text{shift}} \tag{7.30b}$$

Equation (7.30b) is valuable when X is a first-derivative function of G because X_{shift} then vanishes. This is spelled out for the volume V in Eqs. (7.31), but analogous expressions apply to the entropy S and the enthalpy H.

$$V = (V)_\alpha, \qquad (V)_{\text{shift}} = 0 \tag{7.31a}$$

$$V = n_1 V_1 + n_2 V_2 \tag{7.31b}$$

$$V = n_1[(V_1)_\alpha + (V_1)_{\text{shift}}] + n_2[(V_2)_\alpha + (V_2)_{\text{shift}}] \tag{7.31c}$$

$$V = (V)_\alpha = n_1(V_1)_\alpha + n_2(V_2)_\alpha \tag{7.31d}$$

$$(V)_{\text{shift}} = n_1(V_1)_{\text{shift}} + n_2(V_2)_{\text{shift}} = 0$$

$$\therefore \quad \frac{(V_2)_{\text{shift}}}{(V_1)_{\text{shift}}} = -\frac{n_1}{n_2} \tag{7.31e}$$

To show consistency between the separateness of isomolar and molar-shift parts [Eqs. (7.31d) and (7.31e)] and the developments of the preceding sections, we will outline the derivation of Eq. (7.31e) from Eqs. (7.16a) and (7.28).

PROOF. From Eqs. (7.16a) and (7.28),

$$\frac{(V_2)_{\text{shift}}}{(V_1)_{\text{shift}}} = -\frac{n_1}{n_2} = \frac{\partial\alpha/\partial n_2}{\partial\alpha/\partial n_1} = -(\partial n_1/\partial n_2)_\alpha$$

Consider the partial derivative $(\partial n_1/\partial n_2)_\alpha$. As n_2 changes by ∂n_2, α can stay constant only if the simultaneous change in n_1 is controlled so that n_1/n_2 remains constant. Therefore $(\partial[n_1/n_2]/\partial n_2)_\alpha = 0$, which yields the solution

$$(\partial n_1/\partial n_2)_\alpha = n_1/n_2 \qquad \text{Q.E.D.}$$

Equations (7.31) are valuable also because they establish correspondences between the isomolar partial volumes of the *components* and the partial volumes of the descendant molecular species. For composition tree (T-7.1A), for example, the correspondence is between $[(V_1)_\alpha = (\partial V/\partial n_1)_\alpha, (V_2)_\alpha = (\partial V/\partial n_2)_\alpha]$ and $[V_A, V_C, V_T]$, as outlined in Eqs. (7.32): (7.32a) represents V in terms of partial volumes of molecular species, (7.32b) is a memo giving the definitions of these partial volumes, and (7.32c) derives the correspondences.

$$\text{For (T-7.1A):} \quad V = n_1 V_A + n_2(1-\alpha)V_C + n_2\alpha V_T \tag{7.32a}$$

$$V_A = (\partial V/\partial n_A)_{n_C, n_T}, \quad V_C = (\partial V/\partial n_C)_{n_A, n_T}, \quad V_T = (\partial V/\partial n_T)_{n_A, n_C} \tag{7.32b}$$

$$(\partial V/\partial n_1)_\alpha = (V_1)_\alpha = V_A, \qquad (\partial V/\partial n_2)_\alpha = (V_2)_\alpha = (1-\alpha)V_C + \alpha V_T \tag{7.32c}$$

The correspondences of course depend on the nature of the composition tree. For composition tree (T-7.2), a similar procedure arrives at Eqs. (7.33):

$$\text{For (T-7.2):} \quad V = (n_1 - n_2\alpha)V_A + n_2(1-\alpha)V_U + n_2\alpha V_{A\cdot U} \tag{7.33a}$$

$$(\partial V/\partial n_1)_\alpha = (V_1)_\alpha = V_A$$

$$(\partial V/\partial n_2)_\alpha = (V_2)_\alpha = V_U + \alpha(V_{A\cdot U} - V_A - V_U) = V_U + \alpha\Delta V_{A+U\to A\cdot U} \tag{7.33b}$$

The constraints implied by Eqs. (7.31a) and (7.31b) can be expanded to apply to a system with any number of components (f) and any number of independent equilibria, as in Eq. (7.34):

$$V = \sum_{i=1}^{f} n_i V_i = \sum_{i=1}^{f} n_i (V_i)_\alpha, \qquad \sum_{i=1}^{f} n_i (V_i)_{\text{shift}} = 0 \tag{7.34}$$

Mathematically analogous constraints apply to the entropy S and enthalpy H, which (like V) are first-derivative functions whose molar-shift terms $(S)_\alpha$ and $(H)_\alpha$ vanish at equilibrium.

MOLAR-SHIFT TERMS AND TRADITION

Traditional chemical thermodynamics has been productive for a hundred years. Why complicate it? If a few phenomena, like enthalpy–entropy compensation, cannot be explained, is that so awful?

The question is reasonable, but it misses the point. Just as reactions and equilibria are central to chemistry, they should also be central to chemical thermodynamics. They should be included at a very basic level, even at the cost of complexity. This chapter tries to do just that.

Moreover, the results look right. Looking beyond the free energy G, which is the starting point, the most frequently used functions are H (the enthalpy, *not* the *partial* enthalpy), S, V, and the *partial* free energies of the components. These are all first derivatives of G, and therefore they are the same in the α and $y = 0$ representations as in traditional thermodynamics. Molar-shift terms do not arise. The first-derivative functions can therefore serve as benchmarks to indicate success in traditional chemical thermodynamics.

Some Benchmarks

We shall give two examples: (1) standard partial free-energy changes ΔG^0 for equilibria in dilute solutions and (2) those enthalpy changes that are first derivatives in calorimetric titrations.

Thermodynamic equilibrium constants, $K = \exp(-\Delta G^0/RT)$, are based on partial free energies, and the measured data therefore are first-derivative functions. As a test for validity, we shall require that the thermodynamic constants agree with values obtained by accurate nonthermodynamic methods.

Table 7.2 compares the acid dissociation constants K_A for some weak acids in water. The reaction is $HX \rightleftarrows H^+ + X^-$; water is chosen as the solvent because so many properties in water are "anomolous" from the traditional point of view.

The experiments whose results are summarized in Table 7.2 were outstanding at the time, and even now they come close to being state-of-the-art. Ionic-strength effects are treated by established methods of interionic attraction theory (Chapter 12), and the K-values are given at high dilution. The thermodynamic measurements employ reversible electrochemical cells. The comparison values are based on conductance data, or on data for optical absorption. It is clear that the results accord very well.

Table 7.2. Acid dissociation constants in water at 25°C[a]

Acid HX	Method	K_A	Reference
Acetic acid	Electromotive force[b]	$(1.754 \pm 0.01) \times 10^{-5}$	Harned and Owen [1958]
	Conductance	$(1.758 \pm 0.01) \times 10^{-5}$	Harned and Owen [1958]
Similar measurements for other acids show the following discrepancies between electromotive force and conductance (Harned and Owen [1958]): formic acid, 3.2%; propionic acid 0.8%; butyric acid, 0.2%; chloroacetic acid, 1.5%; lactic acid, 1.2%.			
2,6-Dinitrophenol	Conductance	$(1.940 \pm 0.01) \times 10^{-4}$	Kortum and Wilski [1954]
	Optical density	$(1.938 \pm 0.01) \times 10^{-4}$	Kortum-Seiler, cited by Kortum and Wilski [1954]

[a]Corrected for ionic-strength effects.
[b]In reversible cells.

Our second benchmark is calorimetric titration. The experimental amounts of heat evolved (ΔH) are bona fide enthalpies and thus are first-derivative functions. The plot of heat evolved versus formula weights of added titrant must conform to equilibrium theory of titration and yield the same value for the equilibrium constant of the titration reaction as do other methods [Grunwald and Kirschenbaum, 1972].

The traditional scheme for a calorimetric titration is given in Fig. 7.2. Increments of titrant are added with stirring to the substrate solution, and the heat evolved, ΔH, is deduced from the temperature rise in a calorimeter. The overall ΔH, according to Fig. 7.2, has a contribution from physical mixing (which is extraneous to the titration) and also has a major contribution from the heat of reaction. The heat of mixing, which is small and often negligible in dilute solutions, is estimated from data for unreactive molecular analogs and is substracted. ΔH is also corrected for the heat of stirring, for exchange of heat with the surroundings, for the time lag of the thermometer, and for heats contributed by adventitious reactions [Christensen et al., 1972; Gunn, 1971; Leung and Grunwald, 1970]. These corrections are by no means negligible, but with care and

Separate solutions: (solvent + substrate), (solvent + titrant)

↓ ΔH(mixing)

Mixture before reaction: (Combined solvent + substrate + titrant)
(hypothetical)

↓ ΔH(reaction)

Mixture at equilibrium: (Combined solvent +
unreacted substrate and titrant + titration product)

Figure 7.2. Stepwise scheme for a calorimetric titration.

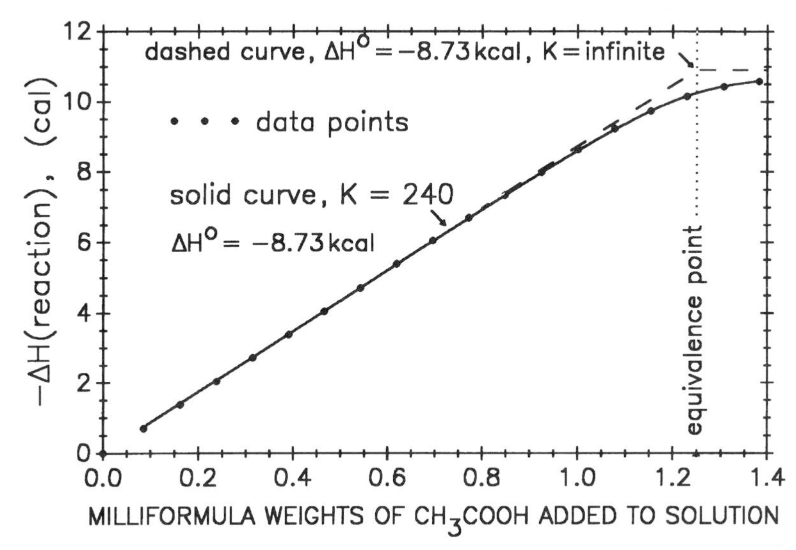

Figure 7.3. Calorimetric titration of 0.01251 volume-formal imidazole with 0.2235 volume-formal acetic acid in water. (Data of Wrathall cited by Eatough et al. [1972].)

precision an accuracy of 1% can be attained. Figure 7.3 shows a titration curve [D. P. Wrathall, cited by Eatough et al., 1972] based on accurate data for the titration of imidazole (a weak base) with acetic acid in water, Eq. (7.35):

$$\begin{array}{ccccccc} CH_3COOH & + & \text{N}\!\!\diagdown\!\!\text{NH (ring)} & \rightleftharpoons & CH_3COO^{\ominus} & + & \text{HN}\!\!\diagdown\!\!\text{NH (ring)}^{\oplus} \\ HA & + & Im & \rightleftharpoons & A^{\ominus} & + & ImH^{\oplus} \end{array} \tag{7.35}$$

According to the traditional theory of titration [Christensen et al., 1972], a calorimetric titration curve is a function of two parameters: the equilibrium constant (K) and the heat of reaction (ΔH^0). The solid curve in Fig. 7.3 is the theoretical titration curve, calculated using parameters obtained by nonlinear least-squares fitting [Wentworth, 1965] of the corrected heat evolved, ΔH_{corr}, versus added titrant. The fit is excellent; the deviations from experiment are so small that one cannot see them on the scale of the figure. The parameters and their standard errors (propagated from the errors of the data) are as follows: $K = 240 \pm 40$, $\Delta H^0 = -8.73 \pm 0.03$ kcal/mol. The ΔH^0 parameter is determined largely by the slope,* while K is determined largely by the difference

*Because the standard enthalpy change ΔH^0 for the titration reaction is calculated from the *slope* of the experimental ΔH_{corr} (which is a first-derivative function) with respect to formula weights of added titrant, the ΔH^0 parameter is a second derivative. This is consistent with the fact that ΔH^0 for reaction (7.35) is equal to $(H^0_{\text{ImH}^+} + H^0_{\text{A}^-} - H^0_{\text{Im}} - H^0_{\text{HA}})$ and thus is on the level of standard partial enthalpies, which are second derivatives of G and *do* in principle involve molar-shift terms.

between the actual titration curve and the dashed titration curve for $K = \infty$. The smallness of that difference accounts for the 20% error in K. Even so, K agrees adequately with an independent potentiometric result of 195 ± 10, based on the acid dissociation constants of HA and ImH^+ in water.

Further Questions

The concept of molar-shift terms is inherent in the thermodynamics of molecular species, but is not a normal part of traditional chemical thermodynamics. Having satisfied ourselves that the traditional approach is accurate on the first-derivative level (where molar-shift terms are zero), we wonder whether it also works well on the second-derivative level, where molar-shift terms in principle are present. The indications are mixed. On the one hand, Chapter 6 describes significant propensities for enthalpy–entropy compensation that cannot be rationalized traditionally, and there are well-documented examples of "anomalous" partial volumes, entropies, and heat capacities for solutes in water and alcohol–water mixtures. [Franks and Desnoyers, 1985; Franks and Ives, 1966; Blandamer et al., 1985]. Also telling is the number of seasoned, thoughtful chemical thermodynamicists who are reluctant to give chemical interpretations, on the grounds that the results have formal significance only.

In spite of these signs, the traditional approach is mostly trouble-free on the level of second and higher derivatives of the free energy. This is especially true for solutions in non-hydrogen-bonding solvents, the kind of liquid systems that fit the regular-solution model. Needed to reconcile the "anomalous" domains and the trouble-free domains are objective estimates of the magnitudes of molar-shift terms, so that one may know when these terms are significant. Such estimates will be made in Chapter 8.

8

MOLAR-SHIFT TERMS IN DILUTE SOLUTIONS

This chapter focuses on partial entropies, enthalpies, and volumes—properties whose equilibrium ($y = 0$) representation includes a molar-shift term. We will also take a glance at partial heat capacities, just long enough to grasp the complicated molar-shift term of Eq. (7.18).

In principle the overall molar-shift term is a sum of independent terms, one for each independent reaction that relaxes to equilibrium on the time scale of the observation. This was proved in Chapter 7. For the partial entropy, enthalpy, and volume the contributing terms all follow the same mathematics. In particular, for any component 2, the molar-shift term due to any independent reaction relaxing to equilibrium with progress variable α is proportional to $(\partial\alpha/\partial n_2)_{n_1,n_3,\ldots}$. In dilute solutions it is convenient to convert to formal concentrations (m) per kilogram of solvent, since m_1 is constant by definition. The molar-shift term then becomes proportional to $(\partial\alpha/\partial m_2)_{m_3,\ldots}$. When α at equilibrium is insensitive to changes in m_2, $\partial\alpha/\partial m_2 = 0$ and the corresponding molar-shift term vanishes.

For dilute solutions, instances in which $\partial\alpha/\partial m_2$ equals 0 are surprisingly common, and many molar-shift terms *do* vanish. In the end, only two molar-shift mechanisms remain: (1) shifts that are based on the Law of Mass Action and (2) solute-induced molar shifts in the solvent.

The impact of the Law of Mass Action is well understood, and the effect of changes in the molality m_2 of a dilute solute on dilute-solution equilibria can be predicted. In spite of this, molar-shift terms to allow for mass action are often overlooked. We will work an example involving the partial volume to call attention to the resulting error.

Solute-induced molar shifts in the solvent are of two kinds. Some shifts may be described as medium or continuum effects inherent in the addition of the solute. Others derive from specific short-range van der Waals interactions that alter the ratios of solvent species around solute molecules, relative to those for the same species in the bulk of the solvent.

A RENEWED LOOK AT THE DILUTE-SOLUTION MODEL

An exemplary "dilute solution" is so dilute that the solute molecules sense only the presence of solvent molecules. Solute–solute interactions at short range are neglected, except in transition states of solute–solute reactions. The *standard* partial free energy of a dilute *molecular species* is therefore independent of the concentration of any solute, throughout the dilute range. By contrast, solute–*solvent* interactions are significant, and the standard partial free energy of a solvent species may vary upon addition of a dilute solute.

Molar-Shift Categories

Composition tree (T-8.1) divides the components of a dilute solution into three categories: the solvent 1, the active solute 2 whose partial enthalpy, entropy, and volume are of interest, and additional solutes represented by a single component 3. Each component is modeled as parent to two interconverting molecular species; the respective progress variables are α_1, α_2, and α_3. The complete molar-shift term for (say) the active solute 2 is therefore a sum of three terms that are proportional, respectively, to $\partial\alpha_1/\partial m_2$, $\partial\alpha_2/\partial m_2$, and $\partial\alpha_3/\partial m_2$.

Composition Tree (T-8.1)

Solvent 1	Active solute 2	Additional solute 3
$A\ (1-\alpha_1) \rightleftarrows B\ (\alpha_1)$	$U\ (1-\alpha_2) \rightleftarrows V\ (\alpha_2)$	$X\ (1-\alpha_3) \rightleftarrows Y\ (\alpha_3)$

Additional Solute

By hypothesis of the dilute-solution model, the additional solute 3 does not sense the presence of the active solute 2. We may therefore state at once that $\partial\alpha_3/\partial m_2 = 0$. Thus *all* partial properties of the active solute 2 ($G_2, S_2, C_{P,2}, \ldots$) lack molar-shift terms attributable to α_3. This is true for *any* added component that is dilute and for any stoichiometry (1 : 1, 1 : 2, or 1 : j) underlying α_3. It remains true for more complex composition models, regardless of the number of added components and regardless of the number of equilibria, all of whose reactants and products are species derived from added components.

Active Solute

The active solute 2 is shown as consisting of two interconverting species, U $\rightleftarrows$ V, in 1 : 1 equilibrium. Because the solution is dilute, G_{U}^{0} and G_{V}^{0} are independent of m_2 (and m_3—see above). At constant T and P, the equilibrium constant $K_2 = \alpha_2/(1 - \alpha_2) = \exp(-[G_{\mathrm{V}}^{0} - G_{\mathrm{U}}^{0}]/RT)$ is therefore independent of m_2, so that at constant T and P, α_2 is constant throughout the dilute-solution range. Hence $(\partial\alpha_2/\partial m_2)_{T,P} = 0$, and all molar-shift terms proportional to $\partial\alpha_2/\partial m_2$ vanish. Thus all second derivatives of G, including particularly S_2, H_2, and V_2, lack molar-shift terms attributable to α_2. On the other hand, partial properties based on third and higher derivatives of G may retain some dependence on α_2. For example, the partial heat capacity $C_{P,2}$ is an intricate sum, which is derived for a single equilibrium in Eq. (7.17). Some terms in that sum do not go to zero for a 1 : 1 equilibrium in the active solute. We shall state without proof that the effect of the U $\rightleftarrows$ V equilibrium in (T-8.1), on $C_{P,2}$, is to add a term that expresses the heat absorbed per degree due to the shift in α_2 with temperature:

$$C_{P,2} = (C_{P,2})_{\alpha_2} + \underbrace{\Delta H_{\mathrm{U}\rightarrow\mathrm{V}}(\partial\alpha_2/\partial T)_{y=0}}_{\text{Molar-shift term}}$$

$$= (C_{P,2})_{\alpha_2} + \underbrace{\alpha_2(1 - \alpha_2)R[\Delta H_{\mathrm{U}\rightarrow\mathrm{V}}/RT]^2}_{\text{Molar-shift term}}$$

The α_2 subscript in $(C_{P,2})_{\alpha_2}$ indicates that only α_2 is constant; other equilibria are allowed to shift.

Departing from the model shown in (T-8.1), molar shifts enter even at constant T and P when the stoichiometry is other than 1 : 1. These shifts are mass-action effects. For definiteness, assume that the solute molecules U (formula-weight fraction $1 - \alpha_2'$) form a dimer: $2\mathrm{U}[m_2(1 - \alpha_2')] \rightleftarrows \mathrm{U}_2[m_2\alpha_2'/2]$. Hence

$$K_2' = (\alpha_2' m_2/2)/(1 - \alpha_2')^2 m_2^2 = \exp(-[G_{\mathrm{U}_2}^{0} - 2G_{\mathrm{U}}^{0}]/RT).$$

Because the solutions are dilute, the standard partial free energies $G_{\mathrm{U}_2}^{0}$ and G_{U}^{0}, and hence K_2', are independent of m_2 and m_3. But mass-action effects cause the progress variable α_2' to vary with m_2, Eq. (8.1):

$$\frac{\partial\alpha_2'}{\partial m_2} = \frac{\alpha_2'(1 - \alpha_2')}{(1 + \alpha_2')m_2} \tag{8.1}$$

In summary, 1 : 1 equilibria among molecules of the active solute component do not contribute molar-shift terms to S_2, H_2, and V_2, but equilibria with other stoichiometries, such as dissociation or self-association, produce molar shifts and contribute molar-shift terms by mass-action effects.

Solvent Species

The solvent 1 in composition tree (T-8.1) consists of two interconverting isomers: $A \rightleftarrows B$; $K_1 = m_B/m_A = \alpha_1/(1 - \alpha_1)$. At equilibrium, $G_A = G_B$; hence

$$G_B - G_A = (G_B^0 + RT \ln m_B) - (G_A^0 + RT \ln m_A) = 0$$

Standard partial free energies of solvent species are specifically sensitive to the presence of solutes. Hence $(G_B^0 - G_A^0)$ and K_1 are functions of m_2:

$$\begin{aligned} -(\partial/\partial m_2)(G_B^0 - G_A^0) &= RT(\partial/\partial m_2)(\ln[m_B/m_A]) = RT(\partial \ln K_1/\partial m_2) \\ &= RT(\partial \ln[\alpha_1/(1 - \alpha_1)]/\partial m_2) \neq 0 \\ \therefore \quad (\partial \alpha_1/\partial m_2) &= \alpha_1(1 - \alpha_1)(\partial \ln K_1/\partial m_2) \neq 0 \end{aligned} \tag{8.2}$$

Equation (8.2) shows that $\partial \alpha_1/\partial m_2$ may be positive, negative, or zero, depending on the sign and magnitude of the solute-induced change in the equilibrium constant K_1. (For further details about solute-induced medium effects, see Chapter 3.)

Although the derivative, $\partial \alpha_1/\partial m_2$, in principle does not vanish, m_2 is small and the actual change, $\delta \alpha_1 = m_2(\partial \alpha_1/\partial m_2)$, is small compared to α_1. We may state, therefore, that even though $\partial \alpha_1/\partial m_2$ is nonzero, α_1 itself is virtually constant. Moreover, for solutes obeying Henry's law, $\ln K_1$ varies linearly with m_2, and the factor $(\partial \ln K_1/\partial m_2)$ in Eq. (8.2) is constant as well. The solute-induced molar shift $(\partial \alpha_1/\partial m_2)$ then is a specific parameter, independent of m_2.

This conclusion may be generalized: Regardless of the stoichiometry of the solvent equilibrium that is being shifted, solute-induced molar shifts in the solvent are specific and virtually independent of solute concentration. This is because the formal concentration of the solvent, in any units, is nearly constant throughout the dilute-solution range. When molal or formal units are used, the solvent formality m_1 ($= 1000/\mathcal{L}_1$) is of course exactly constant.

For definiteness, consider an equilibrium such as the head-to-head dimerization of acetic acid molecules in liquid acetic acid: $2A \rightleftarrows A_2$. Let $1 - \alpha'_1$ denote the formula-weight fraction of the monomer. Then

$$K'_1 = (\alpha'_1 m_1/2)/(1 - \alpha'_1)^2 m_1^2.$$

Since m_1 is constant,

$$\frac{\partial \alpha'_1}{\partial m_2} = \frac{\alpha'_1(1 - \alpha'_1)}{(1 + \alpha'_1)} \left(\frac{\partial \ln K'_1}{\partial m_2} \right)$$

Since ln K_1' varies linearly with m_2, it follows that $\partial\alpha_1'/\partial m_2$ is a specific parameter independent of m_2. Q.E.D.

The fact that derivatives such as $\partial\alpha_1/\partial m_2$ are independent of m_2 is important. We shall find that because of it, solute-induced molar shifts in the solvent affect the *standard* partial properties of the *solute*. For instance, derivatives with respect to m_2 affect S_2^0, H_2^0, and V_2^0; and when the solute-induced shifts differ for reactants and products, the difference causes the addition of specific molar-shift terms to ΔS^0, ΔH^0, and ΔV^0 for the reaction.

The Well-Behaved Dilute Solute

A dilute solute in a liquid is "well-behaved" if its self-association, dissociation, and reaction with other solute components are negligible. Such a solute obeys Henry's law, its colligative properties correspond to one mole of particles per formula weight, and the optical absorption attributable to it follows Beer's law. Equally important, the partial free energy of such a solute is a logarithmic function of its concentration according to Eq. (8.3), which parallels the previously derived Eq. (3.4). (See also Table 3.2.)

$$G_2(\text{M}) = G_2^0(\text{M}) + RT \ln m_2 \tag{8.3a}$$

$$G_2^0(\text{M}) = G_2^0(g) - RT \ln \mathcal{H}_2(\text{M}) \tag{8.3b}$$

The affix (M), which is normally omitted, indicates that the given property varies with the solvent medium. For instance, $\mathcal{H}_2(\text{M})$ is the Henry's law constant of solute 2 in solvent M. Because $G_2(\text{M})$ is linear in the *logarithm* of m_2, it remains sensitive to changes in m_2 even at very high dilutions. $G_2^0(\text{M})$, on the other hand, is independent of m_2; it gathers and sums up all contributions to $G_2(\text{M})$ that are independent of m_2 or that become so at sufficiently low concentrations.

In terms of molecular species, well-behaved solutes are self-contained and consist entirely of isomers, as (for example) U and V in composition tree (T-8.1). The partial free energies of molecular species like U share the logarithmic character of those of the parent component, as shown in Eqs. (8.4), which parallel the previously derived Eqs. (3.6).

$$G_\text{U}(\text{M}) = G_\text{U}^0(\text{M}) + RT \ln c_\text{U} \tag{8.4a}$$

$$G_\text{U}^0(\text{M}) = G_\text{U}^0(g) - RT \ln \mathcal{H}_\text{U}(\text{M}) \tag{8.4b}$$

The concentration dependences of S_2, H_2, V_2, and $C_{P,2}$ and of $(S_2)_\alpha$, $(H_2)_\alpha$, $(V_2)_\alpha$, and $(C_{P,2})_\alpha$ can be derived from that for G_2 through Eqs. (8.5).

$\{n\}$ = constant; $G_2 = (G_2)_\alpha$ in Eqs. (8.5)

In $y = 0$ representation: $S_2 = -(\partial G_2/\partial T)_P$, $\quad V_2 = (\partial G_2/\partial P)_T$

$$H_2 = G_2 + TS_2, \quad C_{P,2} = -T(\partial^2 G_2/\partial T^2)_P \tag{8.5a}$$

In α-representation: $(S_2)_\alpha = -[\partial(G_2)_\alpha/\partial T]_P$, $\quad (V_2)_\alpha = [\partial(G_2)_\alpha/\partial P]_T$

$$(H_2)_\alpha = (G_2)_\alpha + T(S_2)_\alpha, \quad (C_{P,2})_\alpha = -T[\partial^2(G_2)_\alpha/\partial T^2]_P \tag{8.5b}$$

The results are listed in Table 8.1. Note that the two representations yield the same concentration dependences.

MOLAR-SHIFT TERMS BASED ON THE LAW OF MASS ACTION

Dilute solutes fail to be "well-behaved" when the solute molecules associate, dissociate, or react with molecules from other components. When this happens, virtually all solute properties deviate from the "well-behaved" norm, as do colligative properties of the solvent. In particular, we must be on the lookout for molar-shift terms.

Dimerization of the Solute

The general rule is that at equilibrium at constant T and P, molar-shift terms vanish for properties that are first derivatives of the Gibbs free energy, but

Table 8.1. Concentration dependence of partial properties for "well-behaved" dilute solutes

Property	Concentration Dependence
In the $y = 0$ Representation	
G_2	$G_2 = G_2^0 + RT \ln m_2$
$S_2 = -(\partial G_2/\partial T)_{\{n\},P}$	$S_2 = S_2^0 - R \ln m_2$
$H_2 = G_2 + TS_2$	$H_2 = G_2^0 + TS_2^0 = H_2^0$
$V_2 = (\partial G_2/\partial P)_{\{n\},T}$	$V_2 = V_2^0$
$C_{P,2} = (\partial H_2/\partial T)_{\{n\},P}$	$C_{P,2} = C_{P,2}^0$
In the α-Representation	
$(G_2)_\alpha = G_2$	$(G_2)_\alpha = (G_2^0)_\alpha + RT \ln m_2$
$(S_2)_\alpha = -[\partial(G_2)_\alpha/\partial T]_{\{n\},P}$	$(S_2)_\alpha = (S_2^0)_\alpha - R \ln m_2$
$(H_2)_\alpha = (G_2)_\alpha + T(S_2)_\alpha$	$(H_2)_\alpha = (G_2^0)_\alpha + T(S_2^0)_\alpha = (H_2^0)_\alpha$
$(V_2)_\alpha = [\partial(G_2)_\alpha/\partial P]_{\{n\},T}$	$(V_2)_\alpha = (V_2^0)_\alpha$
$(C_{P,2})_\alpha = [\partial(H_2)_\alpha/\partial T]_{\{n\},P}$	$(C_{P,2})_\alpha = (C_{P,2}^0)_\alpha$

molar-shift terms must be considered for properties that are second or higher derivatives. To illustrate this rule when there are molar shifts by mass action, we will consider only one property and one reaction type in detail, because other cases will be analogous. In particular, we will choose the volume V and a reaction in which a dilute solute forms a dimer, according to composition tree (T-8.2). The active reaction is $2M \rightleftarrows D$, where M is the monomer and D ($= M_2$) is the dimer of the solute. The progress variable α is the formula-weight fraction. The equilibrium constant K equals $(\alpha m_2/2)/(1-\alpha)^2 m_2^2$, and the molar shift $\partial\alpha/\partial m_2$ equals $\alpha(1-\alpha)/m_2(1+\alpha)$. Except for the simpler notation, these relationships are equivalent to those leading up to Eq. (8.1).

Composition Tree (T-8.2)

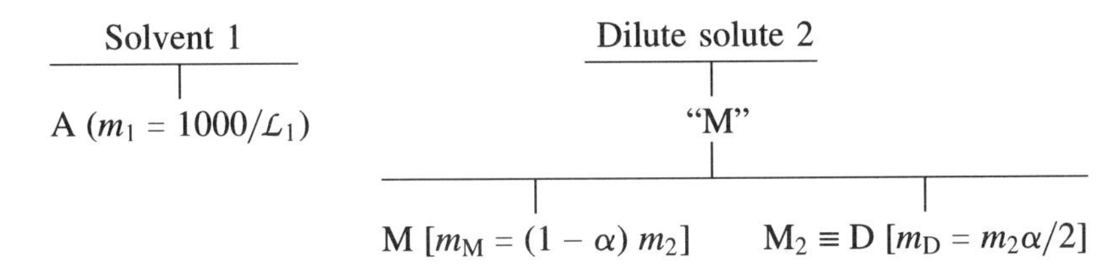

In the present case, the volume V is a first derivative of the Gibbs free energy; hence the molar-shift term vanishes, as stated in Eq. (8.6). [Compare with Eq. (7.31).]

$$V \equiv (\partial G/\partial P)_{y=0} = (V)_\alpha \equiv (\partial G/\partial P)_\alpha, \qquad T, \{n\} = \text{constant} \tag{8.6}$$

On the other hand, the partial volume V_2 is a second derivative: $V_2 = (\partial^2 G/\partial P\, \partial n_2)_{y=0}$ and thus entails a molar-shift term. To arrive at the latter, we shall begin with the volume V and express it as a function of the molar composition according to Eq. (8.7). Since the composition variables here are (n_1, n_2, α), V in Eq. (8.7) belongs to the α-representation.

$$\begin{aligned} V = (V)_\alpha &= n_A V_A + n_M V_M + n_D V_D \\ &= n_1 V_A + n_2(1-\alpha)V_M + (n_2\alpha/2)V_D, \qquad T, P = \text{constant} \end{aligned} \tag{8.7}$$

The partial volumes V_A, V_M, and V_D are shown in Eqs. (8.8a) and (8.8b) to be isomolar properties, by partial differentiation of Eq. (8.7) at constant α. The related expression for $(\partial V/\partial\alpha)$ is obtained in Eq. (8.8c), where ΔV^0 is the standard volume change for the dimerization: $2M \rightarrow D$. The derivations of Eqs. (8.8) employ the Gibbs–Duhem equation.

$$(V_1)_\alpha = (\partial V/\partial n_1)_\alpha = V_A, \quad T, P, \{n \neq n_1\} \text{ constant} \tag{8.8a}$$

$$(V_2)_\alpha = (\partial V/\partial n_2)_\alpha = V_M + \alpha(V_D/2 - V_M), \quad T, P, \{n \neq n_2\} \text{ constant} \tag{8.8b}$$

$$(\partial V/\partial\alpha) = n_2(V_D/2 - V_M) = n_2\Delta V^0/2, \quad T, P, \{n\} \text{ constant} \tag{8.8c}$$

Next, the partial volume V_2 is derived in Eqs. (8.9) by transforming variables from (n_2, α) to (n_2, y) and solving at equilibrium ($y = 0$). As expected, the result includes a molar-shift term.

***T*, *P*, $\{n \neq n_2\}$= constant in Eqs. (8.9)**

$$(\partial V/\partial n_2)_y = (\partial V/\partial n_2)_\alpha + (\partial V/\partial \alpha)_{n_2}(\partial \alpha/\partial n_2)_y \tag{8.9a}$$

$$V_2 = (\partial V/\partial n_2)_{y=0} = (V_2)_\alpha + (\partial V/\partial \alpha)_{n_2}(\partial \alpha/\partial n_2)_{y=0} \tag{8.9b}$$

Finally, substitution from Eqs. (8.1) and (8.8), along with introduction of $(\partial m_2/\partial n_2)_{n_1} = m_2/n_2$, converts Eq. (8.9b) into the working equation (8.10):

$$\begin{aligned} V_2 &= V_M + (V_D/2 - V_M)[\alpha + n_2(m_2/n_2)(\partial \alpha/\partial m_2)] \\ &= V_M + (V_D/2 - V_M)[2\alpha/(1+\alpha)] = V_M + \Delta V^0[\alpha/(1+\alpha)] \end{aligned} \tag{8.10}$$

PROOF OF CONSISTENCY WITH THE GENERAL EQUATION (7.16). The preceding derivation took a shortcut by beginning with the volume V rather than the free energy G. The free-energy based Eq. (7.16) is repeated below.

$$V_2 = (V_2)_\alpha + \left(\frac{-\partial^2 G}{\partial \alpha^2}\right)_{n_2,P} \left(\frac{\partial \alpha}{\partial n_2}\right)_{P,y=0} \left(\frac{\partial \alpha}{\partial P}\right)_{n_2,y=0} \tag{7.16}$$

To show equivalence, we will show that the molar-shift terms in Eqs. (8.9b) and (7.16) are identical. After dividing both by $(\partial \alpha/\partial n_2)_{P,y=0}$, we obtain (at constant T, n_2, and $\{n\}$)

$$\left(\frac{-\partial^2 G}{\partial \alpha^2}\right)_P \left(\frac{\partial \alpha}{\partial P}\right)_y = \left(\frac{\partial V}{\partial \alpha}\right)_P \tag{A}$$

Then, applying the chain rule

$$\begin{aligned} (\partial \alpha/\partial P)_y &= -(\partial y/\partial P)_\alpha/(\partial y/\partial \alpha)_P = -(\partial^2 G/\partial \alpha\, \partial P)/(\partial^2 G/\partial \alpha^2)_P \\ &= -(\partial V/\partial \alpha)_P/(\partial^2 G/\partial \alpha^2)_P \end{aligned}$$

Substitute in (A) and simplify. Q.E.D.

The Curve of V_2 Versus m_2

The volume change ΔV^0 for the combination of two molecules is significant. When two uncharged molecules form a normal covalent bond, the volume contracts because some of the nonbonded van der Waals distances are replaced by more compact covalent distances. In liquids, the accompanying volume change ΔV^0 is typically −20 ml/mol, as shown by the following examples [Riddick and Bunger, 1970]:

Examples

Cyclohexene + Benzene (101.9 ml/mol) (89.4 ml/mol)	$\rightarrow$	Cyclohexylbenzene (170.7 ml/mol)	$\Delta V^0 = -20.6$ ml (−11% at 25°C)
$2CH_3COCH_3$ (2 × 74.04 ml/mol)	$\rightarrow$	$(CH_3)_2COH \cdot CH_2COCH_3$ (124.3 ml/mol)	$\Delta V^0 = -23.7$ ml (−16% at 25°C)

When two oppositely charged ions form an ion pair, the volume may expand. This is because the normal volume decrease on association is opposed by a volume increase due to the relief of *electrostriction* of the solvent. The strong electric fields produced by the dissociated ions interact with dipoles of nearby solvent molecules and compress the medium. When the ions form ion pairs, the ionic fields are replaced by the less effective dipolar fields due to the electric poles in the ion pairs, and the compression is eased. The measured net values of ΔV^0 vary considerably. Typical values are shown below.

Examples

$K^+ + Cl^-$ (6±1 ml/mol) $\rightarrow (K^+, Cl^-)$: $\Delta V^0 = +29$ ml in methanol, $K_{as} = 12 \pm 3$ liter/mol in methanol at 25°C [Grunwald and Brown, 1982]

4-MMP$^+$ + I^- $\rightarrow$ (4-MMP$^+$, I^-): $\Delta V^0 = +16$ ml in acetone, K_{as} = 600 liter/mol in acetone at 20°C [Ewald and Scudder, 1970], 4-MMP$^+$ = 4-carbomethoxy-N-methylpyridinum ion

$CF_3COO^- + H^+(aq) \rightarrow CF_3COOH$: $\Delta V^0 = +2.5 \pm 1.5$ ml in water, $K_{as} = 0.9 \pm .3$ liter/mol in water at 25°C [Grunwald and Haley, 1968]

3-MMP$^+$ + 8-theo$^-$ $\rightarrow$ (3-MMP$^+$, 8-theo$^-$): $\Delta V^0 = -4.0 \pm 0.3$ ml, $K_{as} = 3.0 \pm 0.3$ kg/mol, $I = 0.5$ in water at pH 7.0 and 25°C, 3-MMP$^+$ = 3-carbomethoxy-*N*-methylpyridinium ion, 8-theo$^-$ = 8-chlorotheophyllinate anion [Williams, 1981]

In theory, the volume change due to relief of electrostriction increases directly with the compressibility of the medium, and inversely with the dielectric constant. As T and P approach the critical point of the solvent (where the compressibility becomes infinite), electrostriction of the solvent by dissolved ions increases sharply. The effect is truly dramatic and is well documented in near-critical water [Wood, 1989; Majer and Wood, 1994].

According to Eq. (8.10), the plot of V_2 versus m_2 reflects the behavior of $\alpha/(1+\alpha)$ versus m_2. For the monomer–dimer equilibrium modeled in (T-8.2), we obtain $\alpha/(1-\alpha)^2 = 2m_2K$ and

$$\alpha = 1 + x - [(1 + x)^2 - 1]^{1/2}, \qquad \text{where } x = 1/(4m_2K)$$

Figure 8.1 shows a theoretical plot of V_2 versus $4m_2K$ based on plausible values for V_M and ΔV^0. The figure also shows a plot of $(V_2)_\alpha$ based on the same parameters. The difference, introduced by the molar-shift term, is important. V_2 is halfway between V_M and $V_M + \Delta V_D/2$ when $4Km_2 = 1.5$, while $(V_2)_\alpha$ is halfway when $4Km_2 = 4$. In practice, the molar-shift term is often overlooked.*

Related Matters

As shown in Fig. 8.1, the plot of V_2 versus m_2 can yield a value for the equilibrium constant K, because the half-change point for V_2 occurs when $4Km_2 = 1.5$. However, for experimental reasons, the determination is probably not very accurate. In practice, one deduces V_2 from solution volumes based on measured densities. The differentiation $(\partial V/\partial n_2)$ is performed at constant n_1, and because n_1 remains constant in the operations described in Eqs. (8.9) and (8.10), it is convenient to let the fixed amount of solvent be 1 kg. In this way, $n_1 = m_1 = 1000/\mathcal{L}_1$, and $n_2 = m_2$. The solution volume per kilogram of solvent will be denoted by ${}^kV$, and is related to the density ρ by: ${}^kV = (1000 + m_2\mathcal{L}_2)/\rho$. In this notation, $V_2 = \partial({}^kV)/\partial m_2$. Because the solution is dilute, $m_2 \ll m_1$, and the effect of the solute on the density ρ is small. It follows that ρ must be measured with great precision.

*The formation of ion pairs from free ions in dilute solutions is similar to the dimerization of uncharged molecules, except that interionic effects now need to be included in the equations (see Chapter 12).

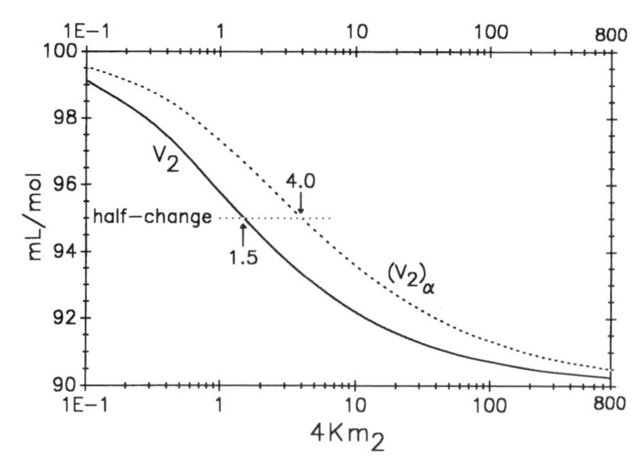

Figure 8.1. Plot of V_2 and $(V_2)_\alpha$ versus $4Km_2$ for dimerization of a dilute solute. $V_M = 100$ ml/mol, $V_D - 2V_M = \Delta V^0 = -20$ ml.

The requirement of high precision is ubiquitous in the analysis of dilute-solution properties. It exists whenever a property of the entire solution is used to derive a partial property for a dilute solute. Physical chemists have developed precise difference methods, or differential methods, to alleviate this problem, but such measurements are seldom routine. If the required solute property can be measured directly, rather than indirectly as the derivative of a property of the entire solution, the result is likely to be more accurate.

Irrespective of experimental errors, major errors result if molar-shift terms are neglected. Let x denote any distinct specific solution property per unit mass or unit volume, which is a function of molar composition. Let X denote the extensive analog of x, and let ${}^{k}X$ denote the value of X for a solution containing 1 kg of solvent. (Examples are ρ, V, and ${}^{k}V$.) The general analog of Eqs. (8.9) then is (8.11):

$$(\partial^{k}X/\partial m_2)_y = (\partial^{k}X/\partial m_2)_\alpha + (\partial^{k}X/\partial\alpha)_{m_2}(\partial\alpha/\partial m_2)_y \tag{8.11a}$$

$$X_2 = (\partial^{k}X/\partial m_2)_{y=0} = (X_2)_\alpha + (\partial^{k}X/\partial\alpha)_{m_2}(\partial\alpha/\partial m_2)_{y=0} \tag{8.11b}$$

According to Eq. (8.11b), any partial property in the $y = 0$ representation requires a molar-shift term. The actual rendition of $(X_2)_\alpha$, $\partial^{k}X/\partial\alpha$, and $\partial\alpha/\partial m_2$ in terms of properties and concentrations of molecular species must be derived from the composition model.

SOLUTE-INDUCED MOLAR SHIFTS IN THE SOLVENT

When the pure solvent consists of two or more species in equilibrium, thermodynamics permits the equilibrium constant to depend on the solute concentration. Discussion of this phenomenon raises questions concerning solvation mechanisms and unavoidably takes us into the realm of speculation. Two questions stand out: (1) At what level of interaction does a pair of interacting molecules become a molecular complex? (2) Is the shell of solvent molecules surrounding a solute molecule a distinct subspecies of the solvent, or are the molecules in that shell part of the existing solvent species?

Brief answers will be given in terms of the cage model of liquids. In first approximation, a molecule in a liquid exists in a potential-energy cage that represents the molecule's time-average interaction with its liquid environment. The cage restricts the translation and rotation of the molecule; but within the bounds imposed by this restriction, the molecule moves as a distinct Brownian entity.

On the other hand, a molecular complex is a kinetic unit made up of two (or more) molecules. The interaction joining these molecules is strong enough and localized enough to be called a bond. Molecular-complex formation leads to the appearance of vibration frequencies representative of the coupled unit, with vibrational coherence times that outlast collisional bombardment for many collisions.

The solvent molecules adjacent to a solute molecule *are a distinct subspecies of the solvent only if they are distinguishable*—either by real observation or through an idealized thought experiment whose design is rooted in primary theory. For liquid environments, it will be shown in Chapter 9 that distinguishability is common. Accordingly, the solvent molecules adjacent to a species of solute molecules may usually be modeled as a separate subspecies of the solvent. When this model is adopted, the molar composition in the bulk of the solvent remains intact, and the existing solvent species *by definition* do not experience molar shifts. The thermodynamic consequences of separate subspecies in solvation shells will be developed in Chapter 11.

Necessary Conditions

Molar shifts can be induced in the solvent only if the solvent molecules in the solvation shells fall into groups that can be correlated one-to-one with the molecular species existing in the pure solvent. In this way, the molecules in the solvation shells can be assigned to the same species that exist in the bulk of the solvent. Molar shifts occur when the mean composition of the solvation shells differs from that in the bulk of the solvent.

The hypothesis of one-to-one correlation is quite plausible when the solvent consists of species with well-apart molecular conformations. For example, consider a dilute solution of catechol (1,2-dihydroxybenzene) in liquid 1,2-dichloroethane. This solvent is known to be an equilibrium mixture of discrete *gauche* and *trans* conformational isomers, shown in Fig. 1.2. Discreteness is important: It means that the fraction of molecules that belong neither to *gauche* nor to *trans* is small. (The population with transitional *eclipsed* conformations is actually less than 0.2%.) On the other hand, when a dichloroethane molecule sits next to a catechol molecule, a nearly *eclipsed* conformation (8.12) might conceivably be favored, because then both Cl atoms can interact with both OH groups in a bimolecular interaction. But if this were important, the bimolecular interaction would be strong enough to create a molecular complex, and the dichloroethane ligand would cease being a solvent molecule.

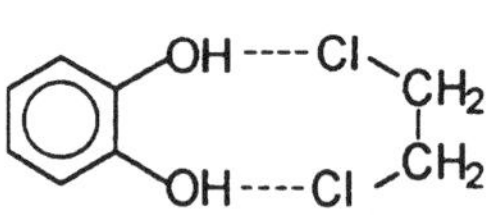

$$\begin{array}{l} \text{OH}\cdots\text{Cl}-\text{CH}_2 \\ \qquad\qquad\quad\;| \\ \text{OH}\cdots\text{Cl}-\text{CH}_2 \end{array} \tag{8.12}$$

The argument is as follows: The energy cost estimated for converting the normal *trans/gauche* mixture to the *eclipsed* conformation and for letting that conformation be competitive with *trans* and *gauche* as neighbors to catechol is >20 kJ/mol. This cost must be paid by the pairwise interaction energy of (8.12), which is therefore more negative than −20 kJ/mol—sufficient for a molecular complex to be more stable than an adjacent pair. (In calculations in Chapter 11, the changeover from an adjacent pair to a molecular complex occurs at −20 kJ/mol of pairwise interaction energy.) The solvent molecules that will remain as uncomplexed molecular neighbors are therefore either discretely *trans* or discretely *gauche*.

The hypothesis of one-to-one correlation is more questionable when the solvent species in the pure solvent are environmental isomers. Then the distinctive property is likely to be the number of linkages that tie the solvent molecule to its pure-solvent cage—the number of hydrogen bonds, or the number of van der Waals interactions that are above average in strength and are centered on specific atoms or groups.

When a linked solvent molecule sits next to a nonlinking solute molecule (such as a water molecule next to a methane molecule), the links between the solvent molecule and its solvent neighbors become perturbed. The linkage geometries become distorted, and the linkage numbers may change. One-to-one correlation between solvent species in the solvation shells and solvent species in the bulk of the solvent then is possible *only if the linkage numbers that are significant in the solvent include all linkage numbers that are significant in the solvation shells*. We may think of this as a necessary condition for molar-shift models when the solvent species are environmental isomers.

Derivation of Molar-Shift Terms

Now let us apply the molar-shift model to derive specific expressions from the general equations of Chapter 7. The composition tree will be (T-8.3). The solvent is a mixture of two isomers, A and B, that interconvert with progress variable α_1. The solute is a single species U. The equilibrium ratio $K_1 = [\alpha_1/(1-\alpha_1)]_{eq}$ varies with m_2 to give a solute-induced medium effect. Because the solution is dilute, G_B^0, G_A^0, and $\ln K_1$ at constant T and P are linear functions of m_2.

Composition Tree (T-8.3)

Solvent 1 (n_1, m_1)		Dilute solute 2 (n_2, m_2)
A $(n_A = n_1[1-\alpha_1])$	B $(n_B = n_1\alpha_1)$	U $(n_U = n_2)$

The derivation begins by expressing the free energy G at constant T and P in terms of mole numbers, Eq. (8.13). This yields the specific expressions (8.14) for $y_1 = (\partial G/\partial \alpha_1)$ and (8.15) for $(\partial^2 G/\partial \alpha_1^2)$. The derivation of Eq. (8.14) employs the Gibbs–Duhem equation.

$$G = n_1(1-\alpha_1)G_A + n_1\alpha_1 G_B + n_2 G_U \tag{8.13}$$

$$\begin{aligned} y_1 &= (\partial G/\partial \alpha_1)_{\{n\},T,P} = n_1(G_B - G_A) \\ &= n_1 RT[\ln(\alpha_1/[1-\alpha_1]) - \ln K_1] \end{aligned} \tag{8.14}$$

$$(\partial y_1/\partial \alpha_1) = (\partial^2 G/\partial \alpha_1^2)_{n_1,n_2,P,T,\mathrm{eq}} = n_1 RT/(\alpha_1[1-\alpha_1]) \tag{8.15}$$

Next we recall Eq. (7.10) for the partial entropy S_2 of the solute:

n_1, P = constant

$$S_2 = \underset{\text{Isomolar}}{(S_2)_{\alpha_1}} + \underbrace{\left(\frac{\partial^2 G}{\partial \alpha_1^2}\right)_{T,n_2} \left(\frac{\partial \alpha_1}{\partial n_2}\right)_{T,y=0} \left(\frac{\partial \alpha_1}{\partial T}\right)_{n_2,y=0}}_{\text{Molar-shift term}} \tag{7.10}$$

To evaluate $\partial\alpha_1/\partial n_2$, we use Eq. (8.2) and the equation $m_2/n_2 = m_1/n_1$:

$$(\partial\alpha_1/\partial n_2)_{n_1,T,\mathrm{eq}} = (m_2/n_2)(\partial\alpha_1/\partial m_2) = (m_1/n_1)\alpha_1(1-\alpha_1)(\partial \ln K_1/\partial m_2)$$

By definition, $\ln K_1 = \ln[\alpha_1/(1-\alpha_1)]_{\mathrm{eq}}$. Therefore

$$(\partial\alpha_1/\partial T)_{n_1,n_2,\mathrm{eq}} = \alpha_1(1-\alpha_1)(\partial \ln K_1/\partial T) = \alpha_1(1-\alpha_1)\Delta H_{\mathrm{A}\to\mathrm{B}}/RT^2$$

Substitution in Eq. (7.10) then yields Eq. (8.16):

$$S_2 = (S_2)_{\alpha_1} + m_1\alpha_1(1-\alpha_1)(\Delta H_{\mathrm{A}\to\mathrm{B}}/T)(\partial \ln K_1/\partial m_2)_{T,P} \tag{8.16}$$

Related expressions for G_2, H_2, and V_2 are given in Eqs. (8.17)–(8.19). The expression for H_2 follows directly from those for G_2 and S_2, with $(H_2)_{\alpha_1} = (G_2)_{\alpha_1} + T(S_2)_{\alpha_1}$. The expression for V_2 makes use of $-RT(\partial \ln K_1/\partial P)_T = \Delta V_{\mathrm{A}\to\mathrm{B}}$.

$$G_2 = (G_2)_{\alpha_1} \tag{8.17}$$

$$H_2 = (H_2)_{\alpha_1} + m_1\alpha_1(1-\alpha_1)(\Delta H_{\mathrm{A}\to\mathrm{B}})(\partial \ln K_1/\partial m_2)_{T,P} \tag{8.18}$$

$$V_2 = (V_2)_{\alpha_1} + m_1\alpha_1(1-\alpha_1)(\Delta V_{\mathrm{A}\to\mathrm{B}})(\partial \ln K_1/\partial m_2)_{T,P} \tag{8.19}$$

The molar-shift terms for third and higher derivatives of the free energy require the full machinery of the general theory of Chapter 7. To illustrate their potential complexity, we shall use Eq. (7.17) to solve for $C_{P,2}$. The result is Eq. (8.20). Some of the required partial derivatives are listed in Table 8.2. Note that $C_{P,2}$ depends not only on $\Delta H_{\mathrm{A}\to\mathrm{B}}$, but also on $\partial\Delta H_{\mathrm{A}\to\mathrm{B}}/\partial m_2$.

$$C_{P,2} = (C_{P,2})_{\alpha_1} + R\alpha_1(1-\alpha_1)m_1(\Delta H_{\mathrm{A}\to\mathrm{B}}/RT)\big[(2/RT)(\partial\Delta H_{\mathrm{A}\to\mathrm{B}}/\partial m_2) + (1-2\alpha_1)(\Delta H_{\mathrm{A}\to\mathrm{B}}/RT)(\partial \ln K_1/\partial m_2)\big] \tag{8.20}$$

Table 8.2. Some partial derivatives for composition tree (T-8.3)

Functions of $(T, P, n_1, n_2, \alpha_1)$

$\partial^2 y_1/\partial\alpha_1^2 = -n_1 RT(1 - 2\alpha_1)/[\alpha_1^2(1 - \alpha_1)^2]$

$\partial^2 y_1/\partial n_2\, \partial T = -m_1 R(\partial \ln K_1/\partial m_2) - (m_1/T)(\partial \Delta H_{A\to B}/\partial m_2)$

$\partial^2 y_1/\partial\alpha_1\, \partial T = n_1 R/[\alpha_1(1 - \alpha_1)]$

Functions of (T, P, n_1, n_2, y_1)

$(n_2/m_2)(\partial^2\alpha_1/\partial n_2\, \partial T)_{y_1=0} = \alpha_1(1 - \alpha_1)(\partial\Delta H_{A\to B}/\partial m_2)/RT^2)$
$\quad + \alpha_1(1 - \alpha_1)(1 - 2\alpha_1)(\partial \ln K_1/\partial m_2)(\Delta H_{A\to B}/RT^2)$

$(\partial^2\alpha_1/\partial T^2)_{y_1=0} = [\alpha_1(1 - \alpha_1)\,\Delta H_{A\to B}/RT^2][-2/T + (1 - 2\alpha_1)\,\Delta H_{A\to B}/RT^2]$

Standard Partial Properties of Solute

Properties of the solvent [such as α_1, $\Delta H_{A\to B}$, and $\Delta V_{A\to B}$ in Eqs. (8.16)–(8.20)] are nearly constant in dilute solutions, even though their derivatives with respect to m_2 (such as $\partial\alpha_1/\partial m_2$ and $\partial\Delta H_{A\to B}/\partial m_2$) are significant. This is because m_2 is so small that such changes as $\delta\Delta H_{A\to B} = m_2(\partial\Delta H_{A\to B}/\partial m_2)$ are negligible fractions of $\Delta H_{A\to B}$. Moreover, the derivatives themselves, such as $\partial \ln K_1/\partial m_2$ and $\partial\Delta H_{A\to B}/\partial m_2$, are practically constant in dilute solutions. The molar-shift terms in Eqs. (8.16)–(8.20) are therefore constant in dilute solutions at constant T and P, being made up of constant factors.

According to the accounting procedure given in Table 8.1, all constant contributions to a partial property $(G_2, S_2, \ldots)$ of a solute are gathered up in the *standard* value $(G_2^0, S_2^0, \ldots)$ of that partial property. The molar-shift term due to solute-induced molar shifts in the solvent therefore is part of the standard value of the respective partial property. The resulting expressions for dilute solutes are summarized in Eqs. (8.21)–(8.25).

Standard partial properties of solute, with solute-induced molar-shift terms

[1 : 1 equilibrium in the solvent, see (T-8.3)]

$$G_2^0 = (G_2^0)_{\alpha_1} \tag{8.21}$$

$$S_2^0 = (S_2^0)_{\alpha_1} + m_1\alpha_1(1 - \alpha_1)(\Delta H_{A\to B}/T)(\partial \ln K_1/\partial m_2)_{T,P} \tag{8.22}$$

$$H_2^0 = (H_2^0)_{\alpha_1} + m_1\alpha_1(1 - \alpha_1)(\Delta H_{A\to B})(\partial \ln K_1/\partial m_2)_{T,P} \tag{8.23}$$

$$V_2^0 = (V_2^0)_{\alpha_1} + m_1\alpha_1(1 - \alpha_1)(\Delta V_{A\to B})(\partial \ln K_1/\partial m_2)_{T,P} \tag{8.24}$$

$$C_{P,2}^0 = (C_{P,2}^0)_{\alpha_1} + R\alpha_1(1 - \alpha_1)m_1(\Delta H_{A\to B}/RT)\big[(2/RT)(\partial\Delta H_{A\to B}/\partial m_2) + (1 - 2\alpha_1)(\Delta H_{A\to B}/RT)(\partial \ln K_1/\partial m_2)\big] \tag{8.25}$$

When the stoichiometry of the solvent equilibrium is not 1 : 1, the molar shift, $\partial\alpha_1/\partial m_2$, stays independent of m_2, as explained in the text following Eq. (8.2). The molar-shift terms then differ in detail from those in Eqs. (8.16)–(8.20) but continue to be part of the *standard* values of the partial properties of the solute. As exemplified by Eqs. (8.22) and (8.23), the molar-shift terms for H_2^0 and $T \cdot S_2^0$ are equal at equilibrium and are compensated in G_2^0.

Standard Changes in Chemical Reactions

Barring constraints imposed by relaxation times and observational time scales, the preceding results apply directly to ΔG^0, $\Delta H^0, \ldots$ of reaction and to $\Delta G^\ddagger, \Delta H^\ddagger, \ldots$ of activation. As explained in Chapter 6, molar-shift terms due to interconversions in the solvent contribute to $\Delta H^0, \Delta S^0, \ldots$ for the primary reaction if the relaxation times in the solvent are less than τ_{obs}. In the case of $\Delta H^\ddagger$, $\Delta S^\ddagger, \ldots$, the relaxation times in the solvent must be less than τ_{kin} for the reaction whose rate is being measured.

Expressions for the standard changes when there are molar-shift terms are given in Eqs. (8.26)–(8.31). The solvent species are in 1 : 1 equilibrium, as modeled in (T-8.3). The relevant solutes (R) are the primary reactants and products. The Δ-operator has the usual significance; that is,

$$\Delta(\partial \ln K_1/\partial m_{\text{R}}) = \sum (\partial \ln K_1/\partial m_{\text{products}}) - \sum (\partial \ln K_1/\partial m_{\text{reactants}})$$

Expressions for standard changes when 1 : 1 equilibrium in the solvent

$$\Delta G^0 = \Delta(G^0)_{\alpha_1} \tag{8.26}$$

$$\Delta S^0 = \Delta(S^0)_{\alpha_1} + m_1\alpha_1(1-\alpha_1)(\Delta H_{\text{A}\to\text{B}}/T)\Delta(\partial \ln K_1/\partial m_{\text{R}})_{T,P} \tag{8.27}$$

$$\Delta H^0 = \Delta(H^0)_{\alpha_1} + m_1\alpha_1(1-\alpha_1)(\Delta H_{\text{A}\to\text{B}})\Delta(\partial \ln K_1/\partial m_{\text{R}})_{T,P} \tag{8.28}$$

$$\vdots$$

$$\Delta G^\ddagger = \Delta(G^\ddagger)_{\alpha_1} \tag{8.29}$$

$$\Delta S^\ddagger = \Delta(S^\ddagger)_{\alpha_1} + m_1\alpha_1(1-\alpha_1)(\Delta H_{\text{A}\to\text{B}}/T)\Delta_{\text{act}}(\partial \ln K_1/\partial m_{\text{R}})_{T,P} \tag{8.30}$$

$$\Delta H^\ddagger = \Delta(H^\ddagger)_{\alpha_1} + m_1\alpha_1(1-\alpha_1)(\Delta H_{\text{A}\to\text{B}})\Delta_{\text{act}}(\partial \ln K_1/\partial m_{\text{R}})_{T,P} \tag{8.31}$$

$$\vdots$$

The molar-shift terms in $T\Delta S^0\,(T\Delta S^\ddagger)$ are equal to those in $\Delta H^0\,(\Delta H^\ddagger)$ and compensate in $\Delta G^0\,(\Delta G^\ddagger)$. Because of the factor $\alpha_1(1-\alpha_1)$, the molar-shift term is small if either A or B is a minor species; both species must be stoichiometrically significant. Furthermore, $\Delta H_{\text{A}\to\text{B}}$ must be substantial, and the reactant-induced shifts of $\ln K_1$ must differ substantially from the product-induced shifts.

WATER AND NONPOLAR SOLUTES

Liquid water consists of two molecular species, as shown by accurate isosbestic points in the OH-stretching region of the spectrum. Figure 8.2 shows the first-overtone absorption by HOD in D_2O in the near-infrared [Worley and Klotz, 1966]. Figure 8.3 shows Raman scattering spectra from pure water in the fundamental region, for three different arrangements of the plane of polarization of the incident beam and the plane of detection of the perpendicularly scattered light [Walrafen et al., 1986a,b]. Each of these physically distinct spectra shows good isosbestic behavior.

A temperature-dependent spectrum typically indicates that two or more species are present whose ratio varies with the temperature. At the isosbestic point the spectral behavior is independent of that ratio because the extinction or scattering coefficients of the species happen to be equal. This is not remarkable when there are two species. Equality at a single point for more than two species is rare, however, so that an isosbestic point probably indicates two species. When the full spectrum shows several isosbestic points (as does the join of Figs. 8.2 and 8.3), this probability becomes near-certainty. In the present case, the

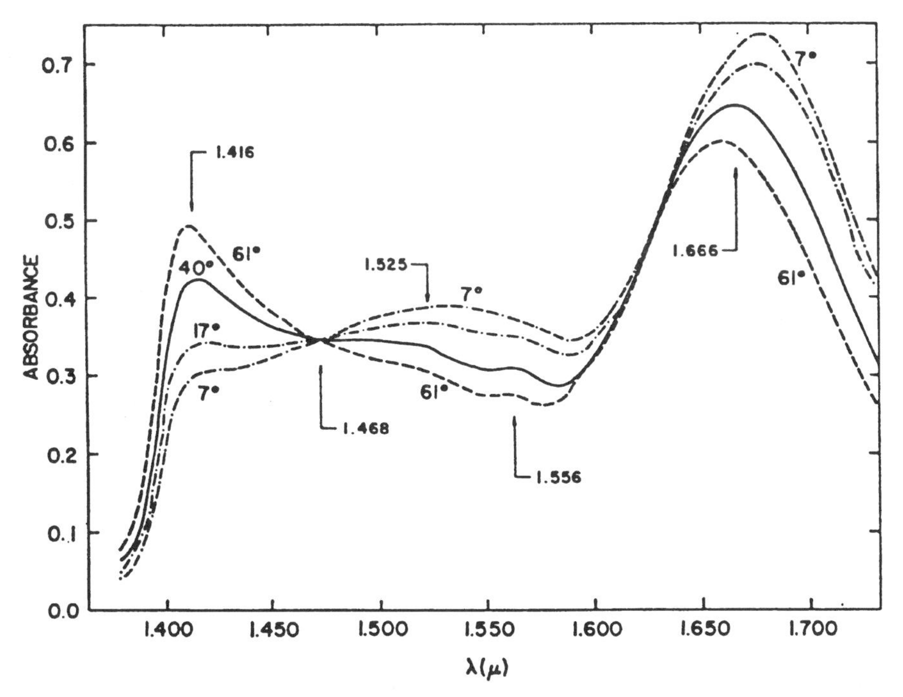

Figure 8.2. Effect of temperature on absorption spectrum for 6.0 volume-formal HOD in D_2O in the region of the first overtone of the OH-stretching vibration. The reference cell contains pure D_2O. (Reproduced from Worley and Klotz [1966], courtesy of Professor I. M. Klotz, with permission of the American Institute of Physics.)

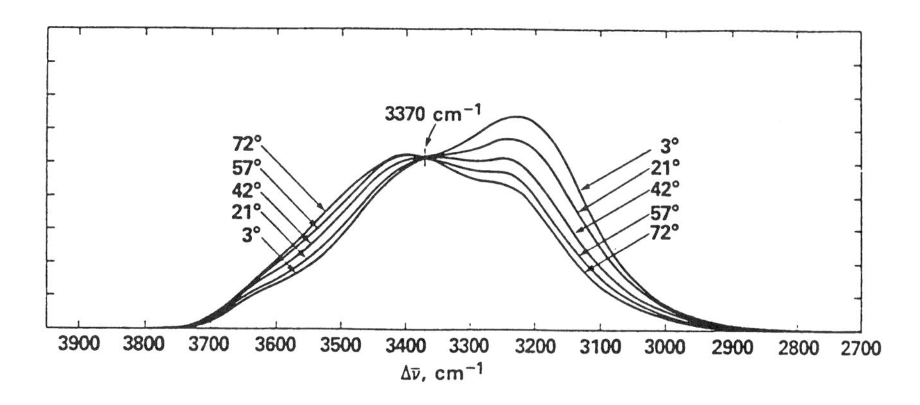

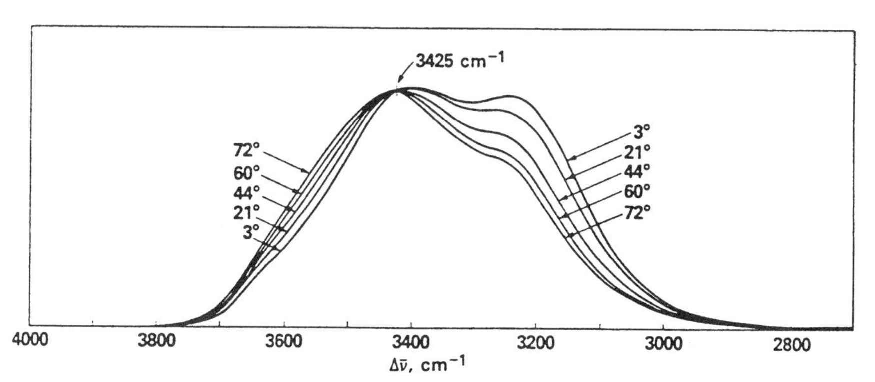

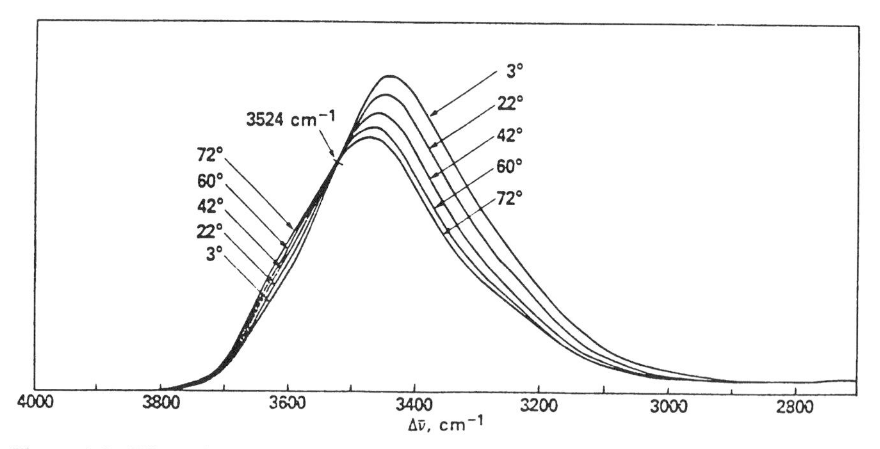

Figure 8.3. Effect of temperature on intensity of Raman scattering from triply distilled water in the fundamental OH-stretching region, for three different positions of the plane of polarization of the incident beam and the plane of detection of the perpendicularly scattered light. (Reproduced from Walrafen et al. [1986a], courtesy of Professor G. E. Walrafen, with permission of the American Institute of Physics.)

OH-stretching spectra reveal every water molecule in the liquid. Water therefore consists of two molecular species, except perhaps for a slight fraction whose departure from isosbestic behavior is undetectable.

Nature of the Solvent Species

The temperature-dependence of the equilibrium ratio of the water species is large enough to affect most properties of liquid water and to yield values of ΔH^0 for interconversion. Walrafen et al. [1986b] were able to cite more than 10 independent values, based on six different properties, with an average ΔH^0 of 10.5 ± 0.5 kJ/formula weight. The unusually high heat capacity of water is particularly well suited to this determination. The relevant equations are (8.32), with the molar-shift term in (8.32b) accounting for the bulk of the heat capacity of liquid water. Here α denotes the formula-weight fraction, because we still need to specify the stoichiometry of interconversion.

$$C_P = \left(\frac{\partial H}{\partial T}\right)_{P,y=0} = \left(\frac{\partial H}{\partial T}\right)_{P,\alpha} + \left(\frac{\partial H}{\partial \alpha}\right)_{P,T}\left(\frac{\partial \alpha}{\partial T}\right)_{P,y=0} \tag{8.32a}$$

$$\therefore \quad C_P = (C_P)_\alpha + \Delta H^0(\partial\alpha/\partial T)_{P,y=0} \tag{8.32b}$$

After estimating $(C_P)_\alpha$ by a combination of theory and propensity rules, Benson [1978] obtained both ΔH^0 and ΔS^0 for the interconversion of the water species by fitting the measured heat capacity C_P as a function of T. Assuming an 1 : 1 model (A $\rightleftarrows$ B), he deduced that $\Delta H^0_{A\to B} = 10.5$ kJ/mol and $\Delta S^0_{A\to B} =$ 28.2 J/mol K. Benson and Siebert [1992] subsequently fitted the heat capacity to an 1 : 2 model, O(ctomer) $\rightleftarrows$ 2T(etramer), where O = $(H_2O)_8$ and T = $(H_2O)_4$. This model recommended itself because the part that is predicted by theory is now quite accurate. They deduced that, per formula weight of water, $\Delta H^0_{O\to 2T}$ = 10.4 kJ and $\Delta S^0_{O\to 2T} = 33.7$ J/K. Note that ΔH^0 is nearly the same for the two models. We shall choose between the models by applying the absorbance data of Fig. 8.2.

The particulars of this test are summarized in Table 8.3. The temperature dependence of absorbance is greatest at 1.416 μ, where the high-temperature species is responsible for most of the absorption. Plots of the predicted α's versus absorbance at 1.416 μ are shown in Fig. 8.4. Both models give linear relationships as expected, but only the 1 : 1 model gives physically acceptable extinction coefficients: In the 1 : 2 model, ε_0 is significantly negative. We shall therefore adopt the 1 : 1 model.

The two water species in 1 : 1 equilibrium are best described as environmental isomers. Let W denote the water molecule, and let \a and \b be two distinct environments. The process W\a $\rightleftarrows$ W\b represents the transfer of a water molecule from one environment to another. It is analogous to the distribution of the species, H_2O-monomer, between two media, as in

Table 8.3. Comparison of 1 : 1 model with 1 : 2 model of the species in liquid water (α = formula-weight fraction)

(1:1) $A(1-\alpha_{11}) \rightleftarrows B(\alpha_{11})$ $\quad K_{11} = \alpha_{11}/(1-\alpha_{11})$

(1:2) $O(1-\alpha_{12}) \rightleftarrows 2T(\alpha_{12})$

Mole fraction, O: $(1-\alpha_{12})/(1+\alpha_{12})$; $\quad$ T: $2\alpha_{12}/(1+\alpha_{12})$

$K_{12} = 4\alpha_{12}^2/[(1+\alpha_{12})(1-\alpha_{12})]$

(1:1) $A = \epsilon_A c_1(1-\alpha_{11}) + \epsilon_B c_1 \alpha_{11}$

At 1.416 μ, $\epsilon_A c_1 = 0.01 \pm 0.04$ and $\epsilon_B c_1 = 1.22 \pm 0.1$

(1:2) $A = \epsilon_O c_1(1-\alpha_{12}) + \epsilon_T c_1 \alpha_{12}$

At 1.416 μ, $\epsilon_O c_1 = -0.27 \pm 0.08$ and $\epsilon_T c_1 = 1.24 \pm 0.1$

T (°C)	α_{11}	α_{12}	A, 1.416 μ[a]
7	0.246	0.377	0.301
17	0.276	0.402	0.344
25	0.300	0.422	—
40	0.345	0.457	0.426
61	0.404	0.504	0.494

[a]Absorbances from Fig. 8.2.

H_2O(benzene) $\rightleftarrows$ H_2O(methanol), and it therefore proceeds with 1 : 1 stoichiometry. W\a and W\b differ in the number of water neighbors linked to the W molecule by hydrogen bonds. The mean number of linked neighbors, inferred from the area under the first peak of the experimental O$\cdots$O radial distribution function, is near 4.4 at 25°C and increases slowly with temperature

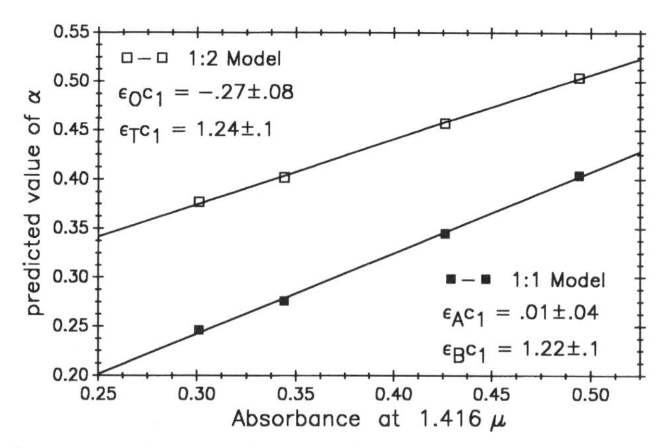

Figure 8.4. Comparison of 1 : 1 model with 1 : 2 model for the two species in liquid water. Absorbances at 1.416 μ from Fig. 8.2.

[Narten and Levy, 1969, 1971; Morgan and Warren, 1938]. One of the two environments is the familiar four-linked environment of the Bernal–Fowler model, with tetrahedrally oriented flexible hydrogen bonds [Bernal and Fowler, 1933]. The other must be five-linked to fit the experimental average. The interconversion may therefore be written as $W\backslash 4 \rightleftarrows W\backslash 5$.

Supporting evidence for the $W\backslash 4 \rightleftarrows W\backslash 5$ model comes from molecular-dynamics simulations of liquid water. In the classic model of Stillinger and Rahman [1974], the water molecule consists of (a) an oscillating dipole of the London-dispersion type and (b) static OH-bond dipoles, each of which is separated into monopoles. The positive monopoles coincide with the water protons; the negative monopoles are located in the regions occupied by the oxygen atom's unshared electron pairs. The directions from the O-atom to the four monopoles are nearly tetrahedral. The magnitudes of the electric charges at the poles and other parameters are adjusted to give overall "best fit" to key properties of liquid water.

Sciortino et al. [1992] simulated an ensemble of such molecules using techniques of molecular dynamics and were able to deduce the distribution of immediate neighbors at configurational equilibrium. "Neighbors" were defined as model molecules with $O \cdots O$ distances belonging to the first peak of the $O \cdots O$ radial distribution function—up to the first minimum. The neighbor configurations sorted themselves so that more than 80% of the model molecules had either four or five neighbors. In addition, the mean energy difference ΔU for the $\backslash 4 \rightarrow \backslash 5$ process was 10 kJ/mol, in agreement with the quasi-experimental ΔH^0 of 10.5 ± 0.5 kJ per formula weight [Walrafen et al., 1986].

One of the key properties of liquid water is its high dielectric constant. In terms of Kirkwood–Onsager theory, the dielectric constant evaluates the time-average dipole moment of the environment around a water molecule—or rather, the ratio $(\mu_{env} \cdot \mu_W)/\mu_W^2$ of the parallel component of that moment to the moment of the central water molecule [Kirkwood, 1939, 1946]. Oster and Kirkwood [1943] showed that the dipole moment is in fair agreement with the model of Bernal and Fowler [1933]. Grunwald [1986a] then improved the fit by treating water as a mixture of two environmental species. The essence of his analysis is shown in Fig. 8.5. The four-linked environment is very polar: $(\mu_{env} \cdot \mu_W)/\mu_W^2 = 2.01 \pm 0.12$, in good agreement with prediction for somewhat flexible near-tetrahedral linkages. The five-linked environment is only half as polar: $(\mu_{env} \cdot \mu_W)/\mu_W^2 = 0.91 \pm 0.18$. One might say that two of the five links are polar as in the four-linked environment, while the three others are essentially bifurcated hydrogen bonds whose dipole directions are such that the time-average moment is small. A possible linkage configuration based on this premise is included in Fig. 8.5, but in fairness a trustworthy model is still unknown.

The simulations of Sciortino et al. [1992] yield instantaneous rather than time-average configurations. Their results indicate that four of the five neighbors are linked by normal hydrogen bonds, while the fifth $O \cdots O$ neighbor distance is longer—long enough to be called a network defect.

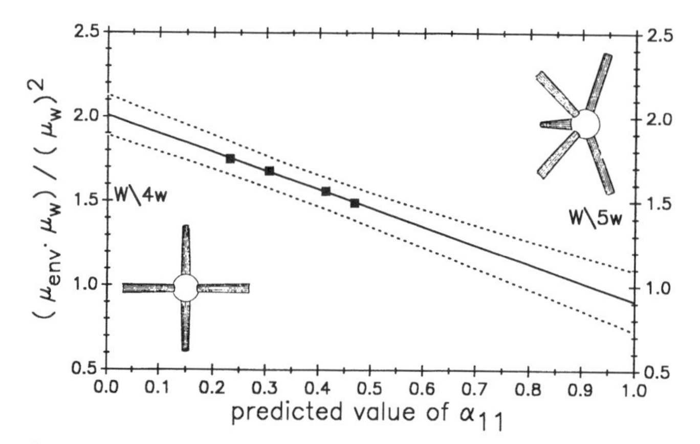

Figure 8.5. Oster–Kirkwood plot based on the dielectric constant, for the dipole moment (parallel component) of the environment of two water species. The result for W\4w fits a tetrahedral linkage geometry. The result for W\5w suggests a half-polar, half-nonpolar linkage geometry [Grunwald, 1986a].

Molar Shifts Induced by Nonpolar Solutes

The model of solute-induced molar shifts in the solvent requires, first, that the solvent species surrounding the solute molecules bear a one-to-one correlation to the solvent species in the bulk of the solvent; and, second, that the occurrence probabilities of the solvent species around the solute molecules differ from those in the bulk. The solvation of nonpolar solutes in water is likely to satisfy these requirements.

The Four-linked (\4w) water environments, even though distorted from the tetrahedral optimum near a solute molecule, are likely to be favored neighbors because tetrahedral links are good cage formers. In organic chemistry, for example, hydrocarbon cages with C–C single bonds are well known, ranging from compact structures with strained bond angles such as cubane, to more spacious and strain-free structures such as 2,2,2-bicyclooctane or 2,3,3-bicyclodecane. The hydrogen bonds to water molecules are more flexible than C–C bonds, and four-linked water molecules can generate form-fitting cages to fit almost any molecular size and shape. Although the four-linked water molecules will be the primary building blocks, fifth links may form when geometry permits.

In addition to being a versatile cage-former, the four-linked water species also offers a relatively low local density, which in turn reduces the work of creating the cavities occupied by the solute molecules. The density is low because at nearest-neighbor distances, up to the first minimum in the O···O radial distribution function, in the case of the four-linked water species the volume contains the centers of five water molecules (one at the origin and four neighbors), while for the five-linked species the same volume contains six. The low local density and ease of cage formation cooperate so that the water in solvation shells around nonpolar solute molecules is enriched in the W\4w species.

Delphic Dissection of the Entropy of Solvation

The analysis of a standard thermodynamic property into an isomolar and a molar-shift term is called *delphic dissection* [Grunwald, 1986b; Grunwald and Comeford, 1988].* For nonpolar solutes in water, the standard entropy of solvation is particularly informative. Solvation, it will be recalled, is the process X(g) $\rightarrow$ X (in solvent M). Accurate standard entropies of solvation are available for noble gases and common inert diatomics and hydrocarbons, both in water and in aprotic solvents ranging from nonpolar benzene to polar dimethylsulfoxide [Wilhelm et al., 1977; Wilhelm and Battino, 1973; Dec and Gill, 1985; Gill et al., 1975, 1976]. In the aprotic solvents, ΔS^0 for solvation is nearly independent of the solvent, as shown by typical data in Fig. 8.6. These aprotic solvents are single species, and solute-induced molar shifts can be ruled out. The horizontal lines in Fig. 8.6 are therefore isomolar lines, and the vertical distance from the line to the point for water estimates the molar-shift term in water.

Regular-solution theory [Hildebrand et al., 1970] supports these assignments. Regular solutions do not distinguish between formal components and molecular species, and therefore they exclude molar shifts: Their processes are isomolar. Two properties of regular solutions are relevant. First, ΔS^0 of solvation varies with the nature of the gas but *not* with the solvent, which fits the horizontal lines in Fig. 8.6. Second, ΔG^0 for the solvation of nonpolar gases (which is isomolar by the Second Law) is reproduced by regular solution theory, in water almost as well as in polar aprotic solvents [Grunwald, 1986b; Grunwald and Comeford, 1988]. This suggests that other predictions of regular-solution theory apply under isomolar conditions in water as well. Accordingly, the horizontal lines in Fig. 8.6 predict the isomolar entropies of solvation for nonpolar gases in water.

Table 8.4 carries out the delphic dissection. The molar-shift term is given by the difference: $(S_2)_{\text{shift}} = \Delta S_2^0(\text{W}) - \Delta S_2^0(\text{M})$, where $\Delta S_2^0(\text{W})$ is the measured value, and $\Delta S_2^0(\text{M})$ is the average for solvation in aprotic solvents, as exemplified by the horizontal lines in Fig. 8.6. The entropies of solvation are more negative in water than in aprotic solvents, and the molar-shift terms are substantial.

Actual molar shifts are calculated by Eq. (8.33), which is derived from Eq. (8.22) with $K_1 = \alpha_1/(1 - \alpha_1)$. Note that $-m_1(\partial\alpha_1/\partial m_2) = (\partial m_{4\text{w}}/\partial m_2)$ and thus evaluates the increase in the molality of four-linked water per unit concentration of the added solute. The calculation is summarized in Table 8.4.

$$-m_1(\partial\alpha_1/\partial m_2) = T(S_2)_{\text{shift}}/\Delta H_{4\text{w}\rightarrow 5\text{w}} = -0.0284(S_2)_{\text{shift}} \quad \text{at 298 K} \qquad (8.33)$$

*When the author first wrote about molar shifts in aqueous solutions, he used the terms *isodelphic*, *lyodelphic*, and *delphic dissection*. To cut down on unfamiliar terminology, *isodelphic* is now replaced by *isomolar*, and the lyodelphic term is renamed the molar-shift term. The term *delphic dissection* remains apt.

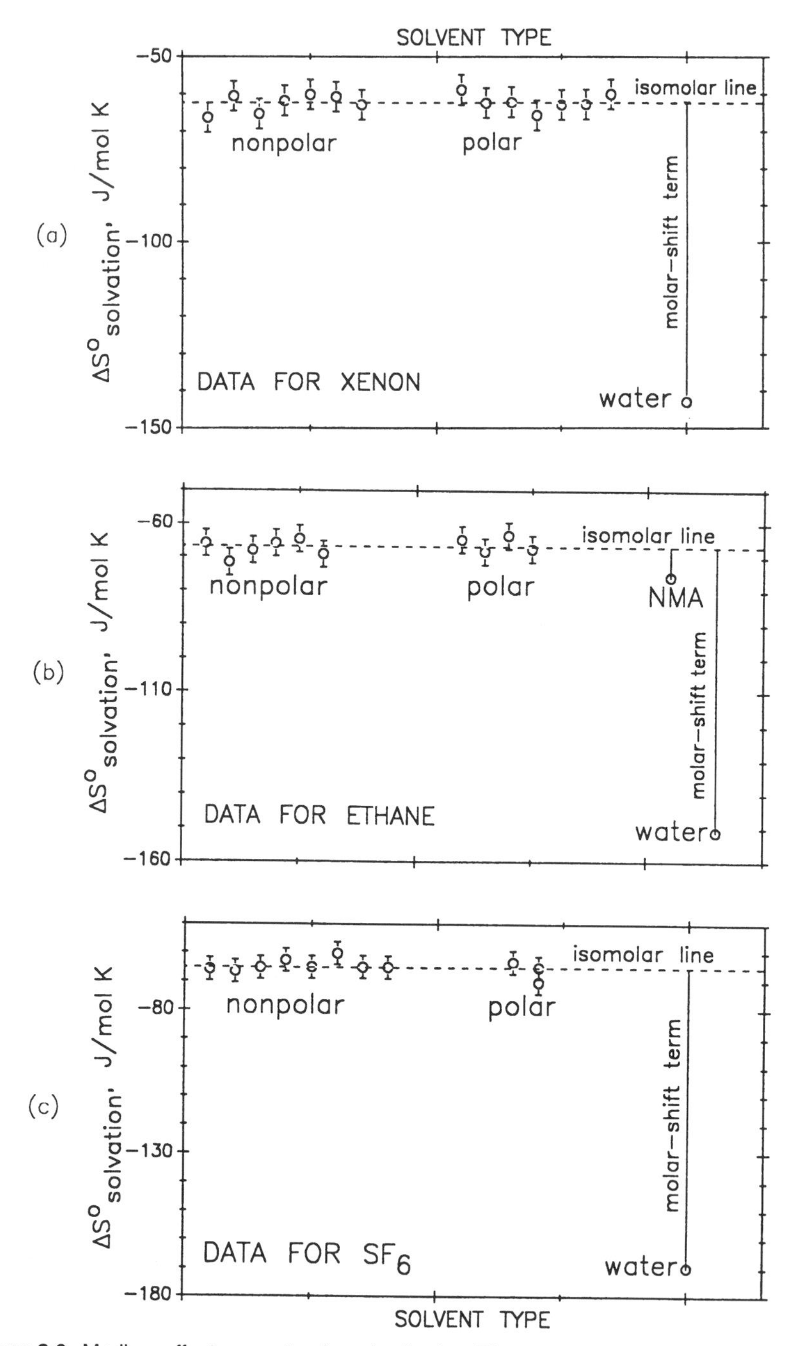

Figure 8.6. Medium effects on entropies of solvation. The nonpolar solvents include hydrocarbons and CCl_4. The polar solvents include halobenzenes, nitrobenzene, acetone, and dimethylsulfoxide [Wilhelm et al., 1977; Wilhelm and Battino, 1973; Dec and Gill, 1985; Gill et al., 1975, 1976].

Table 8.4. Delphic dissection of standard enthalpies of solvation in water at 298 K

Solute	ΔS_2° (J/K)[a] (W)	(M)	$(S_2)_{shift}$	$-m_1(\partial\alpha_1/\partial m_2)$[b]	Block Size[c]
He	−101	−43	−58	1.65	5.4
Ne	−110	−45	−65	1.85	6.1
Ar	−128	−55	−73	2.08	6.8
Kr	−135	−57	−78	2.21	7.2
Xe	−143	−63	−80	2.28	7.5
H_2	−106	−48	−58	1.64	5.4
N_2	−131	−51	−80	2.26	7.4
O_2	−130	−57	−73	2.08	6.8
CO	−128	−55	−73	2.08	6.8
CH_4	−133	−60	−72	2.06	6.7
C_2H_6	−151	−68	−83	2.36	7.7
C_3H_8	−163	−70	−93	2.64	8.6
n-C_4H_{10}	−174	−78	−96	2.73	8.9
n-C_6H_{14}	−202	−92	−109	3.10	10.1
c-C_6H_{12}	−189	−94	−95	2.70	8.8
		Unsaturated Hydrocarbons[d]			
C_2H_2	−109	−63	−46	1.31	—
C_2H_4	−134	−66	−67	1.91	—
C_6H_6	−154	−96	−58	1.64	—
MeC_6H_5	−171	−100	−71	2.01	—
EtC_6H_5	−187	−105	−81	2.31	—
n-PrC_6H_5	−203	−110	−93	2.63	—

[a] (W) = water, (M) = average in aprotic solvents [Wilhelm, et al., 1977; Wilhelm and Battino, 1973; Dec and Gill, 1985; Gill et al., 1975, 1976].
[b] Here $-m_1(\partial\alpha_1/\partial m_2) = \partial m_{4w}/\partial m_2$.
[c] Block size = $(\partial m_{4w}/\partial m_2)/\alpha_1$.
[d] The values of $-m_1(\partial\alpha_1/\partial m_2)$ for the unsaturated compounds are too small, relative to the expected number of water neighbors—as if the pi electrons were accepting hydrogen bonds from water molecules [Grunwald, 1986b; Grunwald and Comeford, 1988].

The values obtained for $-m_1(\partial\alpha_1/\partial m_2)$ are of a serious magnitude. They become intriguing if one assumes that the solute-induced shifts create contiguous blocks of W\4w molecules. Consider a contiguous block of s water molecules. Before the solute is added, $s(1-\alpha_1)$ water molecules are four-linked and $s \cdot \alpha_1$ are five-linked. Now add the solute in such amount that the $s \cdot \alpha_1$ five-linked molecules become four-linked. Then $s \cdot \alpha_1 = -m_1(\partial\alpha_1/\partial m_2)$; and the number ($s$) of water molecules in the block equals $-m_1(\partial\alpha_1/\partial m_2)/\alpha_1$.

Block sizes equal to $-m_1(\partial\alpha_1/\partial m_2)/\alpha_1$ are given in the last column of Table 8.4. The intriguing feature is that the numbers support the formation of closed four-linked water cages around the solute molecules. For example, when gas-kinetic collision radii are used for hard-sphere radii, B. J. Alder's model of random-packing of spheres in hard-sphere liquids [Alder, 1955] predicts 6.5 water neighbors for Ne, compared to a four-linked block size of 6.1. The same model predicts 9.4 water neighbors for *n*-propane, compared to a four-linked block size of 8.6. H. S. Frank, who pioneered the interpretation of entropies of hydration [Frank, 1945; Frank and Evans, 1945; Frank and Wen, 1957], talked about the formation of contiguous "patches of ice" next to a nonpolar molecule in liquid water. Others have visualized contiguous clathrate-like cages. (For information about crystalline clathrates, see Jeffrey and McMullin [1967] and Allcock [1978].) These descriptions invoke poetic license, because they imply a crystalline order rather than the compliant liquid disorder that form-fits a solute molecule. But in essence they envision contiguous blocks of four-linked water molecules. Within this framework, the entropies of solvation are consistent with solute-induced molar shifts that form closed four-linked water cages around the nonpolar solute molecules [Grunwald, 1986b; Grunwald and Comeford, 1988].

In the realm of computer simulations, virtual "snapshots" of hydration shells, based on well-calibrated potential-energy functions, are available for model "solutions" consisting of a nonpolar solute molecule in the midst of a hundred or so water molecules [Rossky and Karplus, 1979; Swaminathan et al., 1978; Geiger et al., 1979; Alagona and Tani, 1980; Ravishanker et al., 1984]. The instantaneous configurations lack clarity because the methods of simulation have built-in Brownian noise, but two elements are unmistakable: First, the packing of water molecules in the hydration shells is dense, with twice as many water molecules as in the "block size" column of Table 8.4. Second, the surface of the nonpolar molecule is surrounded by, or adjacent to pieces of, distorted clathrate-like water structures. For neon in water, Geiger et al. [1979] managed, by mathematical protocol, to suppress the virtual Brownian noise. This revealed that the densely packed water neighbors of the noisy simulations divide themselves into two groups: (1) an inner group of eight water molecules that form a clathrate-like cage around the neon atom and (2) an outer group of six water molecules that are linked to available sites on the outer surface of that cage. While an 8-cage still exceeds the 6.1 water-molecules' "block size" listed for neon in Table 8.4, the two numbers are close enough to be called consistent.

ALCOHOLS AND ALCOHOL–WATER SOLVENTS

Thermodynamic properties in alcohol–water mixtures have long challenged the imagination. Standard free energies of solution and reaction are nearly always monotonic functions of solvent composition, while standard enthalpies and entropies show maxima and minima. Examples for ethanol–water and *t*-butyl alcohol–water were given in Figs. 6.2 and 6.5. The interactions that cause the maxima and minima evidently obey enthalpy–entropy compensation. This and other facts suggest the presence of significant molar shifts. A rudimentary analysis is worthwhile even though the analysis will be speculative, because the nature of the underlying solvent species is not yet fully defined.

We shall begin with some facts. In the water-rich region of the alcohol–water solvent system there can be two extrema, one strong and one weak, as shown in Fig. 6.5. The weak one is missing in many cases, but the strong one is reliable. The formula-weight fraction Z_1^* of water at which the strong extremum occurs is nearly independent of the substrate, reaction, or rate process: In methanol–water, $Z_1^* = 0.75 \pm 0.05$; in ethanol–water, $Z_1^* = 0.85 \pm 0.05$; and in *t*-butyl alcohol–water, $Z_1^* = 0.95 \pm 0.02$ [Lumry and Rajender, 1970; Engberts, 1979]. Each of these ranges includes the composition at which the enthalpy of mixing the solvent components, ΔH^M/formula weight, goes through a minimum.

To illustrate the degree to which Z_1^* is independent of the nature of the substrate or reaction, Fig. 8.7 shows data for five widely different processes in *t*-butyl alcohol–water mixtures. These processes are as follows:

A. Transfer of *p*-tolylsulfonylmethyl benzenesulfonate, *p*-$CH_3C_6H_4SO_2CH_2$-$OSO_2C_6H_5$, from water to *t*-butyl alcohol/water mixtures [Menniga and Engberts, 1976].

B. Change in activation parameters, from water to water–*t*-butyl alco-

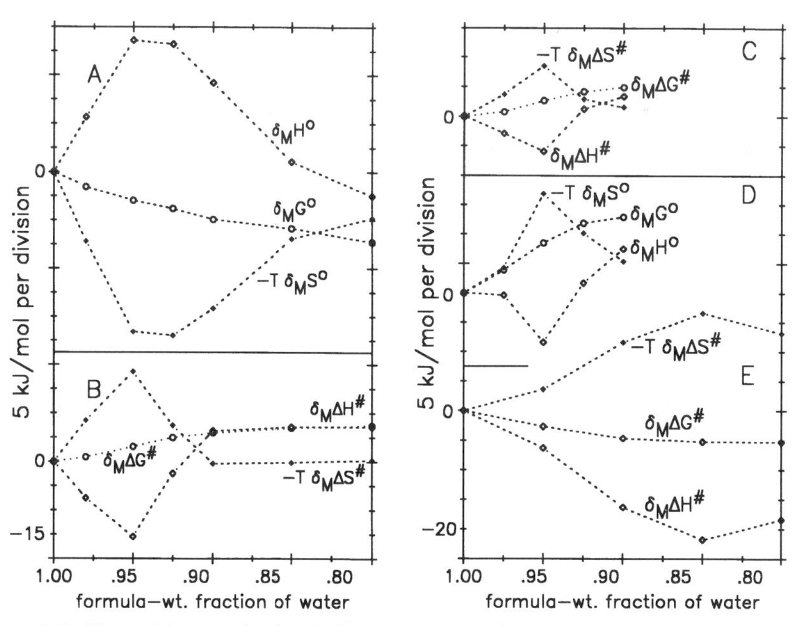

Figure 8.7. Plots of the standard enthalpy, entropy, and free energy for various processes in *t*-butyl alcohol–water solvents at or near 298 K. The processes are described and referenced in the text.

hol mixtures, for the hydrolysis of *p*-nitrophenyldichloroacetate, *p*-$O_2NC_6H_4O(CO)CHCl_2$ [Engbersen and Engberts, 1975].

C. Change in activation parameters for the hydroxide-catalyzed hydrolysis of one arylsulfonate group in *p*-$CH_3C_6H_4SO_2CH_2OSO_2C_6H_5$, from water to water–*t*-butyl alcohol mixtures [Holterman and Engberts, 1979].

D. Extrathermodynamic estimates for the transfer of hydroxide ion (OH^-) from water to water–*t*-butyl alcohol mixtures [Holterman and Engberts, 1979].

E. Change in activation parameters, from water to water–*t*-butyl alcohol mixtures, for the proton removal by water from the following perchlorate ester [Menninga and Engberts, 1976]:

$$p\text{-}O_2NC_6H_4SO_2CH_2OClO_3 + HOH$$
$$\longrightarrow p\text{-}O_2NC_6H_4SO_2CH^{\ominus}\text{—}OClO_3 + H_3O^{\oplus}$$
$$\xrightarrow{+H_2O,\ \text{fast}} p\text{-}O_2NC_6H_4SO_2H + HCOOH + ClO_3^{\ominus}$$

Processes A to D (or their rate-determining steps) are electroneutral. Process E entails the generation of ionic charges. In processes A to D the extremum is close to the normal value of $Z_1^* = 0.95$, while in process E, $Z_1^* = 0.85$. In each process the standard free-energy varies smoothly and monotonically.

Although the solvent species in alcohol–water mixtures are not yet fully defined, it is likely that molar shifts occur. In water, the existence of environmental isomers was suggested by isosbestic points in the infrared and Raman spectrum and was supported by analysis of the dielectric constant. In liquid methanol, with increasing temperature, the OH-stretching spectrum shows a similarly accurate isosbestic point at 3410 cm^{-1} [Luck and Fritzsche, 1992]. The dielectric constants of liquid alcohols show that the cages around the alcohol molecules have significant dipole moments [Oster and Kirkwood, 1943] whose great variability, for the structural isomers of octanol, indicates that the linkage geometry is sensitive to the nature of nonbonded interactions [Dannhauser, 1968]. A conspicuous difference between water and alcohols is that water molecules have two OH hydrogen-bond donor groups while alcohol molecules have only one. Normal alcohols therefore tend to form hydrogen-bonded chains rather than three-dimensional networks. On the other hand, the oxygen atoms in alcohol molecules have two unshared electron pairs and are able to accept two hydrogen bonds, just like the oxygen atoms in water molecules. The excess of acceptor groups over donor groups in the alcohols permits alternative configurations of hydrogen-bonded chains and lends flexibility to the linkage geometries.

There are also significant questions concerning the molecular species that might form. First, do the hydrogen bonds produce distinct molecular complexes? Or do they produce environmental isomers in which the linked solvent molecules act as Brownian neighbors? Second, in the case of environmental isomers, is it correct merely to count the number of links? For example, when the total number of links to an alcohol molecule is constant, but some links are alcohol · water(ROH ··· OH_2) and others are alcohol · alcohol(ROH ··· OHR), will there be more than one environmental isomer?

Even though we do not know the detailed nature of the solvent isomers, there are some molar-shift principles that apply in any case. First, as alcohol is added and Z_1 decreases, the solvent species are bound to change. The environmental equilibrium between W\4w and W\5w in pure water is especially sensitive to the addition of nonpolar solutes, as we have seen. The addition of *t*-butyl alcohol introduces a nonpolar *t*-butyl group, which even at $Z_1 = 0.99$ will decrease the fraction of W\5w very substantially. The [\4w, \5w], molar-shift term is proportional to $\alpha_1(1 - \alpha_1)$ [Eq. (8.22)]. We therefore expect that the [\4w, \5w] molar-shift term will decrease rapidly as alcohol is added.

Second, as alcohol is added to the water-rich solvent, alcohol–water species with progressively higher alcohol content will grow in. Let us suppose that these equilibria are sensitive to molar shifts. As the successive alcohol–water species grow in, reach a maximum, and then diminish, the corresponding molar-shift terms will do the same, because each molar-shift term contains a stoichiometric multiplier [such as $\alpha_1(1 - \alpha_1)$ in Eq. (8.22)] that goes through a maximum. As Z_1 decreases, each molar-shift mechanism therefore generates a *molar-shift band*, whose plot of amplitude versus Z_1 resembles the shape of an optical absorption band. Each successive molar-shift mechanism versus Z_1 generates such a band, and the various bands probably overlap. The full molar-shift term at a given Z_1 is the sum over all mechanisms, just as the full optical density at a given wavelength is the sum over all absorbing species. The chief difference is that molar-shift bands may be negative.

Figure 8.8 shows schematically how these predictions might work out in an alcohol–water mixture. The figure assumes that the molar-shift mechanism in water attenuates rapidly as alcohol is added. It further shows the imagined ebb

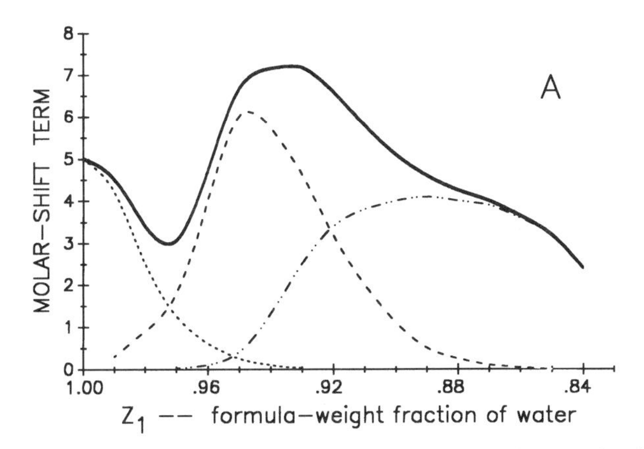

Figure 8.8. Schematic view of individual molar-shift bands and of the total molar-shift term, as envisioned by the author for alcohol–water mixtures.

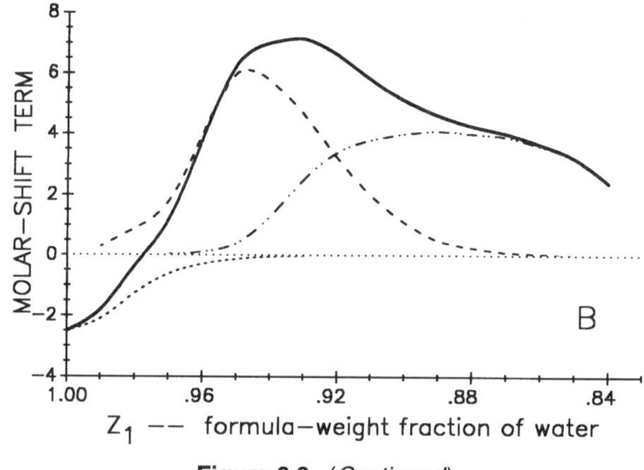

Figure 8.8. (*Continued*)

and flow of two successive alcohol–water bands, both with peaks whose magnitude is comparable to that of the peak in water. In part A all bands are positive. In part B the water band is negative. The model plot of molar-shift terms versus Z_1 readily generates maxima and minima. When molar-shift terms such as these are added to their monotonic isomolar counterparts, the result translates into extrema versus Z_1 for standard enthalpies, entropies, and volumes. On this basis, the reliable extremum at Z_1^* may be identified in Fig. 8.8 with the maximum that is independent of the sign of the water band. The smaller extremum between pure water and Z_1^* may be identified with the weaker extremum that is missing in many real cases. Note that it is predicted only when the water band and the first alcohol–water band have the same sign and comparable peak heights. Further prognostication will have to wait until the solvent species in alcohol–water mixtures have been characterized, but it is encouraging to have a thermodynamic mechanism that takes us this far.

9

HARD-SPHERE AND POTENTIAL-ENERGY CAGES AROUND MOLECULES IN LIQUIDS

In a chemical equation, the molecules are portrayed as individuals, but whether they represent dynamically independent particles depends on the nature of their physical confinement. In this chapter we will consider the cage effect: the confinement, in dense phases, of the molecules within walls consisting of neighboring molecules. When the molecules reside in cages, translation and (in many cases) rotation become hindered motions.

In a low-pressure gas the molecules translate and rotate almost freely, and the only "cage" is the communal volume. This description of a gas as an ensemble of independent molecules is supported by the great success of the ideal-gas model, as well as by the success of quantum theory based on the ideal-gas model in predicting rotational spectra and translational and rotational entropies for molecular species in the gas phase.

At the opposite extreme, in a crystalline solid the molecular motions are confined to cells on a regular lattice. Thermal translation and overall rotation mostly decrease to low-amplitude oscillations or "librations" that, in view of the limited space available per molecule, tend to be concerted. This cage model is supported by the great success of its application in X-ray diffraction, by the slowness of molecular diffusion in solids, and by the existence of crystal vibrational spectra that fit a coupled-phonon model [DiBartolo and Powell, 1976; Brüesch, 1982]. In a solid-state chemical reaction, the entire crystal may be converted in one shot [Cohen et al., 1964; Enkelmann et al., 1993; McBride, 1986].

In a liquid, molecular confinement straddles these extremes. Given the typically high densities, interactions among the molecules are clearly significant, but there is enough empty space to loosen up the motions. Translation and rotation still tend to be hindered, but dynamic memory fades in less than a picosecond. On a time scale of picoseconds or longer, the molecular motions become statistically independent and follow the model of Brownian motion. The molecules then exist in cages that hinder translation and rotation, but within this framework they move as separate kinetic entities and may be treated as individuals. Indeed, our reliance on the concept of separate "molecules" in liquids is so strong that we treat a pair of interacting molecules as a molecular complex only if the interaction destroys their long-term kinetic independence.

Given the freedom of molecular motions in the gas phase, the view of molecules in liquids as separate entities is supported by Henry's law and its corollaries, including, for dilute solutions, the logarithmic concentration dependence of partial free energies, the Law of Mass Action, and the colligative interpretation we give to "colligative" solvent properties. It fits the continuity of gases and liquids at the critical point, as rationalized by the van der Waals and kindred equations of state.

When chemists talk about a cage, they probably mean a material wall built of molecules. When physicists talk about a cage, they probably mean a potential-energy function that exerts restoring forces on the central molecule from all sides. The two concepts differ considerably. When a molecule undergoes long-range Brownian motion such as diffusion, it jumps from the original material cage to a succession of other material cages, but as it jumps, the potential-energy cage effectively travels along. After the molecule's arrival in a new material cage there is a short delay for return to Brownian equilibrium, but the relaxation time for that is substantially shorter than the cage residence time, as will be shown in Chapter 10. In essence, the potential-energy cage and the molecule therefore travel as a unit: The cage is worn by the molecule much as a suit of clothes is worn by a person. And because the potential-energy cage is inseparable from the molecule, the interaction energy of the cage with its molecule becomes part of the energy of the molecular species—just as in methyl chloride, the interaction energy of the CH_3 group with the Cl atom (the C–Cl bond energy) is part of the energy of the methyl chloride species.

The analogy between a cage and a suit of clothes is apt also for another reason. A cage, like a suit of clothes, is a flexible covering, since the molecules in the cage walls are continually in Brownian motion. As a result, the cage potential is noisy and can be analyzed into two parts: (1) a static part, which expresses the long-time average of the potential-energy function, as well as the time-independent standard deviation of the potential-energy noise, and (2) a time-dependent part, which details the Brownian noise. We will find that on the time scale of Brownian fluctuations, cage lifetimes in liquids are long, so that it makes sense to divide the discussion of the cage effects into two parts. In this chapter we will concentrate on the static part and ignore the noise in first approximation. In Chapter 10 we will concentrate on time-dependent phenomena: cage exchange, loss of cage-distinguishability by exchange-averaging, and the influence of Brownian motions.

HARD-SPHERE SIMULATION

In this section we will adopt the chemist's concept of a cage as a material enclosure made of molecules. In the classic work of Rabinowitch and Wood [1936], the cage concept emerged from the computer simulation of a hard-sphere liquid—a liquid in which the molecules resemble incompressible billiard balls in random motion. Attractive forces among the spheres are neglected, and repulsive forces come into play only when the spherical surfaces touch.

A diagram of the analog computer is reproduced in Fig. 9.1. The brass tray (1) is loaded with brass balls of uniform size, and it is shaken so that the balls are set in random motion within the space defined by the fluted border (2). The dark circles in the figure represent balls that are painted with electrical insulating paint. The two light circles (3) represent an unpainted ball in electrical contact with the tray, and a central knob insulated from the tray. The tray and knob are connected to the terminals of the battery. When the unpainted ball hits the central knob, the circuit closes briefly and a voltage pulse is recorded. Inasmuch as the knob and the balls simulate molecules, the voltage pulses mimic the collisions between a specific pair of molecules. The number of voltage pulses divided by the recording time thus represents the collision rate (z) *between two specific molecules*, and the temporal distribution of the voltage pulses expresses the temporal distribution of such collisions. The specific-pair collision rate (z) is distinct from the gas-kinetic *total* collision rate per molecule (Z). In a gas, the total rate Z is proportional to the pressure, while the specific-pair rate z is independent of pressure. In a liquid it is convenient to substitute molecular density for pressure. The molecular density of the simulation is expressed by the packing fraction f of the balls on the tray, where f is the fraction of the fenced area in Fig. 9.1 that is covered by the normal projections of the balls.

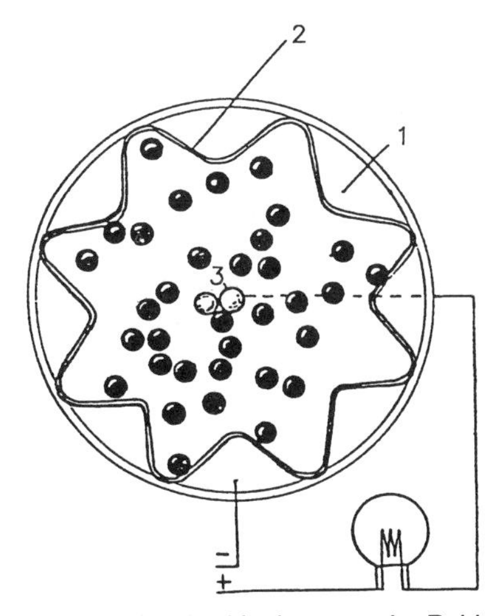

Figure 9.1. The analog computer sketched in the paper by Rabinowitch and Wood [1936].

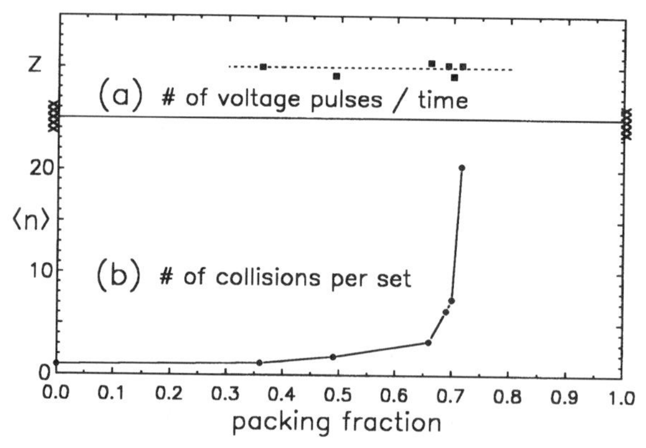

Figure 9.2. Plot of z versus f and of $\langle n \rangle$ versus f, based on computer simulations by Rabinowitch and Wood [1936]. The cage effect becomes marked when $f > 0.68$.

Results for z versus f are given in Fig. 9.2a. As expected from gas-kinetic theory, z is essentially independent of f. The temporal distribution of the collisions is depicted in Fig. 9.3. At low packing fractions (top) the collisions are random, as in a gas, while at high packing fractions (bottom) the collisions occur in long sets. One might say that the unpainted ball is hitting the central knob the way a boxer hits a punching bag: The two remain in close proximity for a finite encounter time rather than escape from each other after each collision. *Ergo* the cage effect!

Let $\langle n \rangle$ be the mean number of collisions per set during an encounter. Results for $\langle n \rangle$ versus f follow a peculiar shape, shown in Fig. 9.2b. At low packing fractions the collisions are gas-like, with $\langle n \rangle \approx 1$. As f increases, $\langle n \rangle$ increases slowly at first. But after f reaches about 0.68, $\langle n \rangle$ increases rapidly—so suddenly in fact that the ensuing cage effect may be regarded as cooperative. Visual observation shows that at high packing fractions, the unpainted ball is penned in by the high surrounding molecular density, until random displacements create a short-lived break in the cage wall to permit escape.

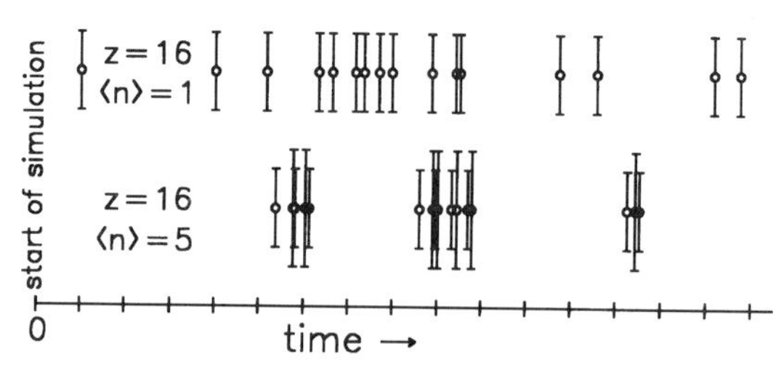

Figure 9.3. (*Top*) Random gas-kinetic collisions between a specific pair of hard spheres at low packing fraction, and (*bottom*) setwise collisions between the same specific pair at high packing fraction. (Based on computer simulations by Rabinowitch and Wood [1936].)

Hard-Sphere Encounters

When $\langle n \rangle$ is large, the two hard spheres hover near each other long enough to be said to exist in a common cage and to form a caged encounter complex. The set length $\langle n \rangle$ is called the *mean number of collisions per encounter*. The collisions of a caged molecule are necessarily limited to the relatively fixed set of its immediate neighbors—"fixed" because the neighbors are also caged.

Let Z be the total hard-sphere collision rate of a caged molecule, in collisions per second; and let s be the mean number of nearest neighbors to the caged molecule. Then Z/s equals z, the pairwise collision rate with a specific neighbor. The mean lifetime τ_{enc} of an encounter complex is equivalent to the time for $\langle n \rangle$ collisions. Hence $\tau_{\text{enc}} = \langle n \rangle/(Z/s)$, which may be rearranged to yield the hard-sphere encounter equation (9.1):

$$\text{Hard-sphere encounter equation:} \quad \langle n \rangle = Z\tau_{\text{enc}}/s = z\tau_{\text{enc}} \qquad (9.1)$$

A MEDLEY OF MODEL-ENCOUNTER RESULTS

In this section we will obtain numerical values for the parameters that describe bimolecular encounters and cage properties in liquids. Some of the parameters will be based on the hard-sphere model, with concepts borrowed from the kinetic theory of gases. Others will be based on the Einstein–Smoluchowsky–Noyes theory of Brownian motion and diffusion in liquids [Einstein, 1956; Smoluchowski, 1916; Noyes, 1961]. Happily, the results obtained by the two approaches mix rather well.

The chief difference between the approaches is in the treatment of collisions. Physically, a collision is any interaction between or among particles that provides an opportunity for the exchange of momentum and energy [Schmidt reference]. In the hard-sphere model of liquids, interactions are significant only at the moment of impact, and gas-kinetic treatments of collisions apply. In real liquids, the molecular packing is dense enough so that significant interactions occur at all times, and the molecules are in a permanent state of collision. Owing to thermal motions, the forces derived from the interactions fluctuate in time and, at random intervals, reach momentary peaks that are strong enough to cause sharp changes in the direction of the molecular Brownian walk.

Unfortunately, there is no unique criterion for the strength that the momentary force peaks must reach. Let l denote the mean length of a step in the Brownian walk, and let r be the mean number of significant changes in direction per unit time. Moderate force peaks are relatively frequent, while strong force peaks are rare. If moderate peaks are sufficient, r will be large and l will be small: The diffusional motion takes place in many short steps. If strong peaks are required, r will be small and l will be relatively long: The diffusional motion takes place in fewer but longer steps. In real life, the speed of diffusion is measured by the macroscopic diffusion coefficient D, which is independent of step size: As l decreases, the number of steps per unit time r increases in precise compensation to keep D constant. If possible, results for Brownian motion are therefore cast in terms of the diffusion coefficient D, so as to make them independent of uncertainties surrounding step size.

Hard-Sphere Collision Rate *Z*

The total number of collisions per second upon a molecule in a liquid will be estimated by extrapolation from the gas phase. At normal T and P, with gas concentration of 0.04 mol liter^{-1}, Z for a typical gas is ~ 1×10^{10} s^{-1}. Thus at a molar concentration of 10 mol liter^{-1}, which is typical of a liquid such as benzene at 298 K, the proportional Z is 2.5×10^{12}. This must be divided by 0.3–0.5, the free-volume fraction in a typical liquid. (Some data are given in Table 9.1.) The hard-sphere collision rate Z in a typical liquid is therefore of order 5×10^{12}–10^{13} s^{-1}.

Estimates for *s* and *z*

The mean number of nearest neighbors, s, to a caged molecule should be somewhat smaller in a liquid than in a crystalline solid. For crystals with cubic or

Table 9.1. Data relevant to the cage effect, for common liquids

Substance	$V(l)$, 25°C (ml/mol)[a]	σ_{hs} (Å)[b]	V_{net}[c] (ml/mol)	Packing fraction f	Viscosity η (cP, 25°C)[a]
n-Hexane	131.6	5.96	66.8	0.507	0.298
Cyclohexane	108.8	5.64	56.6	0.520	0.898
Benzene	89.4	5.29	46.7	0.522	0.603
CCl_4	97.1	5.39	49.4	0.508	0.905
CS_2	60.7	4.52	29.1	0.480	0.347
Pyridine	80.9	5.17	43.6	0.539	0.884
Chlorobenzene	102.2	5.63	56.3	0.551	0.756
Bromoethane	75.1	4.85	36.0	0.48	0.379
Acetone	74.0	4.85	36.0	0.486	0.304
Acetonitrile	52.9	4.24	24.0	0.454	0.706
Chloroform	80.7	5.04	40.0	0.50	0.540
Water	18.07	2.9	7.7	0.426	0.890
Mercury	14.82	3.06	9.03	0.609	1.526
Tetramethylsilane	137[d]	5.91	65.1	0.475	0.227[d]

[a]Riddick and Bunger [1970].

[b]Hard-sphere diameter [Ben-Amotz and Herschbach, 1990].

[c]$V_{net} = L\pi\sigma_{hs}^3/6$.

[d]Parkhurst and Jonas [1975b].

hexagonal closest packing, $s = 12$. For normal liquids we shall adopt the value $s \approx 10$, which will be rationalized later. Thus z, the hard-sphere collision rate between a caged molecule and a specific neighbor, is estimated (through Z/s) as about $10^{12}\ s^{-1}$.

Hard-Sphere Packing Fraction *f*

The liquid state covers a considerable range of densities. For instance, the density of benzene is 0.894 just above the normal melting point of 5.53°C, but decreases to 0.302 at the critical point of 288.94°C and 48.34 atm. To calculate the corresponding packing fractions, we need the macroscopic molar volume of the liquid $V(l)$ and the net molar volume (V_{net}) occupied by the molecules. Kinetic theory lets us deduce an equivalent-sphere diameter σ from any property whose physical mechanism requires that the molecules be of finite size, such as gas viscosity, thermal conductivity, or nonideal compressibility factors [Ben-Amotz and Herschbach, 1990]. For most substances, the σ values inferred from such properties are sufficiently consistent and independent of T and P so that their average may be said to define a hard-sphere diameter, σ_{hs}. Accordingly, $V_{net} = \frac{4}{3}\pi L(\sigma_{hs}/2)^3 = L\pi\sigma_{hs}^3/6$. For benzene, for example, $\sigma_{hs} = 5.29$ Å, so that $V_{net} = 46.7$ ml/mol. Using the densities stated above to calculate $V(l)$, the packing fraction, $f = V_{net}/V(l)$, then is 0.534 at 5.53°C, and 0.181 at the critical point. Similar data for other liquids at 25°C are given in Table 9.1.

On the basis of chemical evidence [such as that inferred from Eq. (9.5) below], the cage effect in benzene is significant up to at least 50°C, where $f = 0.513$. However, as the liquid approaches the critical point and its properties approach those of a dense gas, the cage effect is expected to disappear, probably in a cooperative manner.

Equilibrium for Forming an Encounter Pair

Consider a dilute solution of two solutes, A and B, in a solvent. Let n_a denote the molecules of A per liter, which in turn equals the number of A cages per liter; and let n_b denote the number of B molecules and B cages per liter. Let the "slash" symbol (\) denote "cage" or "environment"; for example, $A\backslash x$ represents an A molecule in an x cage.

In the following, both A and B are dilute solutes. Let $p_{A\backslash b}$ denote the probability that B is a nearest neighbor to A and thus in the cage wall around A, and let $p_{B\backslash a}$ denote the related probability that A is a nearest neighbor and thus in the cage-wall around B. The number of A molecules with a B nearest neighbor then is $n_a \cdot p_{A\backslash b}$, and that of B molecules with an A nearest neighbor is $n_b \cdot p_{B\backslash a}$. Discarding the distinction between the molecule in the cage and the neighbor, the number of A, B encounter pairs, $n_{a,b}$, is given by Eq. (9.2a), in which the prefactor $\frac{1}{2}$ prevents each A, B pair being counted twice:

$$n_{a,b} = \tfrac{1}{2}(n_a \cdot p_{A\backslash b} + n_b \cdot p_{B\backslash a}) \tag{9.2a}$$

In the following, the probabilities p will be given by volume-fraction statistics. Let $\sigma_{a,b}$ denote the mean A, B encounter distance, in centimeters. Then $p_{A\backslash b} = n_b \cdot \frac{4}{3}\pi\sigma^3_{a,b}/1000$, where $\frac{4}{3}\pi\sigma^3_{a,b}\text{cm}^3$ is the volume around an A molecule in which the center of a B neighbor must be located, and 1000 cm^3 is the volume in which there are n_b molecules of B. Likewise, $p_{B\backslash a} = n_a \cdot \frac{4}{3}\pi\sigma^3_{a,b}/1000$. Substitution in Eq. (9.2a) then yields Eq. (9.2b), in which the factor $\frac{1}{2}$ has disappeared.

$$n_{a,b} = n_a n_b \cdot \tfrac{4}{3}\pi\sigma^3_{a,b}/1000 \tag{9.2b}$$

Since A and B are dilute, p is small; hence $n_{a,b} \ll n_a, n_b$, and $c_{a,b} \ll c_a, c_b$. The equilibrium constant $K_{a,b}$ (in liter/mol) therefore takes the form (9.3a).

$$K_{a,b} = c_{a,b}/(c_a c_b) = \tfrac{4}{3}\pi L\sigma^3_{a,b}/1000 \text{ liter mol}^{-1} \tag{9.3a}$$

Later in this chapter we will express the encounter distance $\sigma_{a,b}$ by the equation $\sigma_{a,b} = \frac{1}{2}(\sigma_{hs,a} + \sigma_{hs,b})(1 + \beta)$, where the "cage expansion" β is of order 0.1. If we therefore adopt a value of 6 Å as typical for $\sigma_{a,b}$, the equilibrium constant $K_{a,b}$ is of order 0.5.

Equation (9.3a) represents a rudimentary approximation because it considers only the change in entropy when two originally independent particles are confined in a restricted volume; that is, in Eq. (9.3a), $R \ln K_{a,b} = \Delta S^0_{a,b}$. To obtain a more complete approximation, one allows for a nonzero $\Delta H^0_{a,b}$ and a nonunity statistical factor (statfac). The result is Eq. (9.3b):

$$K_{a,b} = (\text{statfac}) \cdot (4/3000)\pi L\sigma^3_{a,b} \cdot \exp(-\Delta H^0_{a,b}/RT) \text{ liter mol}^{-1} \tag{9.3b}$$

Magnitudes for $\Delta H^0_{a,b}$ for nonelectrolytes will be given in Chapter 10 and are often less than 1 kJ/mol. Electrolytes will be treated explicitly in Chapter 12. Examples of (statfac) are the symmetry factor of $\frac{1}{2}$ when the encounter pair consists of like molecules (A, A), or the spin-statistical factor of $\frac{1}{4}$ when two free radicals $(A\cdot + B\cdot)$ form an encounter pair with singlet (rather than triplet) electron spin.

Dynamics of Forming an Encounter Pair

Rate constants for the formation $[k_f(A, B)]$ and dissociation $[k_d(A, B)]$ of encounter pairs for dilute solutes are predicted most directly by models of Brownian motion and involve the diffusion coefficients D_a, D_b for linear diffusion. The results of different approaches are in essential agreement. Perhaps the simplest derivation is due to Debye [1942]. By making the same approximations as in Eq. (9.3a) and assigning to k_f the same statistical factor (statfac) used for $K_{\mathrm{a,b}}$ in Eq. (9.3b), one obtains Eqs. (9.4). In this way, the statistical factor assigned to k_d is unity, which is usually correct.

$$k_f(\mathrm{A,B}) = (\mathrm{statfac}) \cdot 4\pi L\sigma_{\mathrm{a,b}}(D_{\mathrm{a}} + D_{\mathrm{b}})/1000 \qquad (\text{liter mol}^{-1}\ \mathrm{s}^{-1}) \tag{9.4a}$$

$$k_d(\mathrm{A,B}) = k_f(\mathrm{A,B})/K_{\mathrm{a,b}} = 3(D_{\mathrm{a}} + D_{\mathrm{b}})/\sigma^2_{\mathrm{a,b}} \qquad (\mathrm{s}^{-1}) \tag{9.4b}$$

Geometrical factors in Eqs. (9.4) are unity, and energy gradients or barriers other than those implied by D_{a} and D_{b} are zero. In practice, Eqs. (9.4) work reasonably well.

Coefficients of linear diffusion (D_{a} and D_{b}) in solvents of ordinary viscosity near room temperature are of order 10^{-5} cm^2 s^{-1}. In the absence of experimental data, diffusion coefficients can be estimated by the Stokes–Einstein equation; for example, $D_{\mathrm{a}} \approx kT/(3\pi\eta[\sigma_{\mathrm{hs}}]_{\mathrm{a}})$, where η is the coefficient of viscosity of the liquid in which the diffusion takes place. On letting (statfac) = $1, \sigma_{\mathrm{a,b}} =$ 6 Å, and $D_{\mathrm{a}} + D_{\mathrm{b}} = 2 \times 10^{-5}$ cm^2 s^{-1}, Eqs. (9.4) predict that $k_f(\mathrm{A, B}) = 1 \times 10^{10}$ liter mol^{-1} s^{-1}, and $k_d(\mathrm{A, B}) = 2 \times 10^{10}$ s^{-1}.

Collisions per Encounter. Diffusion-Controlled Reactions

A necessary condition for a cage effect is that $\langle n \rangle$, the mean number of hard-sphere collisions during the lifetime of the encounter pair, be much greater than 1. According to Eq. (9.1), $\langle n \rangle = Z\tau_{\mathrm{enc}}/s$. If we let τ_{enc} equal its conceptual analog $1/k_d$, about 5×10^{-11} s, and use typical values of $s = 10$ and $Z = 1 \times 10^{13}$ s^{-1}, $\langle n \rangle = 50$, which is large enough to represent a substantial cage effect.

A *diffusion-controlled* reaction takes place nearly every time the reactants diffuse together to form an encounter pair. The $\langle n \rangle$ hard-sphere collisions are sufficient to allow the active groups to form a reaction zone and, with high probability, to go on to products. In dilute solutions, the second-order rate constant for a diffusion-controlled reaction is therefore close to the rate constant $k_f(\mathrm{A,B})$ given by Eq. (9.4a)—of order 10^{10} liter mol^{-1} s^{-1} in liquids of ordinary viscosity.

SOME CHEMICAL EVIDENCE

The cage effect was originally proposed to explain the reduced quantum yield when photochemical dissociation occurs in liquid solution rather than in the gas phase [Rabinowitch and Wood, 1937; Lampe and Noyes, 1954]. This is still a major use. But wishing to bypass the inherent intricacies of photochemistry, we will focus on experimental cage effects in thermal reactions.

One consequence of the cage effect is that the molecular fragments produced in a dissociation step remain paired for many collisions and therefore have a chance to recombine before diffusing apart. This kind of chemistry is shown symbolically in Eq. (9.5):

$$\mathrm{A{-}B} \xrightarrow[\text{to caged pair}]{\text{dissociation}} (\mathrm{A,B}) \begin{cases} \xrightarrow{\text{In-cage recombination}} \mathrm{A{-}B} \\ \xrightarrow[\text{diffussion apart}]{} \mathrm{A + B} \end{cases} \tag{9.5}$$

The upper branch symbolizes the cycle A—B → (A, B) → A—B, stoichiometrically a nonreaction. The lower branch yields the dissociated product. The net rate of product formation is therefore smaller than the rate of dissociation; it equals the rate at which the (A, B) pairs separate to form "free" A + B molecules.

To demonstrate a mechanism of this sort, one must choose a reaction in which both the rate of the initial dissociation step [A—B → (A, B)] and the rate of the diffusion apart of the (A,B) caged pair can be measured directly. If the former is greater than the latter and if in-cage recombination can be inferred from complementary evidence, necessary conditions for the cage mechanism are satisfied. An early example that meets these requirements is the thermal dissociation of azo-bis-isobutyronitrile [Eq. (9.6)],

$$(\mathrm{H_3C})(\mathrm{CH_3})(\mathrm{NC})\mathrm{C{-}N{=}N{-}C}(\mathrm{CN})(\mathrm{H_3C})(\mathrm{CH_3}) \longrightarrow \mathrm{N_2} + 2\ (\mathrm{H_3C})(\mathrm{CH_3})\mathrm{C}^{\bullet}{-}\mathrm{CN} \tag{9.6}$$

Azobisisobutyronitrile, R—N=N—R   2-Cyano-2-propyl radical, R•

whose probable mechanism is shown in Eq. (9.7) [Lewis and Matheson, 1949; Overberger et al., 1949; Arnett, 1952; Breitenbach and Schindler, 1952; Walling, 1954; Weiner and Hammond, 1968]. Here as before, molecules residing in a common cage are shown in parentheses.

$$\mathrm{R{-}N{=}N{-}R} \rightarrow (\mathrm{R\cdot, N_2, R\cdot}) \tag{9.7a}$$
$$\mathrm{R{-}N{=}N{-}R} \rightarrow (\mathrm{R{-}N_2\cdot, R\cdot}) \rightarrow (\mathrm{R\cdot, N_2, R\cdot}) \tag{9.7b}$$
$$(\mathrm{R\cdot, N_2, R\cdot}) \rightarrow (\mathrm{R{-}R}) + \mathrm{N_2} \tag{9.7c}$$
$$(\mathrm{R\cdot, N_2, R\cdot}) \rightarrow \mathrm{R\cdot} + \mathrm{R\cdot} + \mathrm{N_2} \tag{9.7d}$$
$$\mathrm{R\cdot} + \mathrm{Scav} \rightarrow \text{distinctive, detectable R—Scav products} \tag{9.7e}$$
$$(\mathrm{R\cdot, N_2, R\cdot}) \not\rightarrow \mathrm{R{-}N{=}N{-}R}$$

The initial dissociation step conceptually has two branches: (1) a concerted process (9.7a) in which the two R—N bonds break simultaneously and (2) a stepwise process (9.7b). In either event, two R· radicals arise in a common cage with a molecule of N_2 and then either combine to form R—R (9.7c) or diffuse apart to become separated radicals (9.7d). The latter are "trapped" and converted to stable detectable products by adding an excess of a free-radical scavenger such as iodine, mercaptan, or diphenylpicrylhydrazyl, $[(C_6H_5)_2N—N\cdot —C_6H_2\text{-}2,4,6(NO_2)_3]$, Eq. (9.7e) [Hammond et al., 1955; Walling, 1957]. Because in the limit of high scavenger concentration the "trapping" of the separated radicals becomes quantitative, any R–R product that manages to be formed under such conditions can only arise from reaction (9.7c) and therefore indicates the intervention of a cage. The reformation of R—N═N—R by reversal of (9.7a,b) can be neglected because of the chemical inertness of N_2, so that the rate of N_2 formation equals the total rate of dissociation.

In fact, the rate of N_2 formation in CCl_4, toluene, and xylene under conditions of quantitative trapping exceeds half the rate of R—Scav formation by 30% to 60%, while the formation of R—R coupling product under such conditions remains significant: No matter how much scavenger is added, one still isolates 10% to 30% of $(CH_3)_2(CN)C—C(CN)(CH_3)_2$ [Walling, 1954, 1957; Hammond et al., 1955; Bevington, 1955]. More recent kinetic evidence shows that the activation barrier for the formation of the R—R coupling product from the singlet-spin radical pair (R·, R·) is small enough to sanction this mechanism [Korth et al., 1983].

THE POTENTIAL-ENERGY CAGE

When seeking "the Truth Even Unto Its Innermost Parts," one risks being confused by the multitude of parts. This risk is certainly present in Fig. 9.4a, which shows the computer simulation of an instant in the life of a two-dimensional "liquid" [Emeis and Fehder, 1970]. The simulation includes attractive as well as repulsive interactions on a realistic scale, and the packing fraction is in the range where the cage effect is significant. There is an interweaving of empty and densely packed regions. Some molecules exist in clusters, others face on channels, and still others are by themselves. And yet, the figure simulates a simple liquid whose particles would all be represented by the same molecular symbol in a chemical equation.

The risk of confusion is reduced by averaging, either over time for a single ensemble or over a set of ensembles. Figure 9.4b shows basically the latter; it is a Fourier transformation of Fig. 9.4a [Warren, 1969]. The original length coordinates are here transformed into reciprocal lengths, in such a way that the stippling represents the probability density of disk matter at the given displacement from the center, averaged over the sample as the center migrates from disk to disk. The good circular symmetry near the center in Fig. 9.4b shows that the disorder seen in Fig. 9.4a averages out to long-range order. To a considerable degree this is also true at shorter ranges, because a vestige of circular symmetry remains even further from the center.

(a)

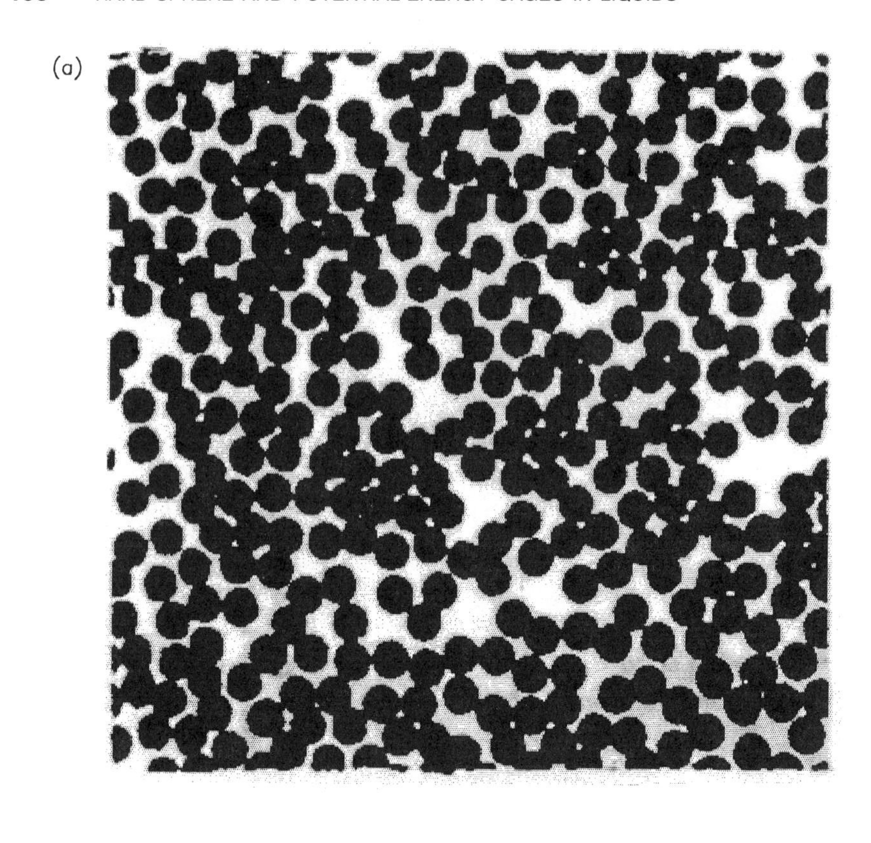

(b)

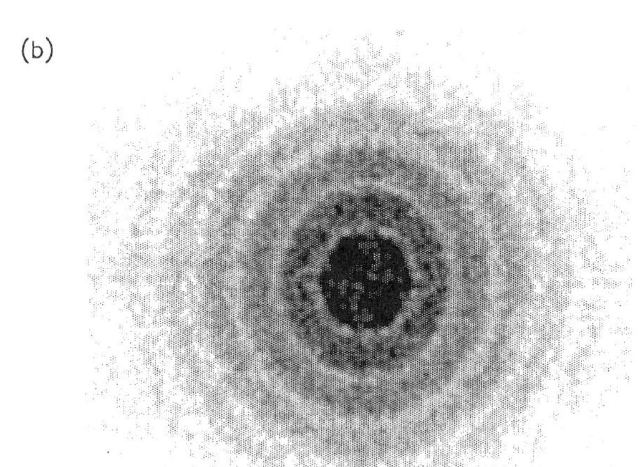

Figure 9.4. (a) Computer simulation of an instant in the life of a two-dimensional Lennard-Jones liquid. (Reproduced with solid circles from Emeis and Fehder [1970], with permission of the American Chemical Society.) (b) Fourier transform of (a), prepared by Dr. Istvan Lengyel using the Optimas program of Bioscan Company, Edmonds, WA. The close-to-circular symmetry shows that the average molecular density in (a) is independent of direction.

The *potential-energy cage* is an abstraction designed to focus on the essence of the interaction between a molecule and its surroundings.* Instead of treating the interaction of the caged molecule with the molecules around it as a mechanical many-body problem, complicated by a multitude of Brownian motions, it simplifies it to a two-body problem: The confined molecule interacts with a single potential-energy cage that represents its specific time-average environment. The molecule exists in a cage because the potential-energy function generates restoring forces all around.

Besides simplifying the molecular mechanics, the potential-energy cage offers another essential advantage: The molecules in the liquid may be regarded as statistically independent, because each molecule interacts only with its own private cage. As a molecule diffuses through the liquid, its potential-energy cage is *not* like a material enclosure whose makeup changes every time the molecule makes a diffusive jump. Rather, it is a companion that accompanies the molecule, because the interaction between the molecule and its environment quickly returns to statistical equilibrium after a diffusive jump—on a time scale set by the Brownian relaxation time that, in liquids, is more than one order shorter than the mean time between diffusive jumps. And because the diffusive motion of the potential-energy cage is coupled to that of the molecule, the interaction energy of the molecule with its cage may be treated as part of the internal energy of the molecule and, thus, of the molecular species.

The space inside a three-dimensional cage is often called the "cavity." In isotropic liquids, cages are pictured as spheres. Spherical cages and "cavities" are part and parcel of familiar models of the liquid state, including Debye's and Onsager's theories of the dielectric constant and the Lorentz–Lorenz theory of molar refraction [Onsager, 1936]. Cages are spherical even when the molecules are not, and even though the electric fields produced by dipoles are quite nonspherical. Models in which the cage containing the dipole is an ellipsoid (rather than a sphere) have fallen into disuse, because the ellipsoidal shape offers no decisive advantage yet requires extra adjustable parameters.

The Cage-Packing Theorem

To characterize a spherical cage in a liquid one needs to specify the long-time average of the cage potential, the cage radius, and the nearest-neighbor number. The cage potential can be constructed from pairwise molecular interaction-energies, and the cage radius accordingly will be defined by molecular center-to-center distances. The cage-packing theorem relates the cage radius (b) and nearest-neighbor number (s) to the packing fraction (f) in the long-time average. These variables are interdependent because at a fixed packing fraction, a larger cage permits more neighbors.

While there are clear differences between the time-average cage structure of

*Early statistical thermodynamic cage models have been described by Fowler and Guggenheim [1956].

a pure liquid and the cell structure of a crystalline solid, there are also some useful analogies. First and foremost, the time-average cages and their "cavities," like unit cells, are all identical. Second, and as a corollary, the packing fraction in a cage, as in a unit cell, equals the packing fraction for the overall pure phase.

The calculation of the packing fraction for a time-average spherical cage uses the cross-sectional geometry shown in Fig. 9.5. The caged molecule (treated as a sphere) is sited at the cage center, and the nearest-neighbors are sited at a distance of one cage radius. These sites do *not* correspond to closest approach. The molecular radius is $a/2$; closest approach is a; but the cage radius $b = a(1 + \beta)$, where $\beta > 0$ is the *fractional cage expansion*. The cage volume is $\frac{4}{3}\pi a^3(1+\beta)^3$. The packed volume inside the cage consists of the full volume of the central molecule, plus that part of the volume of nearest-neighbor molecules that lies inside the cage.

Let γ_{in} denote the fraction of a nearest-neighbor molecule inside the cage, and let s (not necessarily an integer) denote the average number of nearest neighbors. The packing fraction f then is given by

$$f = \tfrac{4}{3}\pi(a/2)^3 \cdot [1 + s\gamma_{in}]/\tfrac{4}{3}\pi a^3 \cdot [1 + \beta]^3$$

Hence

$$s = [8f(1 + \beta)^3 - 1]/\gamma_{in}$$

The calculation of the inside fraction γ_{in} is a straightforward problem in the geometry of figures of revolution. Substitution of the result yields the *Cage-*

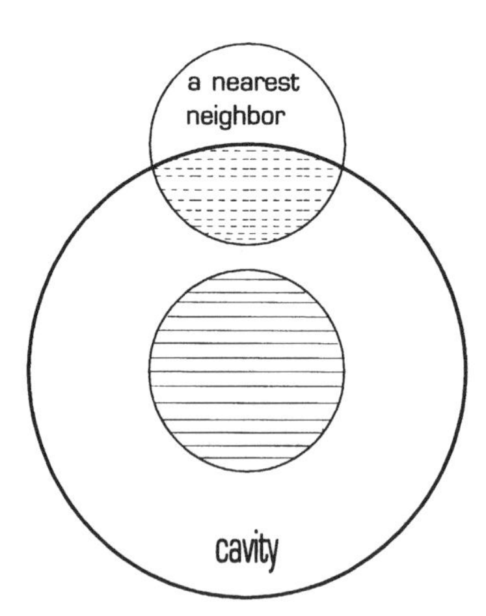

Figure 9.5. Cross section of the time-average spherical cage around a molecule in a liquid, showing one of the s nearest neighbors on the circumference. The cage radius b equals $a(1 + \beta)$.

Packing Equation (9.8):

$$s = \frac{16f(1+\beta)^3 - 2}{1 - 0.1875/(1+\beta)} \tag{9.8}$$

Qualitatively, Eq. (9.8) states that in liquids the neighbor number s is a variable, even at constant packing fraction, because the cage size is variable. But more important, it presents this relationship in the form of a precise equation. In crystalline solids, by contrast, the neighbor number is fixed by the nature of the lattice, and unit-cell size depends solely on the packing fraction.

In the following applications the packing fraction f will be calculated as before, from $f = L\pi\sigma_{hs}^3/6V(l)$, where $V(l)$ is the molar volume and σ_{hs} is the hard-sphere diameter for the given species of molecules. The packing fraction is not a thermodynamic property because σ_{hs} is a gas-kinetic parameter. However, values derived from different properties normally agree within a few percent.

Figure 9.6 shows plots of s versus β for various packing fractions. Each is a plot of Eq. (9.8) for a given packing fraction, but some areas in the s, β plane seem physically implausible. These include the region in which s is greater than 12, because it is doubtful that the number of nearest neighbors in a liquid would be greater than that in a close-packed crystal. They also include the region in which β exceeds about 0.13, because we will find that in this region the cage center is not a potential minimum and concepts of the cage effect begin to blur.

In spite of the difference between unit cells in crystalline solids and average cages in isotropic liquids, the packing fractions predicted for closest packing of spheres agree within 1%. In close-packed cubic or hexagonal crystals, each sphere has 12 nearest neighbors, and the packing fraction is $\sqrt{2}\cdot\pi/6$, or 0.740. When Eq. (9.8) is solved for $s = 12$ and $\beta = 0$, the result for f is 47/64, or 0.734.

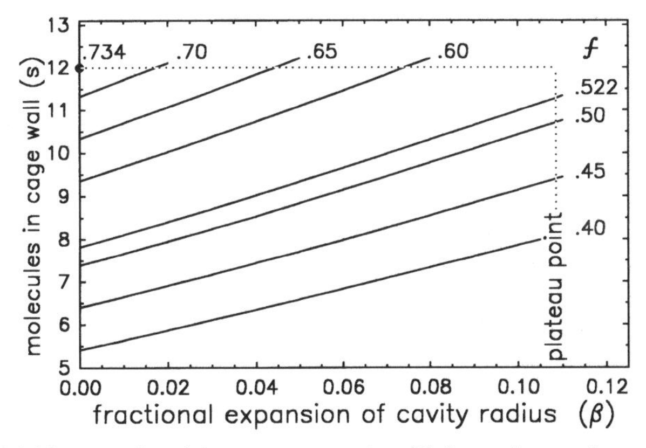

Figure 9.6. Neighbor number (s) versus cage size (β) for various values of the packing fraction f. According to Eq. (9.8).

Containment Controversy and Radial Distribution Function

Equation (9.8) follows directly from basic definitions and involves no assumptions beyond those of Euclidean geometry. Unfortunately it does not give a unique solution: To obtain explicit solutions for s and β, one must introduce additional theory, and that has been a source of controversy.

At issue is the statistical mechanism of the volume expansion at the melting point. Do the effective radii of the spaces between the molecules follow a single distribution function, or is the distribution bimodal, with a substantial fraction of molecule-size holes? The question is not trivial because the answer defines our view of the liquid state.

For instance, Fig. 9.7 shows the molar volume of benzene versus temperature at 1 atm in the range 0–335 K. Between 0 K and the melting point, the number of neighbors is fixed and the 10.5% thermal expansion of the solid takes place because the molecules move apart. The 13.5% expansion on melting, on the other hand, can take place with changes in both β and s. Indeed, because T is constant, one may argue that β should be nearly constant and that s therefore decreases. This would mean that the volume increase on melting introduces a substantial number of molecule-size "holes"—a basic assumption in Henry Eyring's theory of viscosity and diffusion in liquids [Eyring and Jhon, 1969; Glasstone et al., 1941]. A variant of this theory is the *cluster model*, in which melting is viewed as an event in which the solid crystal shatters into small, dynamically interlocking, molecular clusters that are separated by molecule-wide channels.* While it is risky to read too much into Fig. 9.4a because, as in a Rorschach test, one tends to see what one fancies, the figure does suggest such a pattern. The cluster model also gives a natural explanation of the fluidity of liquids: When a shearing stress is applied, the liquid changes its shape because the clusters can slide past one another.

A simple test of the mechanism of melting, based on the increase in fluidity, was suggested by J. H. Hildebrand [Hildebrand, 1971; Hildebrand and Lamoreaux, 1972]. Diffusion in liquids is very much faster than that in solids, with the diffusion rate being nearly proportional to T/η, where η is the coefficient of viscosity, a function of T and P. Hildebrand plotted $V(l)$, the molar volume of the liquid, versus the fluidity $1/\eta$ over the widest possible range of T and P. For benzene, such a plot is shown in Fig. 9.8 [Parkhurst and Jonas, 1975b; McCool et al., 1972; Landolt–Börnstein, NS IV/4]. The data lie close to a smooth line whose intercept at zero fluidity approximates the molar volume of the solid *at the melting point* rather than at 0 K. One may conclude, therefore, that the volume created on melting, and by further expansion of the liquid above the melting point, is specifically richer in the molecule-size cavities that facilitate diffusion. Since the original expansion volume (of the solid between 0 K and the melting point) is still present, one may argue that the total expan-

*For a clear review of molecular clusters, especially of isolated clusters formed in supersonic expansions of gas jets through nozzles, see Jortner [1992].

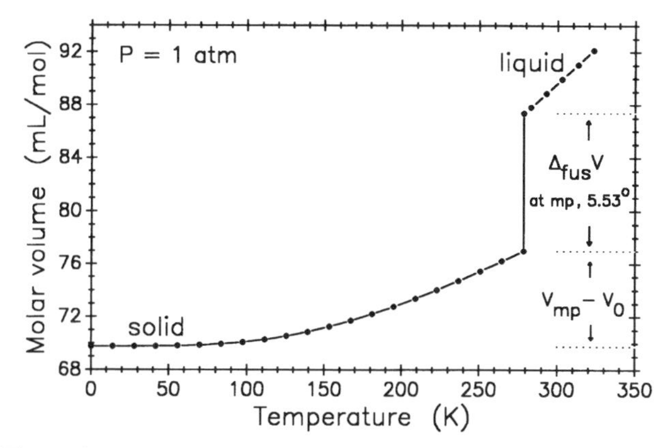

Figure 9.7. Thermal expansion of benzene between 0 K and 350 K. For solid benzene, $(\partial \ln V/\partial T)_P$ is from Eucken and Lindenberg [1942]; the crystallographic cell volume between 80 K and 273 K is from Cox [1958] and Lonsdale [1959]. The solid curve was obtained by integrating the equation $(\partial \ln V/\partial T)_P = 22.56 \times 10^{-11}\ T^3/(1 + 2.555 \times 10^{-7}\ T^3)$, which was obtained by fitting these data. The T^3 dependence in the numerator is based on the Debye T^3 law and the equality $(\partial^2 V/\partial T^2)_P = -(\partial^2 S/\partial P\, \partial T)$. The T^3 dependence in the denominator is based on the fact that near the melting point, $(\partial \ln V/\partial T)_P$ is nearly independent of T. (Data for liquid benzene from Riddick and Bunger [1970].)

sion volume in the liquid consists of two functionally distinct parts. That is, the size distribution of the effective radii of the spaces between the molecules is bimodal.

In spite of this evidence, the majority of scholars are content with a single, rather than a bimodal, distribution function. The reason for this, in the author's view, is that the common approach to the computer simulation of liquids begins,

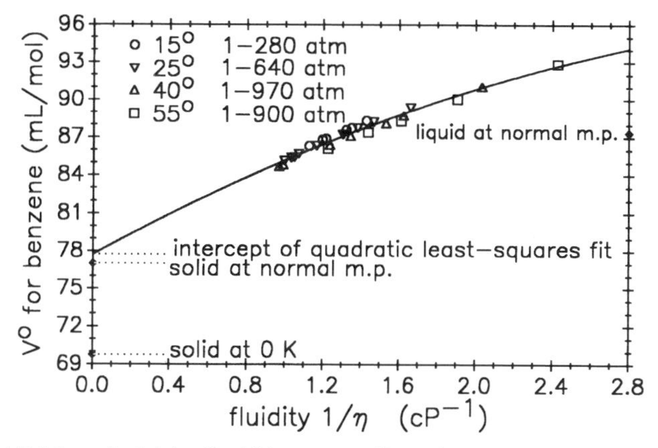

Figure 9.8. Hildebrand plot for liquid benzene. (Data from Parkhurst and Jonas [1975]; McCool et al. [1972]: Landolt–Börnstein [NS IV/4].)

quite reasonably, with pair potentials (i.e., pairwise molecular interactions), and this strategy inherently excludes bimodal distributions. Moreover, the success of the approach has so far discouraged the testing of theoretical alternatives that might produce bimodal distributions. In the following we shall adopt the same strategy and generate cage potentials from pair potentials.

Our pair potentials will be simple, but before going on let us describe a distribution function based on a more refined pair potential—again for benzene. In this simulation, the benzene–benzene interaction contains both London dispersion and electrostatic interactions, the latter from fractional charges on the C and H atoms [Jorgensen and Severance, 1990; Shi and Bartell, 1988]. The resulting distribution function is for the ring center-to-center distances in the liquid, whose definition is more straightforward than that of effective radii for empty spaces.

Results predicted for the radial distribution function $g(r)$ in liquid benzene at 25°C are shown in Fig. 9.9, where r is the benzene ring center-to-center distance [Jorgensen and Severance, 1990]. The ordinate $g(r)$ represents the average distribution function for all relative orientations of the benzene rings. When Fig. 9.9 is interpreted in terms of a cage model, one ring defines the cage center and r is the distance from that cage center.

The radial distribution function can be interpreted in terms of local concentrations. Let c denote the macroscopic molar concentration. Then $g(r)$ is normalized so that $c \cdot g(r) = c(r)$, the average concentration (at statistical equilibrium) in an infinitesimally thin spherical shell at a distance r from the cage center. For two spherical shells with respective radii r_1 and r_2, $g(r_1)/g(r_2)$ therefore equals $c(r_1)/c(r_2)$, the ratio of the equilibrium concentrations, which shows that $g(r)$ is a microscopic kind of free-energy variable. It also follows that when r is

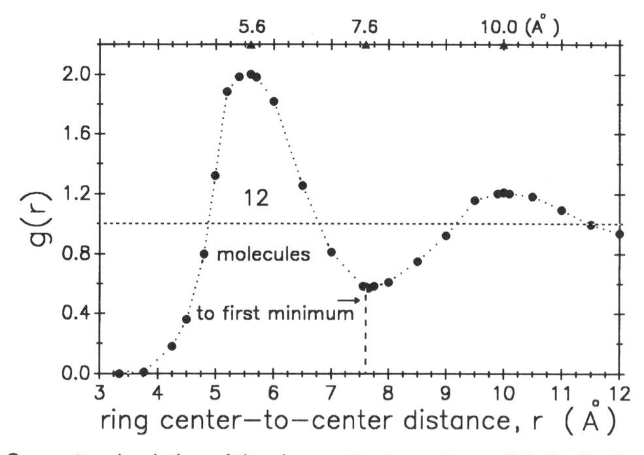

Figure 9.9. Computer simulation of the ring center-to-center radial distribution function $g(r)$ for liquid benzene at 25°C. Integration to the minimum at 7.6 Å gives $s \approx 12$. (Based on Fig. 5 of Jorgensen and Severance [1990].)

expressed in centimeters and c in molecules per cubic centimeter, the integral (9.9) evaluates the number of molecules in the shell (r_1, r_2):

$$\int_{r_1}^{r_2} c \cdot g(r) \cdot 4\pi r^2 \, dr = \text{mean number of molecules in } (r_1, r_2) \tag{9.9}$$

Returning to the cage model: The distance to the first peak in $g(r)$ gives an estimate of the cage radius, and the distance to the first minimum approximates the outer radius of the nearest-neighbor shell. Integration to the first minimum according to Eq. (9.9) thus gives an estimate of s. For benzene according to Fig. 9.9 the resulting estimate is $s \approx 12$, coincidentally the same as the value for close-packed spheres in a crystal.

Finally, let us compare this estimate with the plausible range of s for a liquid with packing fraction $f = 0.522$ (i.e., liquid benzene at 25°C). According to Fig. 9.6, when $\beta = 0$ at the close-packed cage radius, $s = 7.8$; when β grows to 0.13 and the cage model begins to fail, $s = 12.1$. Thus Fig. 9.9 indicates a value for s near the maximum of the plausible range for a cage model.

THE CAGE POTENTIAL FROM PAIR POTENTIALS

The potential energy from the interaction of two molecules in a medium—the *pair potential*—is a function of the center-to-center distance (r_{12}) and the relative orientation. When the medium is a dilute gas, given the pair potential, statistical thermodynamics can predict a host of macroscopic gas properties whose agreement with observation authenticates the pair potential. [Hirschfelder et al., 1954]. When the cage potential in a liquid is constructed from pair potentials, enough relative orientations are involved so that the average, in the absence of linkage bonds, has close to spherical symmetry. The effective pair potential $E^{(2)}$ then may be treated as a function of r_{12} only. (When there are localized linkage bonds, the effective symmetry is lower than spherical.)

In this section we will consider cage potentials that are constructed from spherically symmetric pair potentials. A mathematically simple example is the two-parameter Lennard–Jones potential $E^{(2)}_{\mathrm{LJ}}$, which is often tried as a first approximation:

$$E^{(2)}_{\mathrm{LJ}} = 4\varepsilon_{\mathrm{LJ}}[(\sigma_{\mathrm{LJ}}/r_{12})^{12} - (\sigma_{\mathrm{LJ}}/r_{12})^{6}] \tag{9.10}$$

The two parameters are: $\varepsilon_{\mathrm{LJ}}$, a characteristic energy; and σ_{LJ}, a characteristic distance. A plot of the potential, with parameters that are suitable for benzene over a wide range of gas densities, is shown in Fig. 9.10. $E^{(2)}$ goes to zero as r_{12}, the distance between the molecular centers, becomes large; it reaches a minimum at point a; and it increases sharply when $r_{12} < a$. The minimum is denoted by a because this value of r_{12} can be identified with the closest-approach distance a in the liquid-cage model of Fig. 9.5. By the same token, the distance b in Figure 9.10 might match the cage radius $b = a(1 + \beta)$ in Fig. 9.5, where β is the cage-expansion parameter. Although we shall apply the LJ potential to illustrate qualitative points for benzene cages, the detailed, orientation-dependent potential used in Fig. 9.9 is more realistic.

The Additivity Theorem

In the rest of this chapter, cage potentials ($E^{(C)}$) will be calculated by simply adding pair potentials ($E^{(2)}$). This procedure is approximate, and it is worthwhile to "derive" the underlying Additivity Theorem so that the approximations may be known.

Let E denote the total interaction energy (pairwise and otherwise) in an ensemble of n molecules in a volume V at temperature T. Then $E^{(C)}$ is the contribution to E per molecule. If, hypothetically, E were proportional to n, $E^{(C)}$ would be simply E/n. But proportionality may not be assumed; and, in general, $E^{(C)} = (\partial E/\partial n)_{T,V}$. The partial differentiation is at constant V because E is the *internal* potential energy, so we avoid work in or by the surroundings.

To relate E and $E^{(C)}$ to the pair potential $E^{(2)}$, we let the pairwise interaction have spherical symmetry and define a distance l so that interactions at $r_{12} > l$ may be neglected. Then l expresses the effective range of the pairwise interaction. The molecule at the origin will be called the *caged molecule*, and molecules centered at $a \leq r_{12} \leq l$ will be called *neighbors*. Let s denote the average number of neighbors, and let $E^{(2)}_{\text{av}}$ be the average pairwise interaction energy with a neighbor. Then

$$(\partial E/\partial n)_{T,V} = E^{(C)} = sE^{(2)}_{\text{av}} \tag{9.11a}$$

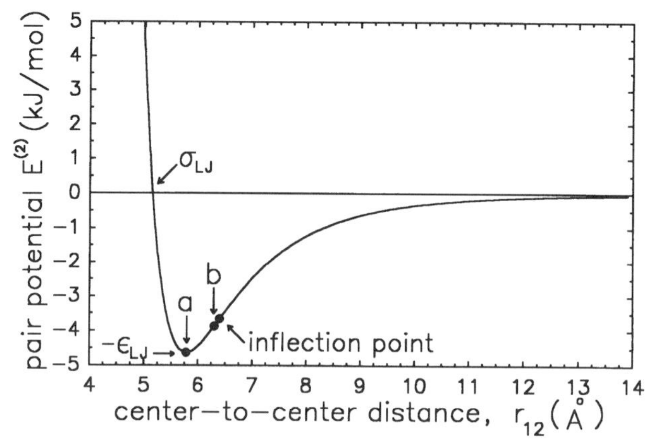

Figure 9.10. Plot of the Lennard-Jones pair potential $E^{(2)}$ with $\sigma_{LJ} = 5.15$ Å and $\varepsilon_{LJ} = 4.63$ kJ/mol, which fits many properties of fluid benzene [Ben-Amotz and Herschbach, 1990].

In order to show that $E^{(2)}_{av}$ is indeed a pairwise interaction, we assume that s is proportional to the molecular density: $s = kn/V$, where k is a specific constant. Then $(\partial E/\partial n)_{T,V} = knE^{(2)}_{av}/V$. We further assume that l, and hence $E^{(2)}_{av}$, is independent of n/V. Integration at constant T and V then yields $E = kn^2E^{(2)}_{av}/2V$. Since E is proportional to n^2, the interaction is pairwise.

In order for the assumptions to succeed, there must be enough empty space between the molecules, in the volume V, so that incremental increases $(\partial n/V)$ in molecular density can be accommodated, and the pairwise interactions must be weak enough so that the van der Waals centers in the interacting molecules are not distorted by changes in molecular density.

It is useful to introduce the definition $E^{(2)}_{av} = (1/s)\ \sum_1^s\ E^{(2)}$. Equation (9.11a) then becomes the *Additivity Theorem*:

$$E^{(C)} = \sum_1^s E^{(2)} \tag{9.11b}$$

Accordingly, the cage potential of a molecule in a liquid is the sum of the pairwise interaction energies of the molecule with its s liquid neighbors.

Potential at the Cage Center

The neighbor number s in Eq. (9.11) is calculated from the cage radius b and the packing fraction f, using Eq. (9.8). In this operation the cage is a sphere, and the molecular centers of the s neighbors all sit on the cage radius: That is, $r_{12} = b = a(1 + \beta)$; and $E^{(2)}_{av} = E^{(2)}(b)$. The time-average distribution of the s neighbors over the cage surface is uniform at $s/4\pi$ neighbors per steradian. On letting $(1 + \beta) = b/a$, Eq. (9.8) takes the form

$$s = [16f(b/a)^3 - 2]/[1 - (0.1875a/b)]$$

In principle, the interaction energy of a pair of molecules depends on the medium in which the molecules reside. However, when the interaction is of short range, the s molecules in Eq. (9.11) are nearest neighbors to the caged molecule, and the intervening space, which transmits the interaction, is empty space. By this reasoning, pair potentials for short-range interactions in liquids are nearly those for the same pairwise geometries in the gas phase. The pair potentials $E^{(2)}$ in the Additivity Theorem for liquid cages then nearly correspond to those in the gas phase. In the following examples we will use this approximation and illustrate the properties of potential-energy cages for a benzene-like liquid, so that we may employ the Lennard-Jones pair potential of Fig. 9.10.

Results for $E^{(C)}$ at the cage center for the benzene-like liquid are shown as a function of β in Fig. 9.11. In these calculations, $\sigma_{LJ} = 5.15$ Å, $\varepsilon_{LJ} = 4.63$

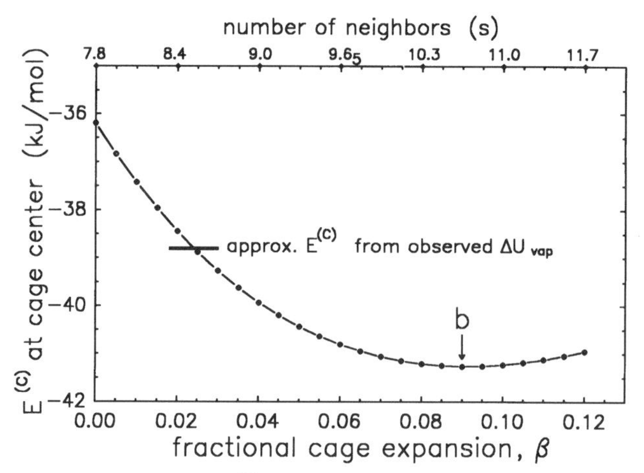

Figure 9.11. The potential energy $E^{(C)}$ at the cage center for a benzene-like liquid. Based on Eqs. (9.8) and (9.11) with $f = 0.522$, as well as on the Lennard-Jones pair potential of Fig. 9.10. The abscissa β equals $(b - a)/a$.

kJ/mol, a (at the pairwise minimum) $= 1.122 \cdot \sigma_{LJ} = 5.78$ Å, and $f = 0.522$. Figure 9.11 of course does not represent real benzene: It represents a virtual fluid with spherical molecules whose pairwise interactions simulate those of gaseous benzene.

The most notable feature of Fig. 9.11 is that the cage potential $E^{(C)}$ at the cage center is rather insensitive to changes in cage size. The reason for this is that the principal variables—the neighbor number s and the pair potential $E^{(2)}(b)$—are interdependent and vary in opposite directions. As the cage radius increases, the *magnitude* of $E^{(2)}$ *decreases*, while s *increases* and the product $s \cdot E^{(2)}$ changes relatively little. A shallow minimum for $E^{(C)}$ occurs at $\beta = 0.09$.

Figure 9.11 also attempts a comparison with real benzene, by including the cage potential predicted from the energy of vaporization at 25°C. In predicting $E^{(C)}$ from $\Delta U_{vap} = 31.4$ kJ/mol, we assume that vibrational energies are equal in the vapor and liquid and apply a correction of $3RT$ for the difference in translational and rotational energies—$\frac{1}{2}RT$ for each mode—since these motions are free in the gas phase and take place in potential-energy cages in the liquid. The result for $E^{(C)}$ intersects the curve predicted from Eqs. (9.8) and (9.11) at $s \approx 8.6$—a remarkable result, which suggests that for pairwise interactions of short range, effective gas-phase pair potentials indeed generate useful cage potentials in liquids.

Displacements from the Cage Center

Up to now we have considered the effect of cage radius and neighbor number, while the caged molecule stays in a fixed position at the cage center. To show that the interaction defines a potential-energy cage, we must allow the caged

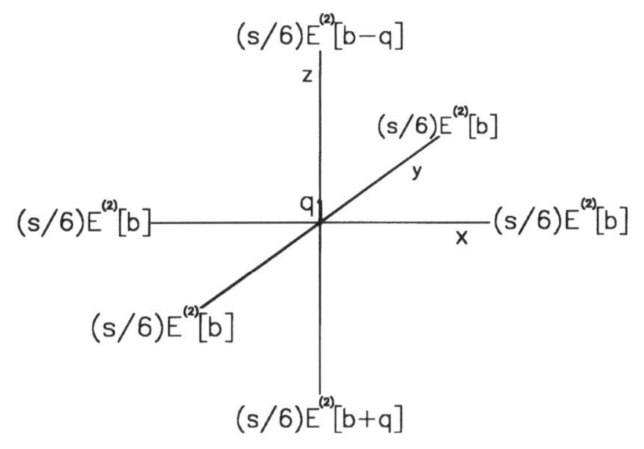

Figure 9.12. Octahedral scheme to calculate the off-center cage potential. $E^{(2)}[b]$ denotes the pair potential when $r_{12} = b$; see Fig. 9.10.

molecule to move off-center and face restoring forces all around. To do so, we shall continue with the example of a benzene-like liquid; the cage is a sphere of fixed radius b with a uniform surface density of identical neighbor centers. Equation (9.11b) and the pair potential of Fig. 9.10 apply.

The individual values of $E^{(2)}$ now are not all the same because the caged molecule has moved off-center. To calculate $E^{(2)}_{\text{av}}$ we shall use the scheme shown in Fig. 9.12: One-sixth of the neighbors each are placed on the cage radius at $\pm x$, $\pm y$, and $\pm z$, and the z-axis is defined so that the off-center displacement q is along $+z$. The pair potentials then are $E^{(2)}(b-q)$ at $+z$, $E^{(2)}(b+q)$ at $-z$, and $E^{(2)}(b)$ at $\pm x$ and $\pm y$. Substitution in Eq. (9.11b) then yields the cage potential given in Eq. (9.12):

$$E^{(C)}(b,q) = (2/3)sE^{(2)}(b) + (s/6)E^{(2)}(b-q) + (s/6)E^{(2)}(b+q) \qquad (9.12)$$

The scheme of Fig. 9.12 makes a common statistical approximation: It substitutes octahedral symmetry for the spherical symmetry of the neighbor density on the cage surface. Analysis shows that this is a good approximation. The factors $(s/6)$ in Eq. (9.12) have fractional errors of order $(q/b)^3$, which will be negligible in the following.

PROPERTIES OF CAGE POTENTIALS

Sample plots of $E^{(C)}(b,q)$ based on Eq. (9.12) and the benzene-like pair potential of Fig. 9.10 are given in Fig. 9.13, for realistic liquid-phase values of $\beta(b)$. We note that $E^{(C)}(b,q)$ is indeed a cage potential, with steeply ascending potential walls as q increases. It is symmetric about the cage center, because it follows from Eq. (9.12) that substitution of $-q$ for q leaves $E^{(C)}$ unchanged. At the cage center, the potential is a minimum for small values of $\beta(b)$. However,

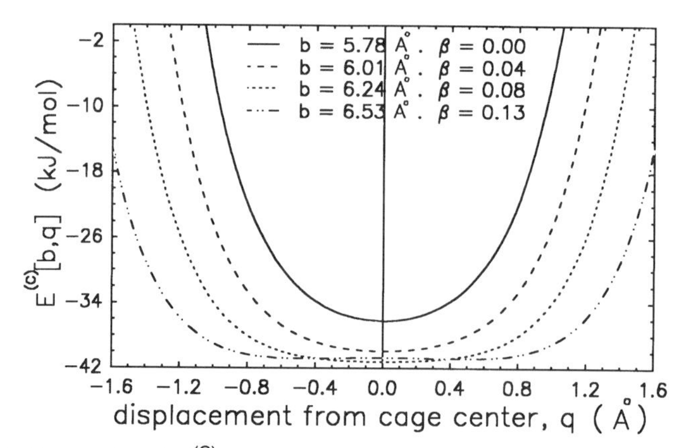

Figure 9.13. Cage potential $E^{(C)}$ versus off-center distance q, for various values of the cage radius b and relative cage expansion $\beta = (b - a)/a$. The pair potential is the Lennard-Jones potential of Fig. 9.10, with parameters that fit the gas-phase properties of benzene.

as β increases, the minimum flattens out and then becomes a maximum. This behavior is shown on an expanded scale in Fig. 9.14. The plateau point is the value of β at which the central curvature is zero.

In the following, we will examine various properties of the cage potential: the half-width, the shape of $E^{(C)}$ at the cage center in relation to the concavity of the pair potential $E^{(2)}(r_{12})$ when $r_{12} = b$, the quasi-harmonic treatment of the libration of the caged molecule, and the thermal noise in $E^{(C)}$ owing to the Brownian motions of the s neighbor molecules.

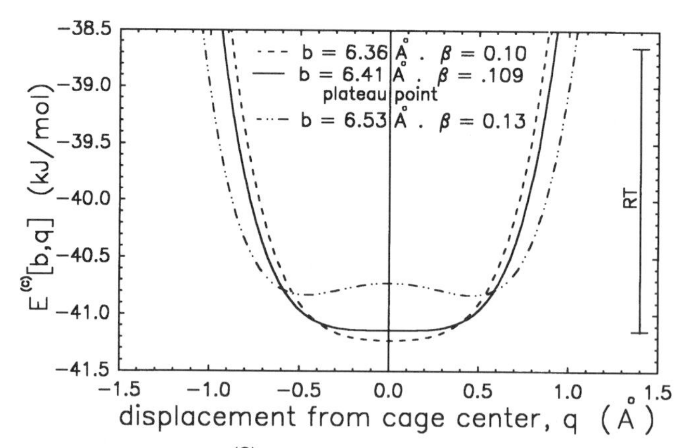

Figure 9.14. The cage potential $E^{(C)}$ of Fig. 9.13 on an expanded scale, to show the change from minimum to maximum at the cage center. For a Lennard-Jones pair potential, the plateau point occurs at $\beta = 0.109$.

Half-Width

As shown in Fig. 9.13, the cage width increases with the radius b and cage expansion β. The half-width is usefully expressed by q_{RT}, the value of q at which $E^{(\mathrm{C})}(b,q) - E^{(\mathrm{C})}(b,q=0)$ equals RT. Since librational frequencies are low, RT is the mean thermal energy per mode of libration. Values of q_{RT} for the benzene-like liquid are listed in Table 9.2. As β increases, the increase in q_{RT} is similar to that in the cage radius b.

Potential Shape at Cage Center

As shown in Fig. 9.14, as the cage expansion $\beta(b)$ increases, the potential energy at the cage center changes from a minimum to a maximum. Since $E^{(\mathrm{C})}$ is constructed from $E^{(2)}$, the change of shape follows from properties of the pair potential. We will show that it follows from the curvature of $E^{(2)}$ at $r_{12} = b$.

The cage center at b of course is not a minimum for $E^{(2)}$; that minimum lies at $r_{12} = a$. We must therefore distinguish between two displacements: (1) the displacement $x = (b - a)$ of the cage center from the pairwise minimum at a and (2) the displacement q of the caged molecule from the cage center at $r_{12} = b$. Let $E^{(2)}(x)$ denote the value of $E^{(2)}$ at the cage center $(a+x)$, and let $E^{(2)}(x+q)$ denote the value at a displacement q from that center. Let $e' = (\partial E^{(2)}/\partial q)_0$ at the cage center, $e'' = (\partial^2 E^{(2)}/\partial q^2)_0$, and so on, for higher derivatives. Expansion in Taylor's series about $q = 0$ then takes the form (9.13), which can be dissected into an odd series (9.14a) and an even series (9.14b). The latter is relevant for the cage potential.

$$E^{(2)}(x+q) = E^{(2)}(x) + e'q + e''q^2/2! + e'''q^3/3! + e''''q^4/4! + \cdots \tag{9.13}$$

Table 9.2. Some cage properties for a benzene-like liquid whose pair potential is that of Fig. 9.10

				$\nu_{q\backslash h}$ [a]		
β	b (Å)	q_{RT} (Å)	Shape of $E^{(\mathrm{C})}(b,q=0)$	$\mathrm{s}^{-1}\cdot 10^{-11}$	cm^{-1}	Thermal Noise[b] $\sigma_{(\mathrm{C})}$(kJ/mol)
0.00	5.78	0.405	Minimum	10	33	0.767
0.04	6.01	0.549	Minimum	7.3	24	1.076
0.08	6.24	0.746	Minimum	5.4	18	2.042
0.10	6.36	0.866	Minimum	4.6	16	2.556
0.109	6.41	0.923	Plateau	4.3	15	2.788
0.13	6.53	1.06(?)	Maximum	[c]	[c]	3.3(?)

[a] Eq. (9.16); $\mathcal{L} = 78$; q_{RT} from third column.
[b] At cage center.
[c] The quasi-harmonic treatment is questionable when the cage potential has a central maximum.

$$E^{(2)}(x+q) - E^{(2)}(x-q) = 2e'q + 2e'''q^3/3! + \cdots \tag{9.14a}$$
$$E^{(2)}(x+q) + E^{(2)}(x-q) = 2e''q^2/2! + 2e''''q^4/4! + \cdots \tag{9.14b}$$

Substitution in Eq. (9.12) then yields the even series (9.15). Note that e'', the *curvature* of the *pair* potential, defines the sign of the coefficient of the q^2 term.

$$E^{(C)}(q) = sE^{(2)}(x) + (s/3)[e''q^2/2! + e''''q^4/4! + \cdots] \tag{9.15}$$

Thus, if $E^{(2)}(r_{12})$ is concave up at $r_{12} = a + x$, e'' is >0 and $E^{(C)}$ is a minimum at the cage center. If $E^{(2)}$ is concave down, $E^{(C)}$ is a maximum. And if $a + x$ is an inflection point for $E^{(2)}$, then $e'' = 0$ and $E^{(C)}$ is a plateau point. The inflection point for a Lennard-Jones potential lies at $x = 0.109a$, and accordingly Fig. 9.14 shows the plateau point for $E^{(C)}$ at $\beta = 0.109$.

Quasi-Harmonic Treatment of In-Cage Libration

The cage potentials shown in Figs. 9.13 and 9.14 are clearly anharmonic, but a quasi-harmonic treatment should be accurate enough to give qualitatively correct libration frequencies. Energy levels for the family of potential functions, $E = z^4 + Bz^2$, with B ranging from -50 to $+100$, show enough similarity between harmonic and anharmonic levels to permit the construction of simple correlation diagrams [Laane, 1970].

The frequency ν of a harmonic oscillator is given by $\nu = (\gamma/\mu)^{1/2}/(2\pi)$. For libration in a liquid potential-energy cage, the reduced mass μ is $\sim \mathcal{L}/L$, where $\mathcal{L}$ is the gram-molecular weight of the caged molecule, and L is Avogadro's number. A representative quasi-harmonic (q\h) value for the force constant γ is obtained by writing $E_{\text{q\h}} = \frac{1}{2}\gamma_{\text{q\h}}q^2$, where q is the off-center displacement, and letting $E_{\text{q\h}} = RT$ and $q = q_{RT}$. On this basis, $\gamma_{\text{q\h}} = 2RT/Lq^2_{\text{RT}}$, and the quasi-harmonic libration frequency is given by Eq. (9.16):

$$\nu_{\text{q\h}} = \frac{\sqrt{(2RT/\mathcal{L})}}{2\pi q_{RT}} \tag{9.16}$$

Quasi-harmonic libration frequencies based on Eq. (9.16) are included in Table 9.2. Based on these frequencies, the spacing $h\nu_{\text{q\h}}$ of the librational energy levels is of order 100–400 J/mol, less than the mean thermal energy RT, and within the potential-energy noise $\sigma_{(C)}$ (Table 9.2) due to Brownian motions—all of which explains why sharp transitions between librational levels are not observed in benzene-like liquids.

Amplitude of Potential-Energy Noise

Given the cage potential $E^{(C)}(b, q)$ and the mean amplitude q_{RT} of the thermal motions, one can calculate the standard deviation, $\sigma_{(C)}$, of the potential energy at the cage center due to the Brownian motions of the s molecules in the cage wall. In Fig. 9.15 the caged molecule 1 is treated as stationary at the center of the primary cage. Let $\sigma_{(2)}$ denote the standard deviation of the potential energy at the primary cage center due to the Brownian motion of molecule 2 in its own, the secondary, cage. The libration of 2 goes on in all possible directions, but only the motion along the line of centers to 1 is relevant to our calculation. The mean amplitude of that motion is $\pm q_{RT}$. Since $E^{(C)}$ equals $s \cdot E^{(2)}$ and since the s neighbors move independently, the desired variance $\sigma^2_{(C)}$ equals $s \cdot \sigma^2_{(2)}$. And because all neighbors are solvent molecules, the s pairwise variances $\sigma^2_{(2)}$ are all equal.

Accordingly, it is sufficient that we calculate $\sigma^2_{(2)}$, the pairwise variance in the 1,2 interaction, *for just one neighbor*. We will let that neighbor be molecule 2, whose off-center displacement in the secondary cage fluctuates in Brownian motion between $+q_{RT}$ and $-q_{RT}$.

To introduce the mathematics, consider a general independent variable q whose statistical weighting function or *distribution function* $f(q)$ is defined in the interval (q_1, q_2). We will assume that $f(q)$ is normalized; that is,

$$\int_{q_1}^{q_2} f(q)\, dq = 1$$

Let $y(q)$ denote a continuous function of q, and let σ_y denote its standard deviation in (q_1, q_2). The variance σ^2_y then is defined in Eqs. (9.17):

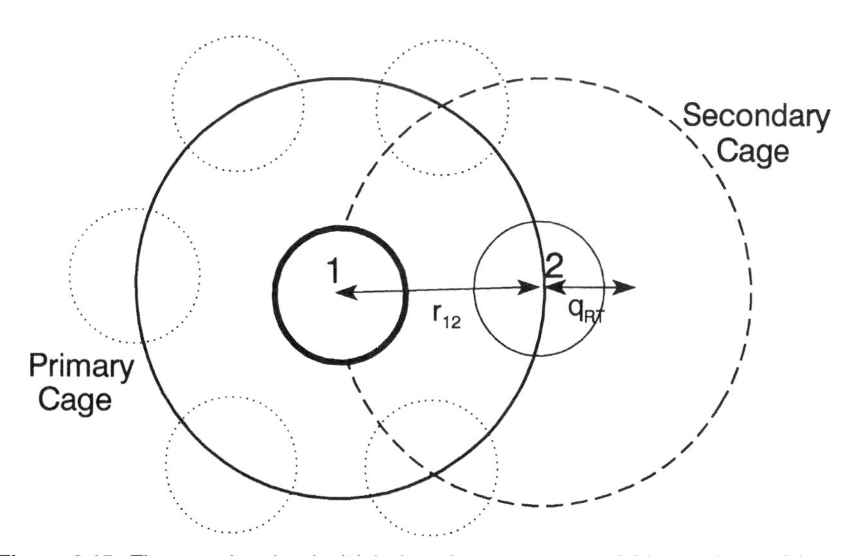

Figure 9.15. The caged molecule (1) in its primary cage; a neighbor molecule (2) in its secondary cage. Brownian motion of 2 affects the potential of 1 through fluctuations in the 1,2 pairwise interaction.

$$\sigma_y^2 = \int_{q_1}^{q_2} f(q) \cdot y(q)^2 \, dq - \left[\int_{q_1}^{q_2} f(q) \cdot y(q) \, dq \right]^2 \qquad (9.17a)$$

In the present case, σ_y^2 corresponds to $\sigma_{(2)}^2$, while $\sigma_{(C)}^2 = s\sigma_{(2)}^2$; $y(q)$ is the pair potential $E^{(2)}(x+q)$; $q_1 = -q_{RT}$, $q_2 = +q_{RT}$; and $f(q)$ is the Boltzmann factor based on the *cage* potential*:

$$f(q) = \exp[-E^{(C)}(b,q)/RT] \Big/ \int_{q_1}^{q_2} \exp[-E^{(C)}(b,q)/RT] \, dq \qquad (9.17b)$$

The calculation of $\sigma_{(C)}$ requires the following input data: any two of the variables (f, β, s) and a suitable pair potential $E^{(2)}$. The integrals in Eqs. (9.17) are readily evaluated by numerical integration.†

Results obtained for the benzene-like liquid by this method are listed in the final column of Table 9.2. The values range from 0.8 to 3.3 kJ/mol. They increase monotonically with β and hence with s, but even at the low end of the range it is clear that $\sigma_{(C)}$ is substantial. Indeed, potential-energy noise is a major factor whenever the concept of a potential-energy cage is employed.

CAGE POTENTIALS AND DISCRETE SOLVATES

It is probable that real potential-energy cages in non-hydrogen-bonding liquids are near the plateau point. As the cage radius increases, the cage potential goes through a minimum somewhat before the plateau point (Fig. 9.11). The quasi-harmonic libration frequency still decreases at the minimum, however, and the librational entropy accordingly still increases (Table 9.2). The Helmholtz free energy $E^{(C)} - TS^{(C)}$ therefore reaches its minimum at a cage radius that is greater than that for the $E^{(C)}$ minimum. But the radius is not *much* greater, because above the plateau point a potential barrier grows in at the cage center. This barrier is slight at first; but when $\beta > 0.13$, its height increases rapidly. The central part of the cage then is effectively walled off, and the librational entropy decreases. It is probable, therefore, that the Helmholtz free energy reaches its shallow minimum right around the plateau point. In this range of cage radius,

*The Boltzmann distribution in terms of potential energy is the equilibrium (i.e., $t = \infty$) solution of the Fokker–Planck equation for Brownian diffusion in a potential well. See, for example, Pathria [1972].

†The results obtained for $\sigma_{(C)}$ are remarkably independent of the form of $f(q)$. If instead of the Boltzmann function one uses a constant $f(q) = 1/(q_2 - q_1)$, the results obtained for $\sigma_{(C)}$ for the benzene-like liquid are within a few percent of those in Table 9.2. When the cage potential is harmonic (i.e., $E^{(C)} = E^{(C)}(0) + \gamma q^2/2$), $\sigma_{(C)} \approx RT(4/5s)^{1/2}$. This harmonic approximation becomes adequate as s becomes small (<4, say) and as the linkage between the cage and its neighbors becomes substantial.

the cage potential is flat at the cage floor and rises steeply at the cage wall, so that the caged molecules exist under conditions that simulate the key assumptions of hard-sphere models of liquids [Hansen and McDonald, 1990]. Given the remarkable success of hard-sphere models, this coincidence is gratifying.

In the formulation of the potential-energy cage, the neighbor number s is the average for a distribution, *not* a stoichiometric integer. Integers are mandatory when solvation leads to molecular-complex formation; and integers are helpful when the interactions produce localized linkage bonds, as in liquid water. For normal van der Waals short-range interactions as in liquid benzene, however, the concept of discrete solvates is an approximation. This is because the outer boundary of the nearest-neighbor shell has no sharp definition. For example, in Fig. 9.9 the outer boundary of the nearest-neighbor shell corresponds formally to the first minimum in $g(r)$, but in fact the nearest-neighbor peak overlaps the next-nearest neighbor peak, so that the neighbor category is ambiguous for a nontrivial fraction of the molecules.

In spite of the want of sharp definition, we will try to analyze the mean neighbor-number s for a benzene-like liquid into discrete solvates and deduce the individual cage potentials at the cage centers as a function of the integral solvation number k. Average properties used in the analysis will be the mean neighbor number s (which may not be an integer), the packing fraction f for the entire liquid, and the cage radius b; a is the distance of closest approach. The corresponding properties for k-solvates will be: k, f_k, and b_k. The cage-packing equation, which is now written in the form (9.18a), is essentially a geometric relationship. Hence it applies analogously to the k-solvates, Eq. (9.19a). The corresponding additivity equations are (9.18b) and (9.19b).

$$s = [16f(b/a)^3 - 2]/[1 - (0.1875a/b)] \tag{9.18a}$$

$$E^{(C)}(b, q = 0) = s \cdot E^{(2)}(r_{12} = b) \tag{9.18b}$$

$$k = [16f_k(b_k/a)^3 - 2]/[1 - (0.1875a/b_k)] \tag{9.19a}$$

$$E_k^{(C)}(b_k, q = 0) = k \cdot E^{(2)}(r_{12} = b_k) \tag{9.19b}$$

In Eq. (9.18a), f and a are fixed: f is the packing fraction for the entire liquid, and $a = r_{12}$ at the *pairwise* potential minimum (Fig. 9.10). Equation (9.18a) therefore expresses s as a function of b, or b as a function of s. We shall choose the second option and rewrite $E^{(C)}(b, q = 0)$ in the form $E^{(C)}(s, f)$. When s is assigned a specific value s_1, we shall write $E^{(C)}(s = s_1, f)$.

In Eq. (9.19a), a is fixed and k is the independent variable (in place of s); f_k and b_k are functions of k. We need two equations to define them, because we may not assume that $f_k = f$ or that $b_k = b$. One of the equations is (9.19a), but there isn't any other. Mathematically, the problem is therefore indeterminate. The cage potential for the k-solvate will be denoted by $E^{(C)}(k, f_k)$.

Although there is no mathematical solution for $E^{(C)}(k, f_k)$, the missing constraint can be supplied by statistics, since the individual solvates are obscured by Brownian noise. We may judge that the difference between f_k and f cannot be great, or else the distinctness of the k-solvate would survive the Brownian noise. This conclusion may be reduced to an inequality in which certain energy differences are less than $\sigma_{(C)}$. There are various ways in which this inequality can be formulated; we will choose (9.20).

$$|E^{(C)}(k, f_k) - E^{(C)}(s = k, f)| \leq \sigma_{(C)} \tag{9.20}$$

Given a value for k, $E^{(C)}(s = k, f)$ can be calculated from Eqs. (9.18)—recalling that f, the average packing fraction for the liquid, is known. The unknown cage potential of the distinct k-solvate, $E^{(C)}(k, f_k)$, then differs from $E^{(C)}(s = k, f)$ by less than the standard deviation $\sigma_{(C)}$ of the Brownian potential-energy noise.

In principle, $\sigma_{(C)}$ is the actual Brownian potential-energy noise, but its value cannot be specified because the mean solvation number s in the actual liquid is not precisely known. We can only say that its value is probably near the plateau point for the benzene-like liquid. We will therefore rewrite Eq. (9.20)—as a matter of choice, not of logical necessity—in the form of Eq. (9.21):

$$E^{(C)}(k, f_k) = E^{(C)}(s = k, f) \pm \sigma_{(C)}(s = k, f) \tag{9.21}$$

Here a known uncertainty, $\sigma_{(C)}(s = k, f)$, takes the place of the prior upper bound. Values of $\sigma_{(C)}(s = k, f)$ for this purpose are available in Table 9.2. Results for distinct solvation numbers k, close to the likely value of s, are listed in Table 9.3. It is clear that when we have k near s, the cage potentials $E^{(C)}(k, f_k)$ are rather insensitive to changes in k.

Table 9.3. Potential energy at the cage center for discrete solvates, for a benzene-like liquid[a]

k (integer)	$E^{(C)}(s = k, f) \pm \sigma_{(C)}(s = k, f)$ (kJ/mol)
8	-37.0 ± 0.8
9	-39.9 ± 1.1
10	-41.1 ± 1.8
11	-41.2 ± 2.6
12	-40.7 ± 3.3

[a]Based on Eq. (9.21). Values from Table 9.2.

10

ENVIRONMENTAL ISOMERS, MOLECULAR COMPLEXES, AND CAGE EXCHANGE IN LIQUIDS

Environmental isomers are subspecies of a molecular species whose molecules exist in distinguishable cages. These cages might differ in molar composition, neighbor number, or packing pattern, and their potential-energy minima are separated by kinetically significant barriers. If the molecules have low symmetry, cage differences can also arise from alternative modes of juxtaposition. In liquids, preferred juxtapositions have important consequences, including molecular recognition, asymmetric synthesis, and the formation of liquid crystals [Evans and Wennerström, 1994; Pirkle and Pochapsky, 1989; Noyori and Kitamura, 1991; Buckingham et al., 1993].

In thermodynamic contexts it is natural to define environmental isomers in terms of potential-energy cages, since the average potential energy at the cage center is part of the energy of the molecular species in the liquid. This follows because a potential-energy cage is intangible and stays with its molecule as the molecule diffuses through the liquid. Moreover, when there are two distinct cages, not only are there two values of the molar energy but there are two complete sets of thermodynamic properties, thus meeting the conditions for defining two isomeric subspecies.

The preceding definitions assume that the environmental subspecies are distinguishable. But this must be proved, because liquid cages are short-lived and alternative cages are in rapid exchange. If the cage-exchange is fast enough, the wave-mechanical energy difference is averaged out. Another subtlety is that the caged molecules will form molecular complexes with their neighbors if the interaction energy is great enough.

This chapter will analyze the exchange averaging of potential-energy cages and arrive at wave-mechanical criteria for distinguishability. Criteria will be obtained for ideal distinguishability, for real distinguishability, and for molecular-complex formation.

TIME SCALES

Four time scales are relevant to cage exchange in liquids: the mean lifetimes of the cages themselves, the correlation times of the Brownian motions, the transit times in cage-switching, and the time scale for relaxation of perturbed electron distributions to stationary states.

Cage Lifetimes

When we regard liquid cages as material walls built of molecules, the kinetic relationships for encounter-pair formation apply [Eqs. (9.2)–(9.4)], and cage lifetimes can be deduced accordingly. In this approach, the caged molecule C together with any molecule E in the cage wall is regarded as an encounter pair. The cage lifetime ends when molecule E leaves the cage wall, or when another molecule leaves or enters or somehow restructures the cage wall. Let $k_d(\mathrm{C}, \mathrm{E})$ denote the rate constant for dissociation, a process in which either C or E departs; then $(1/2)k_d(\mathrm{C}, \mathrm{E})$ is the rate constant for the departure of E only. Let s denote the number of neighbors that constitute the cage wall. The rate constant for the departure of any neighbor then is $(s/2)k_d(\mathrm{C}, \mathrm{E})$. At dynamic steady state, the probability per unit time that a molecule departs from an s-wall is equal to the probability that a molecule enters an $(s-1)$-wall. The combined rate constant k' for leaving or entering the s-wall is therefore $k' \approx 2 \cdot (s/2)k_d(\mathrm{C}, \mathrm{E})$, and the cage lifetime is $\tau' = 1/k' = 1/sk_d(\mathrm{C}, \mathrm{E})$.

The dissociation rate constant k_d for an encounter pair was obtained in Chapter 9 from $k_d = k_f/K_{a,b}$, where k_f is the rate constant for encounter-pair formation, and $K_{a,b}$, the equilibrium constant, is of order 0.5. The theoretical estimate of k_f for encounters among molecules of ordinary mobility in liquids of ordinary viscosity is $1 \times 10^{10}\ \mathrm{s}^{-1}\mathrm{M}^{-1}$. However, this figure is probably too high. As reactivity in bimolecular reactions increases, the second-order rate constant k_2 reaches an upper limit as encounter-pair formation becomes rate-determining. When this limit is reached, $k_2 = k_f \cdot$(statfac), where (statfac) is a predictable statistical factor. A classic example of k_2 approaching encounter-controlled limits is shown in Fig. 10.1. For the forward reaction, (statfac) = 1/2. The k_f values corresponding to the limits are not identical, but both are in the range 10^9–$10^{10}\ \mathrm{s}^{-1}\mathrm{M}^{-1}$, which is consistent with literally hundreds of examples for solutes of ordinary mobility in solvents of ordinary viscosity. If we adopt a median value of $k_f = 3 \times 10^9$ and let $K_{a,b} = 0.5$ and $s = 10$, we obtain $k_d = 6 \times 10^9$ and a median cage lifetime τ' of 2×10^{-11} s.

Correlation/Relaxation Times for Brownian Motions

In the liquid cage model used in this book, each molecule bears an individual potential-energy cage that surrounds it as long as the molecule exists in the liquid medium. In order for this model to be realistic, the time scales, both of the fluctuations owing to Brownian motions and of relaxation to statistical equilib-

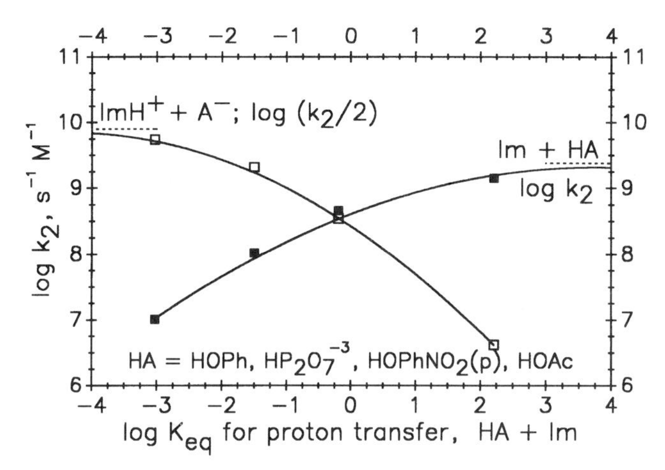

Figure 10.1. Second-order rate constants for reversible proton transfer of imidazole (Im) with a series of acids (HA) of varying strength in water at 25°C. The statistical factor (statfac) is 1 for Im + HA and 1/2 for $ImH^+ + A^-$ owing to two equivalent protons in ImH^+. (Data by G. Maas [Eigen, 1964].)

rium, must be short compared to the cage lifetimes. Lacking experimental data, we shall predict these time scales from statistical theory for Brownian motion in potential-energy cages. The predicted time scales are short compared to the cage lifetimes.

The time scale for noisy fluctuations is given by the correlation time τ_{corr}, which may be thought of as a statistical turnover time or a dynamical memory time. This parameter is so named because it is derived from the noise by a procedure called *autocorrelation*. Brownian motions create a special kind of noise that is compatible with laws governing ensembles of molecules. Thus, when the ensemble exists at statistical equilibrium under given conditions, the distribution functions for its properties (such as the molecular velocities) are fixed and definite. When the ensemble is off equilibrium there are deviations from the equilibrium functions. But if these deviations are small, relaxation to equilibrium occurs by first-order kinetics with a relaxation time denoted by τ_{lax}.

Moreover, it follows from the Principle of Dynamic Balance that rate laws and rate constants at dynamic equilibrium are equal to the corresponding rate laws and rate constants for return to equilibrium. This principle has been validated in chemical kinetics, for example, by comparing rate measurements based on isotopic exchange, spectral line-shape analysis, or fluctuation spectroscopy at equilibrium with independent measurements on the way to equilibrium [Meiboom, 1960; Feher and Weissman, 1973]. It has been validated in physical kinetics, for example, by rate measurements for the return of perturbed rotational distributions to equilibrium in dielectric relaxation [Debye, 1945]. For Brownian motions, therefore, the correlation time τ_{corr} equals the relaxation time τ_{lax}.

ARGUMENT FOR EQUALITY. To argue that $\tau_{\text{lax}} = \tau_{\text{corr}}$, we do a thought experiment analogous to isotopic exchange. We let the system exist at equilibrium and mentally label a set of molecules that all have the same value of a property at $t = 0$. We then let this narrow distribution relax until the labeled molecules reach an effective equilibrium distribution, so that statistical memory of the initial narrow distribution is lost. Since the relaxation takes place with the same rate constant and kinetics as if the system were initially displaced from equilibrium and were returning to equilibrium, the statistical memory time in the thought experiment is equal to the relaxation time. That is, $\tau_{\text{corr}} = \tau_{\text{lax}}$. Note that this thought experiment also measures the statistical turnover time, since the labeled molecules change from their initial position in the distribution to a spectrum of new positions, with probabilities that are governed by the equilibrium distribution function. Q.E.D.

In the present case of molecules in potential-energy cages, the Fokker–Planck equation for the time dependence of average properties in Brownian ensembles has been solved for the kinetics of return to equilibrium in quadratic potential wells [Pathria, 1972]. If we make the quasi-harmonic approximation described for libration [Eq. (9.16)], the available solutions apply directly. The theory yields relaxation times for the mean displacement $\langle q \rangle$ and the mean-square displacement $\langle q^2 \rangle$, which differ by a factor of 2. Of the two, the relaxation for $\langle q^2 \rangle$ is virtually synchronous with that of the cage potential $E^{(\text{C})}$.

PROOF. The key equation is (10.1), which follows from (9.13).

$$E^{(\text{C})} = sE^{(2)}_{q=0} + e' \sum_{s} q_i + \tfrac{1}{2} e'' \sum_{s} q_i^2$$

$$E^{(\text{C})} = s[E^{(2)}_{q=0} + e' \langle q \rangle + \tfrac{1}{2} e'' \langle q^2 \rangle] \tag{10.1}$$

The term involving $\langle q \rangle$ is small since relaxation takes place near equilibrium and $\langle q \rangle_{\text{eq}} = 0$. The fluctuations of $E^{(\text{C})}$ are therefore dominated by those of $\langle q^2 \rangle$. Q.E.D.

The solution of the Fokker–Planck equation for the relaxation time of the ensemble average, $\langle q^2 \rangle$, to equilibrium is $\tau_{\text{lax}} = kT/(2\gamma D)$, where D is the coefficient of linear diffusion, and γ is the force constant of the quadratic potential [Pathria, 1972]. In the quasi-harmonic (q\h) approximation, we let $\gamma = \gamma_{\text{q/h}} = 2kT/q_{RT}^2$, where q_{RT} is the half-width of the potential-energy cage when $E^{(\text{C})} = RT$. [See discussion preceding Eq. (9.16).] The result is Eq. (10.2):

$$\tau_{\text{lax}} = \tau_{\text{corr}} = q_{RT}^2/4D \tag{10.2}$$

Let us calculate some values. For benzene at 298 K, $D = 2.3 \times 10^{-5}$ cm^2/s [Parkhurst and Jonas, 1975a]. If we let $s = 10$, then $q_{RT} = 8.8 \times 10^{-9}$ cm for a benzene-like liquid, according to Table 9.2. Hence $\tau_{\text{corr}} = 8 \times 10^{-13}$ s, about 25 times shorter than the median cage lifetime of 2×10^{-11} s.

For water at 298 K, $D = 2.25 \times 10^{-5}$ cm^2/s [Stejskal and Tanner, 1965]. Because of hydrogen bonding, q_{RT} is substantially smaller than for the benzene-like liquid. According to a cage potential that includes London dispersion interactions, as well as the interaction of the water dipole with the mean dipole-vector sum of its environment, the mean q_{RT} for the two isomers of liquid water is 2.5×10^{-9} cm [Grunwald and Steel, 1994]. Hence $\tau_{\text{corr}} = 0.7 \times 10^{-13}$ s, about 300 times shorter than the median cage lifetime. These results for τ_{corr} are plausible because their magnitude (10^{-12}–10^{-13} s) agrees with that of $1/Z$ (2×10^{-13}–1×10^{-13} s), the mean time between hard-sphere collisions in common liquids.

Transit Times in Cage Switching

A further requirement of the cage model is that the molecules in liquids spend most of their time residing in a cage, and only a small fraction in transit between cages. Assuming that the mean velocity of a molecule in one dimension is of order 0.2 km s^{-1} and that the half-width of the barrier between cages is 0.2 Å, the transit time is of order 10^{-13}s, less than 1/100 of the median cage lifetime.

Relaxation of Electron Distributions

In the wave-mechanical formulation of the exchange of a molecule between liquid cages, one needs to know whether the time scales are such that the Ehrenfest Principle may be applied. This principle states that the energy eigenstate of a system does not change when the system is perturbed at a slow-enough rate that the electronic wave function can follow in a quasi-stationary fashion [Tolman, 1938]. A well-tested corollary is the Born–Oppenheimer approximation, which states that nuclear displacements are slow enough so that electron distributions can follow. In the case of cage-exchange and potential-energy noise due to Brownian motions, the relevant displacements are nuclear displacements. We may therefore assume that in the absence of deliberate electronic excitation, caged molecules, molecules in transit between cages, and molecules exposed to Brownian noise consistently remain in their electronic ground state.

FREQUENCY SWITCHING IN SCHRÖDINGER WAVE TRAINS

We will derive the conditions for distinguishability of cage environments in two parts. In this section we will consider cage exchange. In the next section we will add the effect of Brownian noise. This separation is possible because of the difference in the time scales.

The following derivation is based on Schrödinger wave mechanics. The basic equation for the stationary state of a Schrödinger wave is Eq. (10.3), where $\phi(x)$ is the amplitude as a function of the spatial coordinates x, and t is the time.

$$\psi(x,t) = \phi(x)\exp[-2\pi i(\epsilon/h)t] \tag{10.3}$$

When x is fixed, the equation represents a circularly polarized wave of frequency $\nu = \epsilon/h$, where ϵ is the state energy per molecule. The polarization is circular because $\exp[-2\pi i\nu t] = \cos 2\pi\nu t - i\sin 2\pi\nu t$.

To allow for switching between two cage environments \a and \b, we let ϵ be the energy of the *electronic ground state* and let ϵ_0, ϵ_a, and ϵ_b denote ϵ when the molecule exists in vacuum, at the center of cage \a, or at the center of cage \b, respectively. Since the cage potential ($E^{(C)}$ per mole) is part of the energy of the molecular species, the various energy values are related by Eq. (10.4):

$$\epsilon_a = \epsilon_0 + E_a^{(C)}/L, \qquad \epsilon_b = \epsilon_0 + E_b^{(C)}/L$$
$$\nu_b - \nu_a = (\epsilon_b - \epsilon_a)/h = (E_b^{(C)} - E_a^{(C)})/Lh \tag{10.4}$$

When there is exchange between two cage environments, the frequency of the Schrödinger wave switches between ν_a and ν_b, and the difference between these frequencies is proportional to the difference in the cage potentials, $E_b^{(C)} - E_a^{(C)}$, at the cage centers.

The simple model of switching between two stationary state-frequencies is justified by the time scales: The mean cage residence times, τ'_a and τ'_b, are long compared to the cage-transit times and the Brownian (i.e., cage) relaxation times, while the transit and relaxation times in turn are long enough to allow the electron distribution to keep up in a relaxed, essentially stationary manner.

The exchange of a species of molecules between two liquid cages, with time as variable, is kinetically first-order and statistically stochastic. The term "stochastic" here means that individual behavior is random within the constraints of an exponential lifetime distribution [Grunwald et al., 1995].

Let N_a denote the fraction of molecules in cage \a, and $N_b = 1 - N_a$. Let τ_a and τ_b denote the respective mean residence times, and define a *mean time of exchange* τ according to Eq. (10.5). At dynamic equilibrium these variables follow Eqs. (10.6).

$$\frac{1}{\tau} = \frac{1}{\tau_a} + \frac{1}{\tau_b} \tag{10.5}$$

$$\frac{N_a}{\tau_a} = \frac{N_b}{\tau_b}, \qquad \frac{1}{\tau} = \frac{1}{N_b \tau_a} = \frac{1}{N_a \tau_b} \tag{10.6}$$

In the following we will let $N_a = N_b = \frac{1}{2}$, so that $\tau_a = \tau_b = 2\tau$. When applying Eq. (10.3) we will assume that $\phi_a(x) = \phi_b(x)$; that is, cage effects on the spatial electron distribution are negligible. The illustrations of wave trains versus time will show plane-polarized rather than circularly polarized waves, since the plane-polarized waves exhibit the gist of the action.

It is convenient to define a *distinguishability index* ξ according to Eq. (10.7). The relevance to distinguishability will emerge soon, but for the present the variable $2\pi|\nu_b - \nu_a|$ can serve as a frequency reference for exchange.

$$\xi = 2\pi|\nu_b - \nu_a|\tau \tag{10.7}$$

When $1/\tau$ is smaller than $2\pi|\nu_b - \nu_a|$, exchange is said to be "slow." When $1/\tau$ is greater, exchange is said to be "fast." Figure 10.2 shows an example of each. In slow exchange the frequencies ν_a and ν_b are quite recognizable, while in fast exchange the waveform looks like a jittery sinusoid at the mean frequency $|\nu_b + \nu_a|/2$. In both cases the transit time is short and the phase of the wave motion before and after transit is essentially continuous.

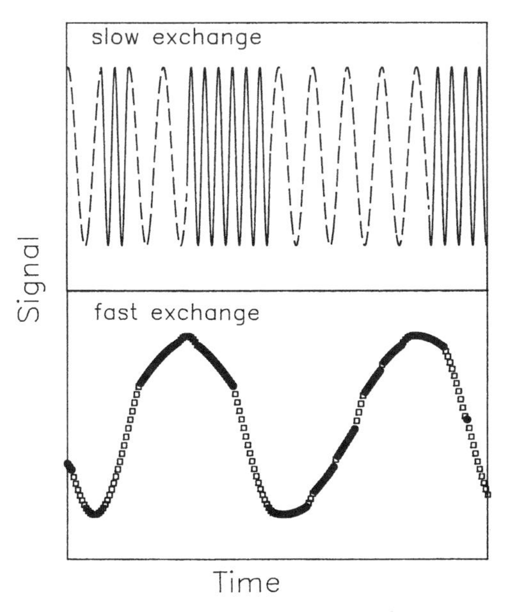

Figure 10.2. Sections of plane-polarized wave trains with frequency switching. *Top*: Slow exchange, $1/\tau = 0.42|\nu_b - \nu_a|$. *Bottom* (expanded time scale): Fast exchange, $1/\tau = 36|\nu_b - \nu_a|$. The full trains yield power spectra consisting of (*Top*) two well-resolved lines and (*Bottom*) a single line.

Mean Time of Exchange and Power Spectrum

Because Schrödinger waves are an abstraction, we will digress briefly to introduce a real example from the field of nuclear magnetic resonance (NMR). Figure 10.3 shows the variable-temperature NMR spectrum at 60 MHz of the methyl protons in 3,3,6,6-tetramethyl-1,2-dithiane (10.8) [Claeson et al., 1961].

(10.8)

The six-membered ring has a near-chair conformation. The protons of two methyl groups have axial chemical shifts, while the other methyl protons have equatorial chemical shifts. At −20°C, exchange is "slow" and the two methyl resonances are distinct. But as the temperature rises and exchange speeds up, distinguishability blurs; above −4°C it vanishes entirely. At 0°C the signal is exchange-averaged and broad. At 25°C it looks just like a single chemical shift.

Although this is magnetic resonance and the McConnell–Bloch equations that describe it [McConnell, 1958] employ NMR variables such as magnetic fields and chemical shifts, Fig. 10.4 demonstrates that exchange-averaging is basically a wave phenomenon. The figure compares steady-state solutions of the

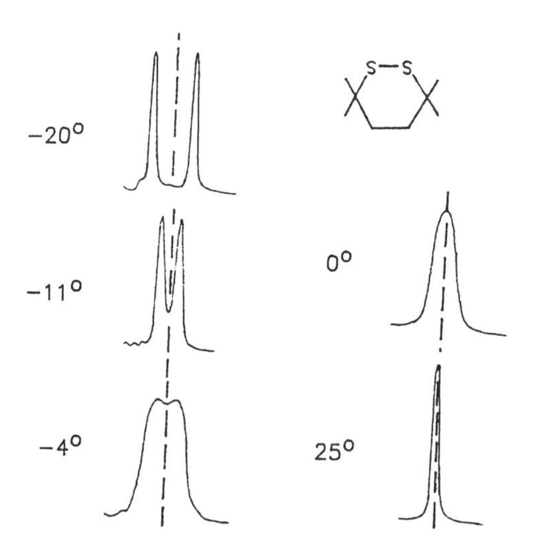

Figure 10.3. Nuclear magnetic resonance of the axial and equatorial methyl protons in 3,3,6,6-tetramethyl-1,2-dithiane (10.8) at 60 MHz. The activation energy derived from $d\tau/dT$ is 67 kJ/mol. (Reproduced from Claeson et al. [1961], with permission of the American Chemical Society.)

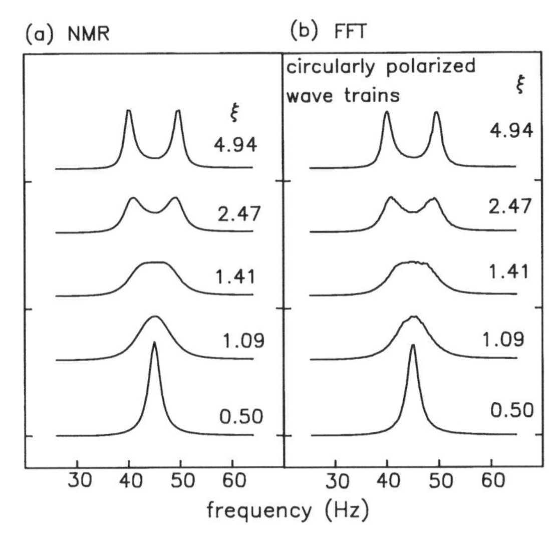

Figure 10.4. (a) NMR absorption mode as a function of ξ, according to steady-state solutions of the McConnell–Bloch equations $\nu_a = 40$Hz, $\nu_b = 50$ Hz, $1/T_2 = 0$, $\tau_a = \tau_b$ [McConnell, 1958]. (b) Fast-Fourier-transform power spectra of the corresponding circularly polarized wave trains, with stochastic frequency switching and exponential lifetime distributions [Grunwald and Steel, 1993a]. The absolute amplitudes in (a) are unknown. After uniform scaling with a single scaling factor, the two panels are superimposable.

McConnell–Bloch equations for NMR with Fourier transforms of frequency-switching wave trains like those of Fig. 10.2 but circularly polarized, that do not involve external magnetic fields [Grunwald and Steel, 1993a,b; Knudsen and Naqvi, 1995]. Fourier transformation, it will be recalled, is a mathematical operation that in this case transforms the frequency-switching wave trains from the time domain to the frequency domain without change of physical content [Bracewell, 1978; Champeney, 1973; Ramirez, 1985]. It is evident that for the same values of $2\pi|\nu_a - \nu_b|\tau$, the Fourier transforms are superimposable on the plots of the McConnell–Bloch equations.

Exchange and Distinguishability

Since Fig. 10.2 illustrates the Schrödinger wave train, without Brownian noise, of a caged molecule when there is exchange between two distinct environments with equal mean residence times and short transit times, panel (b) of Fig. 10.4 applies at once.* In particular, the spectral shapes depend solely on the value of $\xi = 2\pi|\nu_b - \nu_a|\tau$, the distinguishability index defined in Eq. (10.7). In this instance $|\nu_b - \nu_a| = |E_b^{(C)} - E_a^{(C)}|/Lh$; that is, $|\nu_b - \nu_a|$ is governed by the dif-

*Although the transit times are indeed relatively short, the time scales are closer to those for exchange averaging in infrared spectra than in NMR spectra. See, for example, Wood and Strauss [1990].

ference of potential energy at the two cage centers which, in contrast to *absolute* potential energy, is defined without arbitrary zero points. On introducing numerical values for L and h and expressing potential energy in joules per mole and τ in seconds, Eq. (10.7) takes the form (10.9), which will be our key equation. It describes the distinguishability of the electronic ground states of a caged molecule as the molecule exchanges between two cage environments \a and \b with a mean time of exchange τ.

$$\xi = 2\pi |E_{\mathrm{b}}^{(\mathrm{C})} - E_{\mathrm{a}}^{(\mathrm{C})}|\tau/Lh$$

$$\xi = 1.575 \times 10^{10}(\mathrm{mol\ J^{-1}\ s^{-1}}) \cdot |E_{\mathrm{b}}^{(\mathrm{C})} - E_{\mathrm{a}}^{(\mathrm{C})}|\tau \tag{10.9}$$

When $\xi > 5$, the power spectrum in the frequency domain shows two peaks, and distinguishability exists. This is true even when τ_{b} is small compared to τ_{a}. In Fig. 10.4, where $\tau_{\mathrm{a}} = \tau_{\mathrm{b}}$, the two peaks coalesce when $\xi = 1.41$. As τ_{b} becomes small compared to τ_{a}, the coalescence point decreases and reaches a lower limit of $\xi = 1$. Thus, when $\xi < 1$, the power spectrum shows a single peak and distinguishability is lost. In the intermediate range, $1 \leq \xi \leq 5$, simple criteria are equivocal.

If we adopt a median value of 2×10^{-11} s for τ, Eq. (10.9) states that $\xi > 5$ and distinguishability exists when $|E_{\mathrm{b}}^{(\mathrm{C})} - E_{\mathrm{a}}^{(\mathrm{C})}| > 15$ J/mol. It further states that $\xi < 1$ and distinguishability is lost when $|E_{\mathrm{b}}^{(\mathrm{C})} - E_{\mathrm{a}}^{(\mathrm{C})}| < 3$ J/mol.

We will illustrate the significance of these inequalities by calculating some cage effects in non-hydrogen-bonding dilute solutions, where energy differences are relatively small. The specific process is shown in Fig. 10.5. The first step in the figure, A\a $\rightarrow$ A[inX], transfers an A molecule from liquid A to liquid X. ΔE_1 for this step can be predicted from Hildebrand's regular solution model:

$$\Delta E_1 = V_{\mathrm{A}}(\delta_{\mathrm{A}} - \delta_{\mathrm{X}})^2$$

where δ is the Hildebrand solubility parameter and V_{A} is the molar volume of liquid A [Hildebrand and Scott, 1962].

The second step, A[in X] $\rightarrow$ A\x, introduces A neighbors for all but one of the X neighbors. Assuming pairwise additivity of interaction energies and a

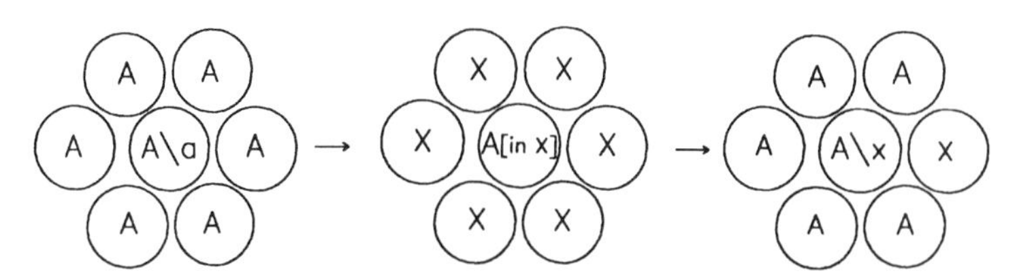

Figure 10.5. Scheme for changing the neighbor shell by substitution.

total of s neighbors, we obtain

$$\Delta E_2 \approx -(s-1)\Delta E_1/s$$

The overall process is A\a $\rightarrow$ A\x. It replaces an A molecule by an X molecule in the nearest-neighbor shell of A. According to the cage model, ΔE for A\a $\rightarrow$ A\x is equal to $E^{(C)}[\backslash x] - E^{(C)}[\backslash a]$, the change in the potential energy of A at the cage center. Thus $E^{(C)}[\backslash x] - E^{(C)}[\backslash a] = \Delta E_1 + \Delta E_2 = V_A(\delta_A - \delta_X)^2/s$.

A series of results, in which A = benzene and in which mixtures of A and X approximately fit the regular solution model, is shown in Table 10.1. On the scale of bond energies, the changes in cage potential are small; yet on the 15 J/mol scale of environmental distinguishability, most changes are substantial.

In Table 10.1 the nature of a neighbor is being changed at constant s. In the earlier Table 9.3, the number of neighbors is changed. In both tables, the cage potentials result from van der Waals interactions in the absence of hydrogen bonds, and we may expect that the magnitudes of any changes are on the small side. Nonetheless, both tables imply a high probability that the addition, subtraction, or change in the nature of a molecule in the neighbor shell changes the cage potential by more than the predicted threshold of 15 J/mol for distinguishability.

How come that observations on liquids give so few clues about this wealth of potential species? The predicted distinguishability is based on Schrödinger wave mechanics, a primary scientific theory. By our definitions the distinguish-

Table 10.1. Sample results for $[E^{(C)}[\backslash x] - E^{(C)}[\backslash a]]$, calculated according to Fig. 10.5 using Hildebrand's regular solution model[a]

X	δ_X	$[E^{(C)}[\backslash x] - E^{(C)}[\backslash a]]$ (J/mol)	ξ	Comment
CCl_4	17.6	13	4	Ambiguous
$(CH_3)_4Si$	12.7	330	100	Distinguishable
CH_2I_2	24.1	250	80	Distinguishable
$Cl_2C{=}CCl_2$	19.0	0.4	0.1	Averaged
CS_2	20.5	26	8	Distinguishable
$(C_4F_9)_3N$	12.1	400	130	Distinguishable
Diethyl ether	15.1	120	40	Distinguishable
n-Hexane	14.9	140	45	Distinguishable
Cyclohexane	16.6	43	14	Distinguishable
C_6F_{14}	12.1	400	130	Distinguishable
Toluene	18.2	3	1	Averaged
Naphthalene	20.3	20	6	Distinguishable

[a] A = benzene, $V_A = 89$ cm^3/mol, $\delta_A = 18.8$ J$^{1/2}$ cm$^{-3/2}$, $s = 10$, and $\tau = 2 \times 10^{-11}$ s.
Source: Hildebrand and Scott [1962].

ability is ideal: We may count on its existence, but observation may be obscured by potential-energy noise due to the Brownian motions of the molecules. This explanation is not *ad hoc*. The standard deviation of the noise, $\sigma_{(C)}$, was found to be of order 1–3 kJ/mol. (*Kilo*joules! See Table 9.2.) By contrast, differences in cage potentials, according to Tables 10.1 and 9.3, are often less than 1 kJ/mol. In the gas phase, Doppler-free two-photon absorption from opposing laser beams is able to cancel out this mode of Brownian noise [Bloembergen and Levenson, 1976], and one may hope that similar cancellation will eventually be feasible for liquids. In the meantime, real distinguishability is limited by the signal-to-noise ratio [Grunwald and Steel, 1994]. The following section will analyze this fact.

REAL DISTINGUISHABILITY. EFFECT OF BROWNIAN MOTIONS

A Schrödinger wave whose energy fluctuates about a mean value ϵ_m with a standard deviation $\sigma_{(C)}$ is in some ways like an FM radio wave, with carrier frequency $\nu_m = \epsilon_m/h$ and randomly fluctuating frequency modulation with standard deviation $\sigma_{(C)}/h$. The fluctuation may be due to cage exchange [as in Eq. (10.4)] or due to Brownian noise, or due to both. In either case, it is a fluctuation of the cage potential $E^{(C)}$; the intrinsic state energy ϵ_0 is fixed. In the absence of cage exchange, the distribution of the potential energy at the cage center is a convolution based on adding the contributions from s Brownian neighbors. Almost always, s is large enough for the convolution to produce a Gaussian distribution, according to the Central Limit Theorem of statistics [Bracewell, 1978].

Figure 10.6a shows a sample of real noise, generated by an accurate Gaussian random-noise generator. The spiky appearance is typical not only of the familiar electrical noise one sees on an oscilloscope, but also of the zigzag Brownian paths of micron-size particles suspended in a liquid and observed under a microscope.

Figure 10.6b shows Gaussian potential-energy noise as simulated by the "zigzag" model, which may be regarded as a Brownian analog of the strong-collision model of kinetic theory [Gilbert and Smith, 1990]. The potential energy is assumed to vary linearly with time until the next Brownian "collision" directs it toward a new value chosen at random from a Gaussian probability distribution. The standard deviation of that distribution is identified with $\sigma_{(C)}$, the standard deviation of $E^{(C)}$ due to Brownian noise at the cage center. The mean value of the Brownian noise is taken as zero. The time between Brownian "collisions" follows an exponential lifetime distribution with a mean time equal to the Brownian correlation time τ_{corr}. As shown in Fig. 10.6, the resemblance of zigzag noise to real noise is quite good.

In the absence of cage exchange, the inclusion in the Schrödinger wave train of Brownian frequency modulation according to the zigzag model causes the Fourier transform to be a Gaussian band rather than a sharp line. The band

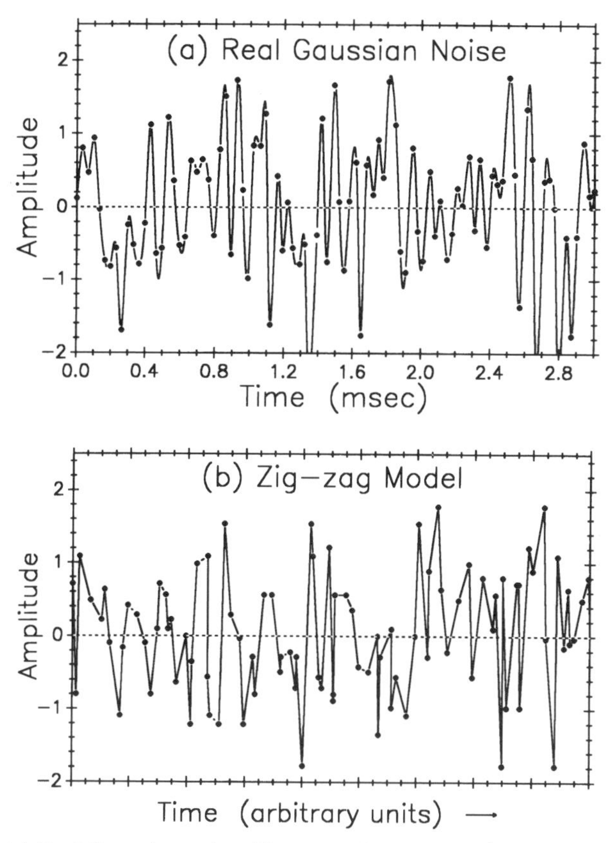

Figure 10.6. (a) Real Gaussian noise. The mean frequency is 20 kHz, and the standard deviation of the random fluctuations is of order 5 kHz. The amplitude was sampled digitally, at the rate of 30,000 samples/s. (b) Comparable noise simulated by the zigzag model.

center is at the mean frequency ϵ_m/h. The half-width decreases somewhat with the Brownian correlation time τ_{corr} but remains of order $\sigma_{(\text{C})}/h$, decreasing from slightly less than $\sigma_{(\text{C})}/h$ when $\tau_{\text{corr}} = 0.8$ ps, to $\sigma_{(\text{C})}/2h$ when $\tau_{\text{corr}} = 0.07$ ps. The width, in the frequency domain, of a Schrödinger wave train in the absence of cage exchange is reminiscent of the natural line width of an NMR spectral line in the absence of chemical exchange. Both result from fluctuations caused by Brownian motions, and both can be broad enough to mask the effect of exchange. The chief difference is that the natural width of an NMR spectral line decreases in proportion to τ_{corr} and is often quite narrow.

Figures 10.7 and 10.8 show power spectra that simulate the combined effects of cage exchange and Brownian potential-energy noise [Grunwald and Steel, 1994]. Figure 10.7 uses values of $\sigma_{(\text{C})}$ and τ_{corr} typical of benzene-like liquids under normal conditions. The mean time for cage exchange (denoted by τ) is 3×10^{-11}s. The distinguishability index ξ is > 40 in each case. In the absence

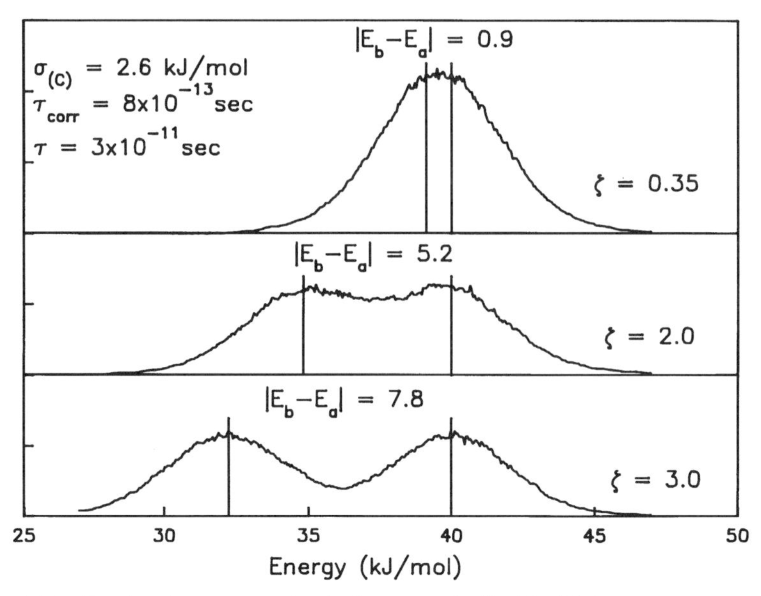

Figure 10.7. Simulated power spectra for benzene-like liquids. Stick spectra are expected in the absence of Brownian potential-energy noise.

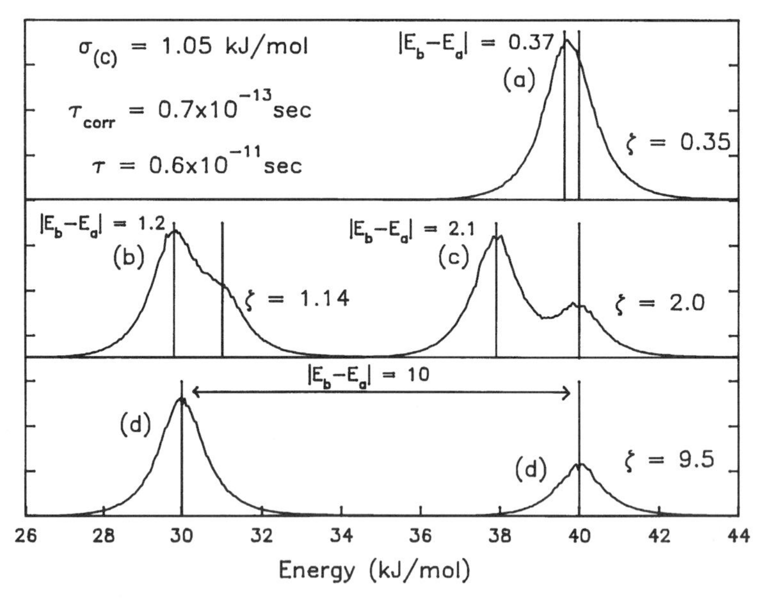

Figure 10.8. Simulated power spectra for water-like liquids. Panel (d) represents values of $|E_b - E_a|$ actually inferred for the environmental isomers described in Chapter 8 for liquid water. Stick spectra are expected in the absence of Brownian potential-energy noise.

of the Brownian noise, each power spectrum would consist of two sharp, discrete lines, as indicated by the sticks. By contrast, in the presence of Brownian noise the lines become broad bands, and some overlap remains even when the simulated difference in cage potential (denoted by $|E_b - E_a|$) is as great as 3.0 $\sigma_{(C)}$, or 7.8 kJ/mol. The figure suggests that real distinguishability depends on the ratio $\zeta = |E_b - E_a|/\sigma_{(C)}$ and that the threshold for real distinguishability is at $\zeta \approx 2$. The corresponding threshold for $|E_b - E_a|$ is about 5 kJ/mol, substantially greater than a benzene-like liquid is likely to produce. This threshold also fits the author's own results for dilute solutions. The latter consist of differences of power spectra for the solvent and for solutions in the presence of Brownian noise, along with specialized applications of information theory [Haber, 1974; Pierce, 1980].

In the case of benzene there is credible theoretical evidence for two major pairwise configurations, with the benzene rings either tilted orthogonal or displaced parallel. $\Delta E^{(2)}$ for these configurations is about 0.9 kJ/mol [Jorgensen and Severance, 1990]. It is clear from Fig. 10.7 that a difference of 0.9 kJ/mol is masked by Brownian noise. If one wishes to observe the two configurations in a real experiment, one must find a technique that defeats the Brownian noise.

Figure 10.8 uses values for $\sigma_{(C)}$ and τ_{corr} and for the mean time of exchange τ that are more typical of water-like liquids. The relative cage populations simulate those inferred for liquid water: \a = 0.7, $s_a = 4$; and \b = 0.3, $s_b = 5$. $\sigma_{(C)}$ is substantially smaller than for the benzene-like liquid, largely because the cage potential is more nearly harmonic and $\sigma_{(C)} \approx RT(4/5s)^{1/2}$. (See second footnote on p. 214.) For water, the harmonic approximation with $s = 4.3$ yields $\sigma_{(C)} = 1.07$ kJ/mol. A better approximation based on a detailed model yields a mean value for $\sigma_{(C)}$ of 1.05 kJ/mol [Grunwald and Steel, 1994, calculation (b)].

Figure 10.8 shows some distinguishability already when $|E_b - E_a|/\sigma_{(C)}$, or ζ, is 1.14. It shows clear distinguishability when $\zeta = 2.1$. And distinguishability is predicted without reservation when $|E_b - E_a| = 10.0$ kJ/mol, which is the [\b – \a] energy difference inferred for water, where the bands stand far apart. Experimental evidence for the existence of two environmental isomers in liquid water was described in Chapter 8.

Figure 10.7 suggests that real distinguishability is rare for environmental isomers in non-hydrogen-bonding liquids, because $\sigma_{(C)}$ is relatively large and the cagewise differences of $E^{(C)}$ tend to be small. In hydrogen-bonding solvents, on the other hand, the linkages are stronger and $\sigma_{(C)}$ is smaller; distinguishable environmental isomers should be quite common.

ENCOUNTER PAIRS, LINKAGE PAIRS, AND MOLECULAR COMPLEXES

The interaction of a caged molecule with its liquid neighbors varies considerably in strength and specificity. In this section we consider how strong it must be to produce a molecular complex.

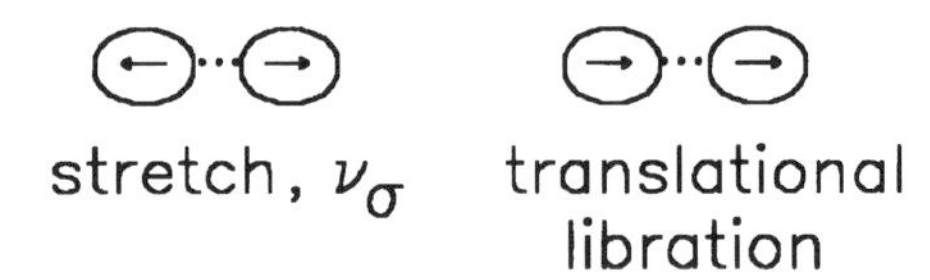

Figure 10.9. When the two modes of translational libration along the A–X axis become coherent, the in-phase motion is translational libration of [A–X] as a unit, and the 180° out-of-phase motion is the stretching mode ν_σ.

It is useful to classify adjacent pairs of molecules into three categories: encounter pairs [A, X], linkage pairs [A; X], and molecular complexes [A–X]. In encounter pairs and linkage pairs, A and X remain separate molecules, with separate centers of mass whose individual motions show little cross-correlation. In molecular complexes, by contrast, the interaction is so strong that A and X form a single kinetic unit, with a single center of mass about which A and X move in concerted motion. For example, as shown in Fig. 10.9, when the two modes of motion along the A··X axis become concerted, A and X move either in-phase, to form a mode of hindered in-cage translation, or 180° out-of-phase, to form the stretching mode ν_σ of the A–X bond.

In encounter pairs the van der Waals interaction is of "normal" strength, without marked directional character. In linkage pairs this interaction is above average strength. A and X have complementary centers of van der Waals attraction that are predisposed to move into juxtaposition, so that the interaction is directional. When the complementary centers are in juxtaposition, the A··X interaction is strong enough to perturb somewhat the internal vibrational modes of A and X. These perturbations cause spectral frequency shifts resembling those for molecular-complex formation—so much so that without special techniques one cannot tell whether the source of the perturbation is a linkage pair [A; X] or a molecular complex [A–X]. Yet the distinction matters because, in the design of composition models, the A and X entities in [A; X] are separate molecular species, while the molecular complex [A–X] is a single species.

The following analogy may dramatize this issue: The distinction between linkage pairs and molecular complexes resembles that between (a) a troop of boy scouts walking as a group and (b) a company of soldiers marching in step. There is no doubt that the motions of the boy scouts are linked, because they head for a common destination and travel at a common average speed; yet their steps are the uncorrelated steps of separate individuals. By contrast, the steps of the marching soldiers are concerted, both in length and in frequency. One cannot help but feel that this difference is major and qualitative.

One would of course like to know some kind of a linkage threshold—how great the linkage energy must be to cause a changeover from linkage pairs to molecular complexes. Unfortunately, that threshold is currently beyond reach. On the other hand, there does exist suitable information for encounter pairs [A, X] and weak molecular complexes [A–X], enough information for the calculation of the magnitude of the [A, X] to [A–X] threshold. We shall therefore calculate that threshold. The information is relevant because the encounter pair is less stable than the linkage pair. The threshold for [A, X] $\rightarrow$ [A–X] therefore sets a lower limit to that for [A; X] $\rightarrow$ [A–X]. We find that the threshold for [A, X] $\rightarrow$ [A–X] is about 20 kJ/mol. Unpublished calculations by the author place the desired threshold for [A; X] $\rightarrow$ [A–X] in the 20–30 kJ/mol range.

Modes of Motion for [A, X] and [A–X]

In estimating the threshold for the changeover from encounter pairs [A, X] to molecular complexes [A–X], our approach is to compare the standard free energy of [A, X] with that of [A–X], the latter as a function of the bond dissociation energy E[A–X]. The threshold is defined as the value of E_J at which $G^0[\text{A–X}] - G^0[\text{A, X}])$ becomes negative. The calculations employ methods of statistical thermodynamics. For simplicity, A and X are represented as physically rigid bodies; that is, internal vibrations within A and X are neglected.

On that basis, the caged species A and X in [A, X] each have three modes of translational libration (tl) and three modes of rotational libration (rl). Since by hypothesis, [A, X] is an ordinary encounter pair without directional linkage, the modewise partition functions are nondirectional and will be denoted simply by z_{tl} and z_{rl}—hence Eq. (10.10):

$$G^0[\text{A, X}] = \text{constant} - RT(6 \ln z_{tl} + 6 \ln z_{rl}) \tag{10.10}$$

When a caged molecular complex [A–X] forms in a liquid, the original translational and rotational librations of A and X—a total of 12 modes—change into three modes of tl and three modes of rl of the complex. The other six modes are converted into internal vibrations, as shown in Fig. 10.10. To derive these

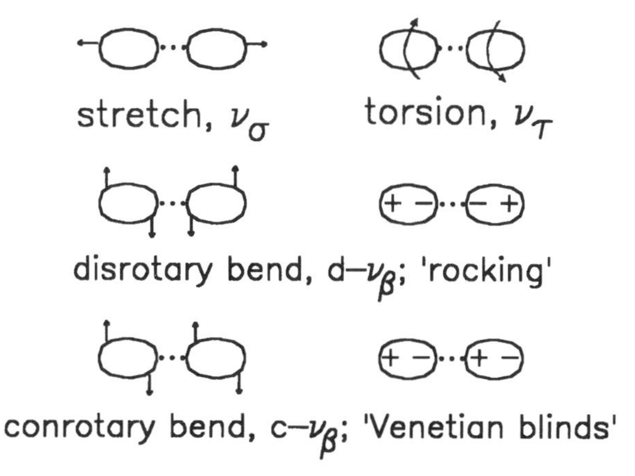

Figure 10.10. The six vibrational modes of a molecular complex formed from two rigid three-dimensional bodies (shown as ovals).

internal modes, A and X are allowed to translate and rotate coherently—either in-phase or 180° out-of-phase—and the motions that are not translations or rotations of the entire complex are sorted out. The ovals in Fig. 10.10 represent the original three-dimensional rigid bodies, but are not intended to suggest axial symmetry.

The six derived modes are: a bond-stretching mode ν_σ, a torsional mode ν_T of opposing oscillations about the A–X bond axis, and four bending modes ν_β consisting of a disrotary pair of rocking motions (d-ν_β) and a conrotary pair reminiscent of the closing/opening motions of slats of Venetian blinds (c-ν_β). Accordingly, G^0 [A–X] will be given by Eq. (10.11), whose constant is the same as that in (10.10). The energy of bond dissociation E[A–X] is a positive number.

$$G^0[\text{A–X}] = \text{constant} - E[\text{A–X}] - RT(3\ln z_{\text{tl}} + 3\ln z_{\text{rl}} + \ln z_\sigma + 4\ln z_\beta + \ln z_T) \tag{10.11}$$

The partition functions ($z_{\text{tl}}, z_{\text{rl}}, z_\sigma, \ldots$) will be approximated by the equation for a harmonic oscillator, (10.12); ν_i is the (actual or quasi-harmonic) frequency of the ith librational or vibrational mode.

$$z_i = \frac{1}{1 - \exp(-h\nu_i/kT)} \tag{10.12}$$

To the accuracy required for a propensity calculation, the partition functions in Eq. (10.11) are accessible, because some values are known for ν_σ and for the average ν_β from high-resolution rotational and vibrational spectra of supersonically expanded gas mixtures. A few results are given in Table 10.2 [Legon and Millen, 1986; Legon et al., 1982; Petersen and Klemperer, 1984]. The measurements of the stretching frequencies are fairly direct, but those of the bending frequencies are indirect and do not show a difference between disrotary and conrotary bending. The bond dissociation energy was actually measured only for HCN··HF. The other values are estimates based on stretching force constants and $r_{\text{B,A}}$ distances.

Both stretching (ν_σ) and bending (ν_β) frequencies vary considerably with the nature of the complex. Based on the few data in the table, $\nu_\beta/\nu_\sigma \approx 0.3$ and, according to theory, ν_σ is proportional to $E[\text{A–X}]^{1/2}$. The distances $r_{\text{B,A}}$ in the complexes are shorter than the corresponding encounter distances by 0.5–1 Å.

Approximate Threshold for [A, X] → [A–X]

G^0 [A, X] and G^0 [A–X] will be calculated from Eqs. (10.10) and (10.11). The threshold for molecular-complex formation is identified with the value of the bond dissociation energy E[A–X] at which (G^0[A–X] – G^0[A, X]) becomes negative.

The only numerical information we have for libration frequencies in liq-

Table 10.2. Results for weak molecular complexes in the gas phase, based on a review by Legon and Millen [1986]

Complex B··HA	Conformation at Potential Minimum	ν_σ (cm^{-1})	$\nu_\beta{}^a$ (cm^{-1})	E[A–X]^b (kJ/mol)	$r_{B,A}$ (Å)
HCN··HF	Linear	155	70	26	2.80
HCN··HCl	Linear	110	—	14	3.40
HCN··HCN	Linear	120	—	17	3.26
HCN··HCF_3	Symmetric top	55	—	—	3.49
H_3CCN··HF	Symmetric top	170	45	—	2.76
HC≡CCN··HF	Linear	140	30	—	2.79
H_2O··HF	Pyramidal O-atom (inversion barrier at 0-atom 2.5 kJ/ mol)	170			2.66
N≡N··HF	Linear	90	—	—	3.12
Ar··HF	Linear	40	—	1.5	2.62
H_3P··HCN	Symmetric top	70	—	—	3.91
| = C_2H_2, acetylene					
|··HCl	T-shaped	85	—	8	3.13
▯ = C_6H_6, benzene					
▯··HCl	C_{6v}	110	—	14	3.18

^{a}Average value.

^{b}The dissociation energy of the molecular complex in the gas phase is here identified with E[A–X]. The value for HCN·HF was measured. The other values were estimated by Legon et al. [1982] from the force constant and $r_{B,A}$.

uid cages are the quasi-harmonic frequencies for translational libration listed in Table 9.2, ~ 15 cm^{-1} for benzene-like cages. Frequencies for rotational libration in nondirectional cages should be somewhat smaller—probably less than 10 cm^{-1} for molecules that rotate with normal friction; in the following, a benzene-like average of $\langle \nu_l \rangle = 10$ cm^{-1} will be adopted for all librations. We shall also adopt the value of 10 cm^{-1} for the torsional frequency ν_T in the molecular complex. On that basis, (G^0[A–X – G^0[A, X]) simplifies to Eq. (10.13), where $\langle z_l \rangle$ denotes the partition function (10.12) based on $\langle \nu_l \rangle$:

$$(G^0[\text{A–X}] - G^0[\text{A, X}]) = -E[\text{A–X}] + RT(6 \ln \langle z_l \rangle - \ln z_\sigma - 4 \ln z_\beta - \ln z_T) \tag{10.13}$$

For ν_σ and ν_β we will assume linear increases with the square root of E[A–X], with intercepts of 10 cm^{-1} and slopes fitting the data in Table 10.2:

$$\nu_\sigma / c = 10 + 27.5\, E[\text{A–X}]^{1/2}\ \text{cm}^{-1}; \qquad E[\text{A–X}] \text{ in kJ/mol} \tag{10.14a}$$

$$\nu_\beta / c = 10 + 7.9\, E[\text{A–X}]^{1/2}\ \text{cm}^{-1}; \qquad E[\text{A–X}] \text{ in kJ/mol} \tag{10.14b}$$

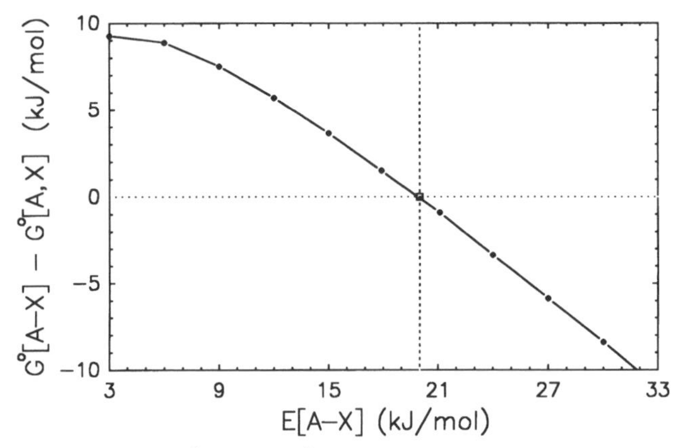

Figure 10.11. Plot of (G^0[A–X] – G^0[A, X]) versus E[A–X], as described in text.

The results for (G^0[A–X] – G^0[A, X]) versus E[A–X] are shown in Fig. 10.11. The threshold occurs at 20 kJ/mol.

Calculations such as those in Fig. 10.11, where the input data are nonspecific propensities, are vulnerable to specific exceptions, but the magnitude of the results is credible. The deduced threshold of 20 kJ/mol is large compared to RT (2.5 kJ/mol at 298 K), showing that the energy gained by bond formation must be substantial before the constraints on internal motions imposed by bond formation become tolerable. When the bond formation involves linkage pairs rather than encounter pairs, the threshold for bond formation will be somewhat higher.

11

SOLVATION IN DILUTE SOLUTIONS

When designing a composition model we try to avoid unnecessary complexity, but sometimes we commit too much simplicity. This risk is greatest for the solvent component, because complexity of solvation is hard to prove. Colligative properties do not define the solvent species (Chapter 3), and distinctions between wave-mechanically distinguishable cage environments can be obliterated by Brownian noise (Chapter 10).

In the absence of evidence to the contrary, the conservative tendency is to represent the solvent as a single molecular species. When there *is* evidence, the tendency is to limit the solvent species to those that are clearly present. When solute-induced medium effects occur (as they do, for example, in acetic acid and water), the model of solute-induced molar shifts combines the solvent species that exist in the solvation shells with logically related solvent species in the bulk of the solvent—even though there are bound to be differences (Chapter 8). The only requirement is that the species in the solvation shells must have distinct properties that can be correlated one-on-one with parallel properties of distinct species in the bulk of the solvent.

The economy of model that results from merging logically related groups of molecules into single species can, however, be false economy. The reason is that, in terms of Schrödinger wave mechanics, the molecules in solvation shells are usually distinguishable from their logical counterparts in the bulk of the solvent, since the underlying criteria for distinguishability of cage environments are undemanding (Chapter 10). For liquids of ordinary viscosity, the difference in the cage potentials at the cage centers, $|E_b^{(C)} - E_a^{(C)}|$, only needs to exceed about 15 J/mol, a condition often met. This criterion defines *ideal* distinguishability and allows us to introduce the corresponding caged species into composition models as environmental subspecies. Owing to the Brownian noise, whose standard deviation $\sigma_{(C)}$ in liquids near room temperature is of order 1–3 kJ/mol, these subspecies may not be directly observable. The criterion for real distinguishability is that $|E_b^{(C)} - E_a^{(C)}| \gtrsim 2\sigma_{(C)}$, or 2–6 kJ/mol—much more demanding!

In this chapter we take the position that ideal distinguishability is sufficient and include the environmental subspecies of the solvent in composition models. The thermodynamic consequences are significant [Grunwald and Steel, 1995]. Solvation emerges as a mechanism for enthalpy–entropy compensation. Guidelines appear for the design of thermodynamic models of solvation. And constraints arise that distinguish genuine solvation in liquids from the limited "solvation" stemming from intermolecular interactions in small clusters in the gas phase.

NOMINAL AND ENVIRONMENTAL EQUATIONS

The conservative equation for the solvation of a gaseous species (X) in a liquid (A) is simply

$$X(g) \rightleftarrows X(\text{in liquid A})$$

The corresponding mass-action expression is $K = P_X/m_X$, which is the expression for a simple 1 : 1 equilibrium. For some purposes this equation is unduly austere because it glosses over the simultaneous changes in the environments of some solvent molecules. For definiteness, assume that the solution of X in A is dilute, and that the possible caged species or subspecies are limited to the four species shown in Fig. 11.1. A\a denotes an A molecule adjacent only to A neighbors—essentially an all-A environment. A\x denotes an A molecule with one X neighbor and the rest A neighbors. X\a denotes an X molecule adjacent only to A neighbors. And [X–A]\a denotes an [X–A] molecular complex adjacent only to A neighbors. This model assumes that A\a is distinguishable from A\x—at least ideally—and that X\a is distinguishable from an [X–A] molecular complex.

Within this framework, the solvation of gaseous X in liquid A *in the absence of molecular complex formation* is represented by Eq. (11.1), in which s_X is the mean number of A neighbors to an X molecule:

$$X(g) + s_X A\backslash a \rightleftarrows X\backslash a + s_X A\backslash x \qquad (11.1)$$

Equation (11.1) states that as 1 mole of X(g) dissolves, s_X moles of A are transferred from \a environments to \x environments. It implies that the distinction between A\a and A\x is consequential, because one does not clutter up equations with nothings; and, by showing X\a rather than X–A, it implies that molecular-complex formation is negligible.

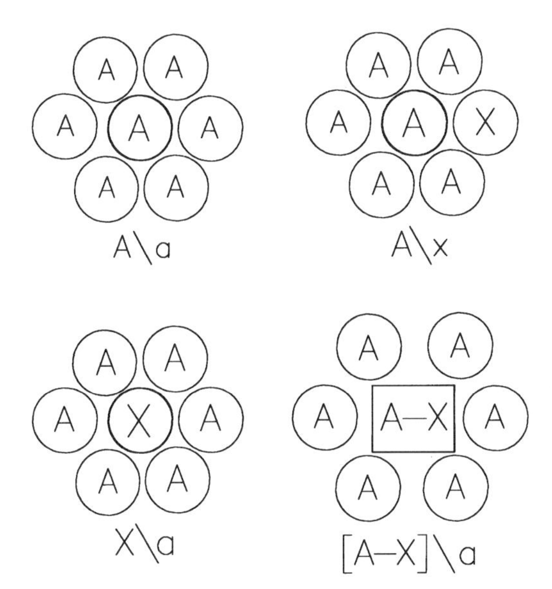

Figure 11.1. Possible molecular species or caged subspecies in dilute solutions of a solute X in a liquid solvent A.

Equation (11.1) may be separated into two stoichiometrically balanced parts: (11.1n) represents nominally the solvation of X(g), and (11.1e) shows the simultaneous environmental change for s_X solvent molecules. Neither equation implies molecular-complex formation.

$$\mathrm{X}(g) \rightleftarrows \mathrm{X\backslash a} \tag{11.1n}$$

$$s_{\mathrm{x}}\mathrm{A\backslash a} \rightleftarrows s_{\mathrm{x}}\mathrm{A\backslash x} \tag{11.1e}$$

Equations such as $s_{\mathrm{x}}\mathrm{A\backslash a} \rightleftarrows s_{\mathrm{x}}\mathrm{A\backslash x}$ are *not* chemical equations, because the A molecules do not change. They are *environmental* equations, because the cage environments change. The separation of Eq. (11.1) into Eqs. (11.1n) and (11.1e) illustrates another generality: The overall equation for a solution process can always be separated into a nominal equation and an environmental equation. The nominal equation expresses the goal of the process, and the environmental equation expresses the associated *solvent reorganization.*

Similarly, the formation of an encounter pair between dilute X and dilute Y in the solvent A is given by Eq. (11.2), where X\y denotes an X molecule with one Y neighbor and the rest A neighbors, and Y\x denotes an Y molecule with one X neighbor and the rest A neighbors. The related solvent subspecies are A\a, A\x, and A\y. As X and Y become neighbors, mA\x and nA\y solvent molecules are displaced and transferred into the bulk of the solvent.

$$\mathrm{X(solv) + Y(solv)} \rightleftarrows \mathrm{[X, Y]\ encounter\ pair\ (solv) + extra\ solvent}$$

$$\underbrace{\mathrm{X\backslash a} + s_{\mathrm{x}}\mathrm{A\backslash x}} + \underbrace{\mathrm{Y\backslash a} + s_{\mathrm{y}}\mathrm{A\backslash y}}$$

$$\rightleftarrows \underbrace{(s_{\mathrm{x}} - m)\mathrm{A\backslash x} + [\mathrm{X\backslash y, Y\backslash x}] + (s_{\mathrm{y}} - n)\mathrm{A\backslash y}} + (m + n)\mathrm{A\backslash a}$$

$$\mathrm{Net:}\ \mathrm{X\backslash a + Y\backslash a} + m\mathrm{A\backslash x} + n\mathrm{A\backslash y} \rightleftarrows [\mathrm{X\backslash y, Y\backslash x}] + (m + n)\mathrm{A\backslash a} \quad (11.2)$$

When Eq. (11.2) is separated into two stoichiometrically balanced parts, the nominal Eq. (11.2n) represents the formation of the encounter pair and the environmental Eq. (11.2e) shows the simultaneous cage reorganization for $(m + n)$ solvent molecules. Again, neither equation implies molecular-complex formation.

$$\mathrm{X\backslash a + Y\backslash a} \rightleftarrows [\mathrm{X\backslash y, Y\backslash x}] \quad (11.2\mathrm{n})$$

$$m\mathrm{A\backslash x} + n\mathrm{A\backslash y} \rightleftarrows (m + n)\mathrm{A\backslash a} \quad (11.2\mathrm{e})$$

The formation of a molecular complex from ligands that are present as an encounter pair is shown in Eq. (11.3). Once again, the overall equation is usefully separated into (a) a nominal part (11.3n) for the complex formation and (b) an environmental part (11.3e) for the solvent reorganization. The nominal equation in this example represents a chemical reaction.

$$\mathrm{[X, Y]\ encounter\ pair\ (solv)} \rightleftarrows \mathrm{[X\text{–}Y]\ (solv) + extra\ solvent}$$

$$\underbrace{(s_{\mathrm{x}} - m)\ \mathrm{A\backslash x} + [\mathrm{X\backslash y,\ Y\backslash x}] + (s_{\mathrm{y}} - n)\ \mathrm{A\backslash y}}$$

$$\rightleftarrows \underbrace{[\mathrm{X\text{–}Y}]\mathrm{\backslash a} + s_{\mathrm{xy}}\mathrm{A\backslash xy}} + (s_{\mathrm{x}} - m + s_{\mathrm{y}} - n - s_{\mathrm{xy}})\ \mathrm{A\backslash a} \quad (11.3)$$

$$[\mathrm{X\backslash y, Y\backslash x}] \rightleftarrows (\mathrm{X\text{–}Y})\mathrm{\backslash a} \quad (11.3\mathrm{n})$$

$$(s_{\mathrm{x}} - m)\mathrm{A\backslash x} + (s_{\mathrm{y}} - n)\mathrm{A/y} \rightleftarrows s_{\mathrm{xy}}\mathrm{A\backslash xy} + (s_{\mathrm{x}} - m + s_{\mathrm{y}} - n - s_{\mathrm{xy}})\mathrm{A\backslash a} \quad (11.3\mathrm{e})$$

We will find that environmental equations obey exact enthalpy–entropy compensation, so that solvent reorganization does not contribute to ΔG^0 for solution processes at constant T and P. In this respect ΔG^0 is unique. In principle, solvent reorganization contributes to all other thermodynamic properties, and the contributions can be considerable [Grunwald and Steel, 1995].

THERMODYNAMIC FORMULATION

Composition tree (T-11.1) traces the levels at which composition is specified for the simplest possible liquid solution: A dilute solute component (2) in a solvent component (1). Each component is a single molecular species. Chemical reactions, including molecular complex formation, are negligible. The solvent A may be divided into two subspecies, A\a and A\x, as in Fig. 11.1; that is, the \a and \x environments are wave-mechanically distinguishable. The solute X is all X\a.

The concentrations of X\a and A\x are related by stoichiometry. Let s_{x} denote the mean number of A neighbors in X\a. Then each X\a molecule is associated with s_{x} A\x molecules, and $n_{\mathrm{A\backslash x}} = s_{\mathrm{x}} \cdot n_{\mathrm{X\backslash a}}$.

Composition Tree (T-11.1)

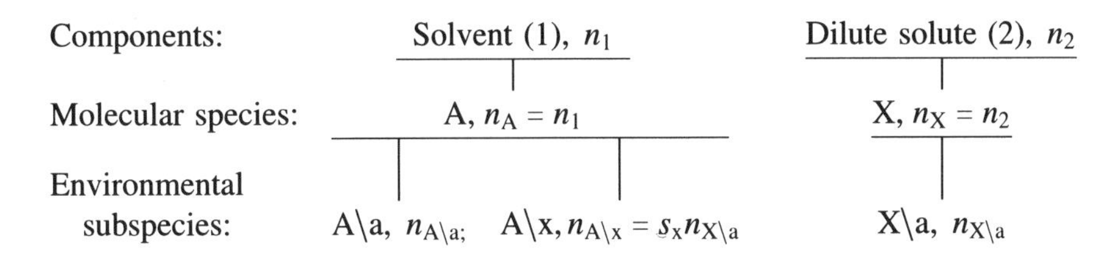

We are interested in relationships among thermodynamic properties on the three levels shown in (T-11.1). On the level of environmental subspecies, the partial properties are defined in terms of the mole numbers $n_{\mathrm{A\backslash a}}$, $n_{\mathrm{A\backslash x}}$, and $n_{\mathrm{X\backslash a}}$. For example, $G_{\mathrm{A\backslash a}} = (\partial G/\partial n_{\mathrm{A\backslash a}})_{n_{\mathrm{A\backslash x}}, n_{\mathrm{X\backslash a}}}$ or $S_{\mathrm{A\backslash x}} = (\partial S/\partial n_{\mathrm{A\backslash x}})_{n_{\mathrm{A\backslash a}}, n_{\mathrm{X\backslash a}}}$. It is convenient, however, to transform the composition variables, letting two of them be the formula-weight numbers n_1 and n_2 and letting the third be the mean number s_{x} of A-neighbors to an X molecule:

$$n_{\mathrm{A\backslash a}} = n_1 - s_{\mathrm{x}} n_2, \qquad n_{\mathrm{A\backslash x}} = s_{\mathrm{x}} n_2, \qquad n_{\mathrm{X\backslash a}} = n_2 \tag{11.4}$$

In this notation, the Gibbs free energy, enthalpy, and entropy of the liquid phase are expressed by Eqs. (11.5)–(11.7):

$$G = (n_1 - s_{\mathrm{x}} n_2) G_{\mathrm{A\backslash a}} + s_{\mathrm{x}} n_2 G_{\mathrm{A\backslash x}} + n_2 G_{\mathrm{X\backslash a}} \tag{11.5}$$

$$S = (n_1 - s_{\mathrm{x}} n_2) S_{\mathrm{A\backslash a}} + s_{\mathrm{x}} n_2 S_{\mathrm{A\backslash x}} + n_2 S_{\mathrm{X\backslash a}} \tag{11.6}$$

$$H = (n_1 - s_{\mathrm{x}} n_2) H_{\mathrm{A\backslash a}} + s_{\mathrm{x}} n_2 H_{\mathrm{A\backslash x}} + n_2 H_{\mathrm{X\backslash a}} \tag{11.7}$$

It will be recalled that in the Cage Packing Equation (9.8) for liquids, neighbor numbers such as s_{x} are variables. Indeed, we may think of s_{x} as an environmental (rather than chemical) progress variable, since it indicates how many moles of solvent are transferred from \a to \x, per mole of solute. When s_{x} is not at equilibrium, $(s_{\mathrm{x}} - s_{\mathrm{x,eq}})$ is a strain, analogous to the molar strain $(\alpha - \alpha_{\mathrm{eq}})$, and the corresponding stress y_{x} equals $(\partial G/\partial s_{\mathrm{x}})_{T,P,n_1,n_2}$. At equilibrium at constant T and P, the value of s_x adjusts itself so that G is at a minimum. That is, at environmental equilibrium, $y_\mathrm{x} = \partial G/\partial s_\mathrm{x} = 0$. When this condition is applied to Eq. (11.5), there results the environmental analog of a Correspondence Theorem:

$$y_\mathrm{x} = (\partial G/\partial s_\mathrm{x})_{T,P,n_1,n_2} = n_2(G_{\mathrm{A\backslash x}} - G_{\mathrm{A\backslash a}}) = 0 \quad \text{at equilibrium}$$
$$\therefore \quad G_{\mathrm{A\backslash x}} = G_{\mathrm{A\backslash a}} \quad \text{at equilibrium} \tag{11.8}$$

Since $s_\mathrm{x}(G_{\mathrm{A\backslash x}} - G_{\mathrm{A\backslash a}})$ is the free energy change ΔG_e for the environmental part of Eq. (11.1), we have here an example that ΔG_e for solvent reorganization is zero.

Because s_x is analogous to a chemical progress variable α, certain conclusions follow at once.

(i) At equilibrium in dilute solutions at constant T and P, s_x (in common with any molar property of a well-behaved dilute solute) is independent of m_2:

$$(\partial s_\mathrm{x}/\partial m_2)_{\mathrm{eq}} = (\partial s_\mathrm{x}/\partial n_2)_{n_1,\mathrm{eq}} = 0 \tag{11.9}$$

It follows from Eq. (11.9) and from $(\partial s_\mathrm{x}/\partial n_1)_{n_2} = -(\partial s_\mathrm{x}/\partial n_2)_{n_1}(\partial n_2/\partial n_1)_{s_\mathrm{x}}$ that $(\partial s_\mathrm{x}/\partial n_1)_{n_2,\mathrm{eq}}$ is also zero.

(ii) The equations derived in Chapter 7 apply to any kind of molar shift. The progress variable α may therefore be identified with s_x. When this is done in Eqs. (7.10) and (7.13), it follows from Eq. (11.9) that the molar-shift terms for S_2 and H_2 vanish. Thus, at constant T and P,

$$S_2 = (\partial S/\partial n_2)_{n_1,y_\mathrm{x}=0} = (\partial S/\partial n_2)_{n_1,s_\mathrm{x}} \tag{11.10a}$$
$$H_2 = (\partial H/\partial n_2)_{n_1,y_\mathrm{x}=0} = (\partial H/\partial n_2)_{n_1,s_\mathrm{x}} \tag{11.10b}$$

Differentiation of Eqs. (11.6) and (11.7) at constant n_1, s_x yields the following for the solute component:

$$S_2 = S_{\mathrm{X\backslash a}} + s_\mathrm{x}(S_{\mathrm{A\backslash x}} - S_{\mathrm{A\backslash a}}) \tag{11.11}$$
$$H_2 = H_{\mathrm{X\backslash a}} + s_\mathrm{x}(H_{\mathrm{A\backslash x}} - H_{\mathrm{A\backslash a}}) \tag{11.12}$$

The same method gives the following for the solvent component:

$$S_1 = S_{\mathrm{A\backslash a}}, \qquad H_1 = H_{\mathrm{A\backslash a}} \tag{11.13}$$

As a corollary,

$$G_1 = H_1 - TS_1 = H_{\mathrm{A\backslash a}} - TS_{\mathrm{A\backslash a}} = G_{\mathrm{A\backslash a}} \tag{11.14}$$

(iii) We may now obtain two key equations. On combining Eq. (11.8) with

Eq. (11.14) we obtain Eq. (11.15), which equates the partial free energy of the formal solvent component to that of either of the solvent's environmental subspecies. Equation (11.15) is a key to solvation mechanisms. At constant T and P and for $y_X = 0$,

$$G_1 = G_{A\backslash a} = G_{A\backslash x} \tag{11.15}$$

Since $G = H - TS$, it follows from Eqs. (11.11), (11.12), and (11.15) that at equilibrium at constant T and P we obtain

$$\begin{aligned} & G_{A\backslash x} - G_{A\backslash a} = H_{A\backslash x} - H_{A\backslash a} - T(S_{A\backslash x} - S_{A\backslash a}) = 0 \\ \therefore\ & H_{A\backslash x} - H_{A\backslash a} = T(S_{A\backslash x} - S_{A\backslash a}) \quad \text{at equilibrium} \end{aligned} \tag{11.16}$$

Equation (11.16) implies that the solvent reorganization attending the solvation of a dilute solute X—any dilute solute—proceeds with enthalpy–entropy compensation. This implication will now be explored.

STANDARD ENTHALPY–ENTROPY COMPENSATION

The term *enthalpy–entropy compensation* normally refers to changes in *standard* partial properties, such as ΔH^0 and ΔS^0 for reaction or solvation in dilute solutions. To see that Eq. (11.16) implies *standard* compensation, we note that A\x and A\a are subspecies of the solvent, so that $(S_{A\backslash x} - S_{A\backslash a})$ and $(H_{A\backslash x} - H_{A\backslash a})$ are solvent properties and thus virtually independent of m_2 in dilute solutions. By conventional standards of reckoning (Table 8.1), therefore, $(S_{A\backslash x} - S_{A\backslash a})$ and $(H_{A\backslash x} - H_{A\backslash a})$ become part of the *standard* entropy and enthalpy of X. These conclusions, as well as $(G_{A\backslash x} - G_{A\backslash a}) = 0$, are expressed in Eqs. (11.17)–(11.19):

$$G_2^0 = G_{X\backslash a}^0 + s_X(G_{A\backslash x} - G_{A\backslash a}) = G_{X\backslash a}^0 \tag{11.17}$$

$$H_2^0 = H_{X\backslash a}^0 + s_X(H_{A\backslash x} - H_{A\backslash a}) \tag{11.18}$$

$$S_2^0 = S_{X\backslash a}^0 + s_X(S_{A\backslash x} - S_{A\backslash a}) \tag{11.19}$$

Furthermore, we can claim enthalpy–entropy compensation only if there is something to be compensated—that is, if $(H_{A\backslash x} - H_{A\backslash a})$ and $T(S_{A\backslash x} - S_{A\backslash a})$ not only are equal but individually are nonzero. Equation (11.16) alone does not prove this because the solution "zero = zero" would satisfy the mathematics. We will show, therefore, that $(S_{A\backslash x} - S_{A\backslash a})$ in general is nonzero. The following proof is based on the Second Law.

PROOF. When equating $(S_{A\backslash x} - S_{A\backslash a})$ to $-[\partial(G_{A\backslash x} - G_{A\backslash a})/\partial T]$ for a closed system, one must be careful about the variables that remain inactive. If the temperature is changing with maintenance of equilibrium, the partial derivative $[\partial(G_{A\backslash x} - G_{A\backslash a})/\partial T]_{\{n\},y_x=0}$ vanishes because Eq. (11.8) is true at all temperatures. But this partial derivative does not express $(S_{A\backslash x} - S_{A\backslash a})$. For the environmental species in composition tree (T-11.1), the composition variables are $\{n_{A\backslash a}, n_{A\backslash x}, n_{X\backslash a}\}$ or their transforms, $\{n_1, n_2, s_x\}$. Thus, to derive $(S_{A\backslash x} - S_{A\backslash a})$, the temperature derivative must be taken with either $\{n_{A\backslash a}, n_{A\backslash x}, n_{X\backslash a}\}$ or $\{n_1, n_2, s_x\}$ constant.

In the following, $\{P, n_1, n_2\}$ will be constant. The active variables will be either (1) T and s_x or (2) T and y_x. Choosing T and s_x, we obtain

$$(S_{A\backslash x} - S_{A\backslash a}) = -[\partial(G_{A\backslash x} - G_{A\backslash a})/\partial T]_{s_x}$$

Transforming to T and y_x, we obtain

$$[\partial(G_{A\backslash x} - G_{A\backslash a})/\partial T]_{y_x} = [\partial(G_{A\backslash x} - G_{A\backslash a})/\partial T]_{s_x} + [\partial(G_{A\backslash x} - G_{A\backslash a})/\partial s_x]_T(\partial s_x/\partial T)_{y_x}$$

When $y_x = 0$, the term on the left-hand side vanishes. Therefore,

$$(S_{A\backslash x} - S_{A\backslash a}) = [\partial(G_{A\backslash x} - G_{A\backslash a})/\partial s_x]_T(\partial s_x/\partial T)_{y_x=0}$$

where by Eq. (11.8), $\partial(G_{A\backslash x} - G_{A\backslash a})/\partial s_x = (\partial^2 G/\partial s_x^2)/n_2$. This is the curvature of the plot of G/n_2 versus s_x and is positive, because G is at a minimum in a closed system at equilibrium. At the same time, $\partial s_x/\partial T$ at equilibrium is a finite number which vanishes only in rare cases when s_x happens to be independent of T. Thus $(S_{A\backslash x} - S_{A\backslash a})$ is normally nonzero and finite.

Q.E.D.

What do Eqs. (11.17)–(11.19) tell us? First, they tell us [through the terms $s_x(H_{A\backslash x} - H_{A\backslash a})$ and $s_x(S_{A\backslash x} - S_{A\backslash a})$] that the introduction of the dilute solute causes solvent reorganization according to $s_x\text{A}\backslash\text{a} \rightarrow s_x\text{A}\backslash\text{x}$. These terms add on to the nominal terms, $H^0_{X\backslash a}$ and $S^0_{X\backslash a}$, and are usually nonzero and finite.

Second, Eq. (11.17) tells us that solvent reorganization has *no effect* on the standard partial free energy G^0_2 of the solute component. As a corollary, nominal equations are perfectly adequate for representing free-energy changes in reactions.

Third, because Eqs. (11.17)–(11.19) are derived by purely thermodynamic methods, they apply regardless of the microscopic nature of the cage environments. Of course, the magnitudes of $s_x(H_{A\backslash x} - H_{A\backslash a})$, and hence the amounts that are being compensated, *do* depend on microscopic particulars.

Magnitudes

The results based on the regular-solution model listed in Chapter 10 (Table 10.1) indicate the magnitude of $(H_{A\backslash x} - H_{A\backslash a})$ for non-hydrogen-bonding solutes in benzene-like solvents. The mean value for 12 solutes is 140 J/mol, and the standard deviation is 150 J/mol. Adopting a value of 10 for s_x, a typical magnitude for $s_x(H_{A\backslash x} - H_{A\backslash a})$, and for the compensating $s_x \cdot T(S_{A\backslash x} - S_{A\backslash a})$, is therefore 1.4 kJ/mol.

From the basic differential $d(uv) = u\ dv + v\ du$, the probable range, $\delta[s_x(H_{A\backslash x} - H_{A\backslash a})]$, to be expected in a reaction series is

$$\delta[s_x(H_{A\backslash x} - H_{A\backslash a})] = s_x \cdot \delta(H_{A\backslash x} - H_{A\backslash a}) + (H_{A\backslash x} - H_{A\backslash a}) \cdot \delta s_x \qquad (11.20)$$

For $\delta(H_{A\backslash x} - H_{A\backslash a})$ we may adopt the standard deviation, 150 J/mol, in Table 10.1. For δs_x we will use ± 1. The mean $(H_{A\backslash x} - H_{A\backslash a})$ is 140 J/mol, and $s_x \approx 10$. Hence $\delta[s_x(H_{A\backslash x} - H_{A\backslash a})]$, the probable range of compensation in a non-hydrogen-bonding reaction series, is expected to be of order 1.6 kJ/mol.

The magnitudes are greater in hydrogen-bonding media. In Pimentel and McClellan's review on hydrogen bonding [1971], the range of ΔH^0 for hydrogen-bonding between more than 20 acceptors and phenol and alcohol donors is 30 kJ/mol, with a standard deviation of 3 kJ/mol. In applying Eq. (11.20), we will therefore adopt 3 kJ/mol as a measure of $\delta(H_{A\backslash x} - H_{A\backslash a})$, and a similar 3 kJ/mol for the mean of $|H_{A\backslash x} - H_{A\backslash a}|$ in a reaction series when the solvent is hydrogen-bonding. The symbol $\langle s \rangle$ will denote the mean, and δs the probable range, of the number of solvent molecules reorganized. On that basis, Eq. (11.20) predicts that the range of compensation due to solvent reorganization in hydrogen-bonding solvents is

$$\delta[s(H_{A\backslash x} - H_{A\backslash a})] = (3\langle s \rangle + 3\delta s)\ \text{kJ/mol}$$

This is rarely less than a few kilojoules per mole and may be substantially greater. Estimates of $\langle s \rangle$ are of order 50 for biological processes such as the opening and closing of ion channels in cell membranes or the conversion of hemoglobin T to hemoglobin R [Zimmerberg et al., 1990; Colombo et al., 1992]; δs could easily be of order 10. Solvent reorganization effects in excess of 100 kJ/mol are therefore possible.

Unfortunately, predictions about the magnitude of environmental effects are hard to verify, because one does not measure thermodynamic properties for solvent reorganization directly: One measures directly only the overall ΔH^0 and ΔS^0, which includes ΔH_n^0 and ΔS_n^0 for the nominal process. If, nonetheless, one succeeds in showing that compensation is significant, that does not guarantee a mechanism of solvent reorganization. Compensation might follow from solute-induced molar shifts in the solvent (considered in Chapter 8) or from circumstances as yet unidentified. That there are compensation mechanisms other than solvent reorganization and molar shifts is clear, for example, from the data for surface reactions described in Chapter 6.

Difficult as the characterization of compensation mechanism may seem, it is a problem that can be solved—perhaps by using the kind of approaches that succeed in characterizing reaction mechanisms.

The compensation expressed by Eqs. (11.17)–(11.19) is readily extended to chemical equilibria. Upon applying the Δ-operator, we obtain Eqs. (11.21)–(11.23), in which the letters R and P denote reactants and products.

$$\Delta G^0 = \sum G^0_{\mathrm{P}\backslash\mathrm{a}} - \sum G^0_{\mathrm{R}\backslash\mathrm{a}} \tag{11.21}$$

$$\Delta H^0 = \sum [H^0_{\mathrm{P}\backslash\mathrm{a}} + s_{\mathrm{P}}(H_{\mathrm{A}\backslash\mathrm{p}} - H_{\mathrm{A}\backslash\mathrm{a}})] - \sum [H^0_{\mathrm{R}\backslash\mathrm{a}} + s_{\mathrm{R}}(H_{\mathrm{A}\backslash\mathrm{r}} - H_{\mathrm{A}\backslash\mathrm{a}})] \tag{11.22}$$

$$\Delta S^0 = \sum [S^0_{\mathrm{P}\backslash\mathrm{a}} + s_{\mathrm{P}}(S_{\mathrm{A}\backslash\mathrm{p}} - S_{\mathrm{A}\backslash\mathrm{a}})] - \sum [S^0_{\mathrm{R}\backslash\mathrm{a}} + s_{\mathrm{R}}(S_{\mathrm{A}\backslash\mathrm{r}} - S_{\mathrm{A}\backslash\mathrm{a}})] \tag{11.23}$$

The solvent-reorganization terms show enthalpy–entropy compensation, but the nominal terms $[\sum H^0_{\mathrm{P}\backslash\mathrm{a}} - \sum H^0_{\mathrm{R}\backslash\mathrm{a}}]$ and $[\sum S^0_{\mathrm{P}\backslash\mathrm{a}} - \sum S^0_{\mathrm{R}\backslash\mathrm{a}}]$ do not, and ΔG^0 is nonzero.

Analogous expressions apply to reaction series. The solvent-reorganization terms again show compensation, and the nominal terms do not.

Examples

The plots of ΔH^0 versus ΔS^0 and of ΔG^0 versus ΔS^0 in Figs. 11.2 and 11.3 cover spectacular ranges of ΔS^0. They are nearly linear and have slopes close to the experimental temperature, thus exemplifying major enthalpy–entropy compensation. The compensation is much too extensive for more than a minor fraction to come from the nominal reaction series. Moreover, since the solvents are hydrogen-bonding, solvent-reorganization effects are expected to be relatively large and very probably account for most of the compensation.

Figure 11.2 shows data for the 1 : 1 association of a series of α-amino acids with 18-crown-6 ether (11.24) or cryptand-2,2,2 (11.25) in methanol and ethanol

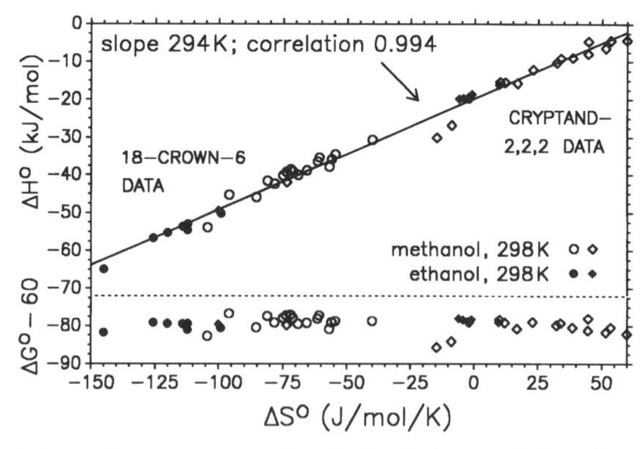

Figure 11.2. Enthalpy-entropy compensation in the 1 : 1 association of a series of α-amino acids with 18-crown-6 ether or cryptand-2,2,2 in methanol and ethanol. Note that ΔH^0 varies by more than 60 kJ/mol [Danil de Namor et al., 1991a].

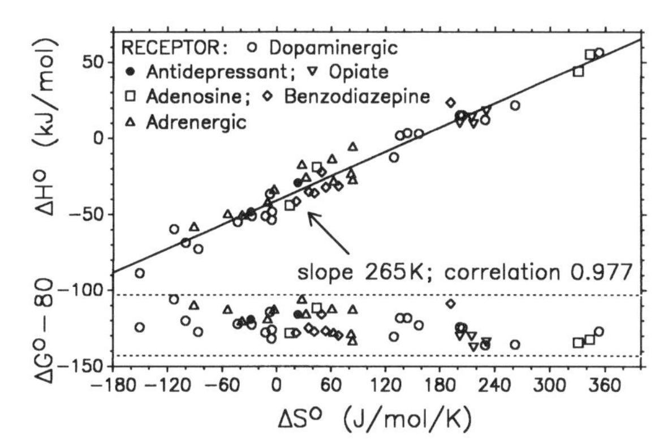

Figure 11.3. Enthalpy–entropy compensation in the 1 : 1 binding of drugs with their respective receptors; $T \approx 290$ K. Based on measurements of the binding constant $K(T)$ at three to five temperatures. Assuming errors of 0.1° in T and 10% in $K(T)$, the error propagated into ΔH^0 is 2–3 kJ/mol [Gilli and Borea, 1991].

[Danil de Namor et al., 1991a]. The amino acids are Gly, Ala, Arg, Asn, Asp, Cys, Glut, His, Ile, Leu, Met, Phe, Ser, Thr, Trp, Tyr, and Val and thus cover a broad spectrum of both hydrophobic and hydrophilic R-groups.

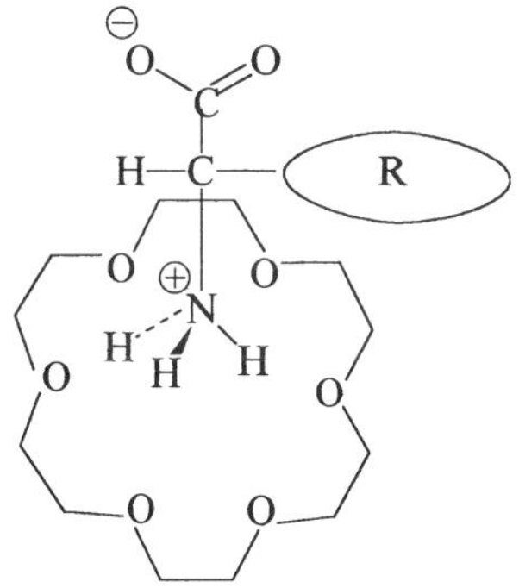

(11.24)

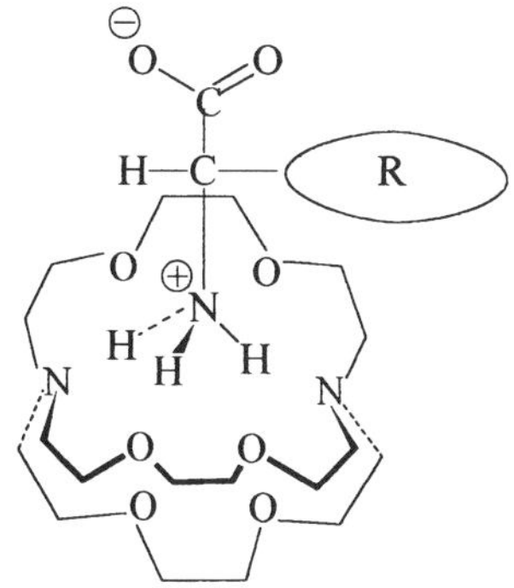

(11.25)

ΔG^0 is nearly invariable; the standard deviation for all ΔG^0 points is only 1.8 kJ/mol. By contrast, the standard deviation of the ΔH^0 points is 17 kJ/mol, and the range is over 60 kJ/mol. The slope is 294 K and $T = 298$K.

The ΔH^0 points in Fig. 11.2 group themselves by host, being more negative for binding by 18-crown-6, and tend to be more negative in ethanol than in methanol. For a given solvent and host, ΔH^0 varies considerably with the nature of the R-group, as much as 35 kJ/mol for association with cryptand-2,2,2 in ethanol. This effect follows no obvious trend. The effect of the R-group on ΔH^0 for binding to 18-crown-6 shows no correlation with that for binding to cryptand-2,2,2, nor does it correlate with pK_a for the α-NH_3^+ group of the amino acids in water. It appears that solvent reorganization is highly specific, with a structural dependence that, at present, still seems chaotic. Indeed, the individual who reviewed the related manuscript by Grunwald and Steel [1995] for the *Journal of the American Chemical Society* was so unhappy with both the size of the R-group effect and the seeming lack of correlation that he suggested that experimental errors might be greater than reported. Yet the results were obtained with good equipment, buttressed by good control experiments, and are based on the established method of calorimetric titration. (Compare Fig. 7.3.)

Figure 11.3 represents the binding of a series of drugs dissolved in water by receptor sites in membrane proteins [Gilli and Borea, 1991; Gilli et al., 1994]. This is not a reaction series but comes close to being a collection of all data available to the cited authors. The binding triggers conformational changes that are thought to resemble the opening or closing of membrane channels [Gilli and Borea, 1991]. The attendant solvent reorganization might involve as many as 50 water molecules [Zimmerberg et al., 1990].

In this case, the values of ΔG^0 spread out over the full 40-kJ/mol range accessible to measurement (indicated by the dashed lines), and the plot of ΔG^0 versus ΔS^0 is a scatter diagram. But the 150-kJ/mol spread of ΔH^0 is so much greater than that of ΔG^0 that enthalpy–entropy compensation is major. Accordingly, the plot of ΔH^0 versus ΔS^0 is close to linear, with a slope that is within 10% of the experimental temperature.

ENTHALPIES OF WATER SUBSPECIES IN PRESENCE OF NONPOLAR SOLUTES

The molar-shift model for water introduced in Chapter 8 gives numerically precise results for the transfer of water molecules into the solvation shells of nonpolar solutes. Such results are as good as the underlying model (the thermodynamic development is exact), and the translation of the results into the language of a solvation model is straightforward. We shall therefore build on the results of Table 8.4 to obtain the analog of $(H_{A\backslash x} - H_{A\backslash a})$ for solutions of nonpolar solutes in water. The procedure is circuitous, in the sense that we first apply the molar-shift model and then translate the results into the solvation model, but it produces real numbers, including algebraic signs, for precisely the sort of envi-

pure water — aq. solution, m_2

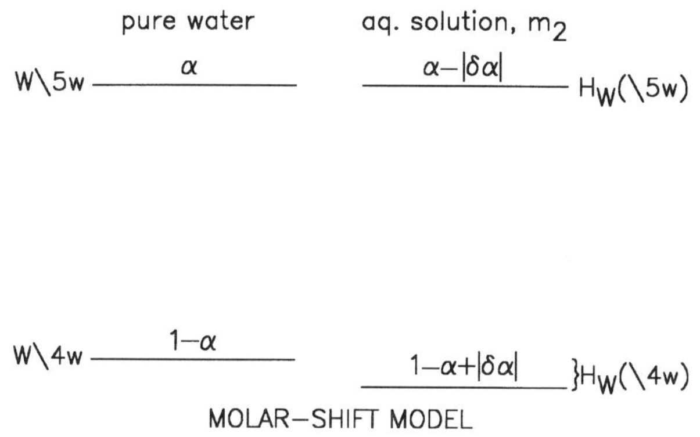

W\5w — α — $\alpha - |\delta\alpha|$ — $H_W(\backslash 5w)$

W\4w — $1-\alpha$ — $1-\alpha+|\delta\alpha|$ — $\}H_W(\backslash 4w)$

MOLAR–SHIFT MODEL

Figure 11.4. Enthalpy–mole-fraction–environmental-species diagram. Molar-shift model for water and an aqueous solution of a nonpolar solute.

ronmental changes that concern us. We shall see that within this framework, the results for nonpolar solutes in water are both substantial and specific.

The essence of the previously used molar-shift model for nonpolar solutes (X) in water (W) is displayed in the form of an enthalpy–mole-fraction–environmental-species diagram in Fig. 11.4. In pure water, the two environmental isomers W\5w and W\4w are present at mole fractions α and $(1-\alpha)$, respectively. In the presence of a nonpolar solute at m_2 formal, a fraction $\delta\alpha$ shifts from W\5w to W\4w. (In the subsequent solvation model, this shift is interpreted as a transfer into solvation shells; see Fig. 11.5.) In Chapter 8, the molar-shift model was developed to derive $\partial\alpha/\partial m_2$ from the molar-shift term of the partial entropy S_2 of the solute. We will now use Eq. (8.25) to derive $\partial\Delta H_{4w\rightarrow 5w}/\partial m_2$ from the values obtained for $\partial\alpha/\partial m_2$, as well as from the molar-shift term of the partial heat capacity $C_{P,2}$. Equation (8.25) is now rewritten in the specialized form (11.26):

$$C_{P,2} - (C_{P,2})_\alpha = (C_{P,2})_{\text{shift}} = Rm_1(1-2\alpha)\left[\frac{\Delta H_{4w\rightarrow 5w}}{RT}\right]^2\left(\frac{\partial\alpha}{\partial m_2}\right) + 2\alpha(1-\alpha)m_1\frac{\Delta H_{4w\rightarrow 5w}}{RT^2}\left(\frac{\partial\Delta H_{4w\rightarrow 5w}}{\partial m_2}\right) \qquad (11.26)$$

On introducing numerical values for pure water at 298 K, as listed for the 1 : 1 model in Table 8.3 (noting that $\alpha \equiv \alpha_1$), and applying Eq. (8.33), we obtain the practical equation (11.27):

$$(\partial\Delta H_{4w\rightarrow 5w}/\partial m_2) = -4.92(S_2)_{\text{shift}} + 2.99(C_{P,2})_{\text{shift}} \qquad (11.27)$$

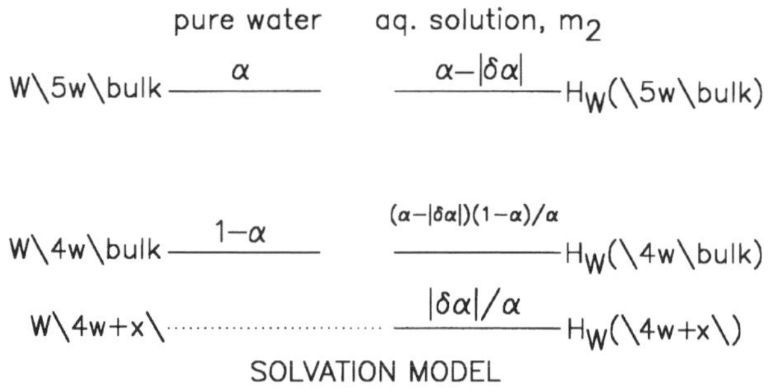

Figure 11.5. This is Fig. 11.4 translated to fit the solvation model.

The calculations according to Eqs. (11.26) and (11.27) are outlined in Table 11.1. The molar-shift term of $C_{P,2}$ is estimated, in analogy to that for S_2, by the equation: $(C_{P,2})_{\text{shift}} = \Delta C_{P,2}(\text{W}) - \Delta C_{P,2}(\text{M})$, where $\Delta C_{P,2}(\text{W})$ is the experimental $\Delta C_{P,2}$ for solvation in water, and $\Delta C_{P,2}(\text{M})$ is that for solvation in regular solvents. $\Delta C_{P,2}(\text{M})$ is approximated ΔC_P for the liquefaction of the pure

Table 11.1. H_w(\4w + x\) − H_w(\4w\bulk) **for nonpolar solutes X in water at 298 K**[a]

Solute X	$\Delta C_{P,2}$[b] (W)	(M)	$(C_{P,2})_{\text{shift}}$[c]	$\lvert\partial\alpha/\partial m_2\rvert$	$\dfrac{\partial \Delta H_{4\text{w}\to 5\text{w}}}{\partial m_2}$	H_w(\4w + x\) − H_w(\4w\bulk)
He	118	33	85	0.0298	0.54	−3.8
Ne	149	33	116	0.0336	0.67	−4.2
Ar	186	33	153	0.0376	0.81	−4.6
Kr	210	33	177	0.0400	0.91	−4.9
H_2	142	38	104	0.0296	0.60	−4.3
N_2	189	38	151	0.0408	0.84	−4.4
O_2	192	38	154	0.0376	0.82	−4.6
CH_4	218	38	180	0.0372	0.90	−5.1
C_2H_6	285	42	243	0.0427	1.14	−5.7
C_3H_8	331	46	285	0.0477	1.31	−5.9
n-C_4H_{10}	389	46	343	0.0493	1.50	−6.5
n-C_6H_{14}	490	54	436	0.0561	1.84	−6.9
c-C_6H_{12}	410	50	360	0.0487	1.54	−6.7
Acetylene	154	38	116	0.0236	0.58	−5.2
Ethylene	236	42	194	0.0346	0.91	−5.6
Benzene	276	54	222	0.0296	0.95	−6.8

[a]Heat capacities are given in J/mol-K; enthalpies are in kJ/mol.

[b](W) = ΔC_P of solvation in water; (M) = ΔC_P of liquefaction of the pure gaseous solute [Wilhelm et al., 1977; Dec and Gill, 1985; Wilhelm and Battino, 1973; Shaw, 1969; Grunwald, 1986b].

[c]$(C_{P,2})_{\text{shift}} = \Delta C_{P,2}(\text{W}) - \Delta C_{P,2}(\text{M})$.

gaseous solute. Note that $\Delta C_{P,2}(\mathrm{M})$ is relatively small, so that high accuracy is not required. The results for $(\partial \Delta H_{4\mathrm{w}\to 5\mathrm{w}}/\partial m_2)$ hover around 1 kJ/mol per unit m_2.

The translation of the results to the corresponding solvation model follows the scheme outlined in Fig. 11.5. There are now three environmental species, denoted as W\5w\bulk, W\4w\bulk, and W\4w + x\. The index after the first slash indicates the nearest-neighbor shell, as before; the index \bulk indicates that the more remote environment is, really or in essence, bulk water. This is worth showing because the equilibrium ratio for W\5w\bulk:W\4w\bulk is $\alpha/(1-\alpha)$, not only in pure water but also in any solution. $|\delta\alpha|$ in Fig. 11.5 is identical to $|\delta\alpha|$ in Fig. 11.4; the sum of the mole fractions of W\4w\bulk + W\4w + x\ in Fig. 11.5 is equal to the mole fraction $(1-\alpha+|\delta\alpha|)$ in Fig. 11.4. The mole fractions entered in Fig. 11.5 follow directly from these guidelines.

In order to relate $H_\mathrm{W}(\backslash 4\mathrm{w}+\mathrm{x}\backslash) - H_\mathrm{W}(\backslash 4\mathrm{x}\backslash \mathrm{bulk})$ in the solvation model to $(\partial \Delta H_{4\mathrm{w}\to 5\mathrm{w}}/\partial m_2)$ as derived by the molar-shift model, we note that in either model, the enthalpy level of W\5w in Fig. 11.4 (W\5w\bulk in Fig. 11.5) is not affected by the addition of the nonpolar solute. On introducing the constraint $\partial H_\mathrm{W}(\backslash 5\mathrm{w})/\partial m_2 = 0$ and solving for $-\partial H_\mathrm{W}(\backslash 4\mathrm{w})/\partial m_2$ (where $H_\mathrm{W}(\backslash 4\mathrm{w})$ is the average of $H_\mathrm{W}(\backslash 4\mathrm{w}+\mathrm{x}\backslash)$ and $H_\mathrm{W}(\backslash 4\mathrm{w}\backslash \mathrm{bulk})$, we arrive at Eq. (11.28):

$$H_\mathrm{W}(\backslash 4\mathrm{w}+\mathrm{x}\backslash) - H_\mathrm{W}(\backslash 4\mathrm{w}\backslash \mathrm{bulk}) = -\alpha(1-\alpha)(\partial \Delta H_{4\mathrm{w}\to 5\mathrm{w}}/\partial m_2)/|\partial\alpha/\partial m_2| \tag{11.28}$$

Results for $H_\mathrm{W}(\backslash 4\mathrm{w}+\mathrm{x}\backslash) - H_\mathrm{W}(\backslash 4\mathrm{w}\backslash \mathrm{bulk})$ are listed in the last column of Table 11.1. For these nonpolar solutes the values are negative and amount to several kilojoules per mole, which is considerable. Equally important, the values vary by several kilojoules per mole with the nature of the solute, which implies that the enthalpy of solvent reorganization can vary significantly in a reaction series. On the whole, the magnitudes and standard deviations in the last column of Table 11.1 support the previous estimates for hydrogen-bonding solvents based on Eq. (11.20).

SOLVATION COMPLEXES

When the solvent (A) and solute (X) molecules have complementary linkage sites, molecular-complex formation (11.29) is a distinct possibility:

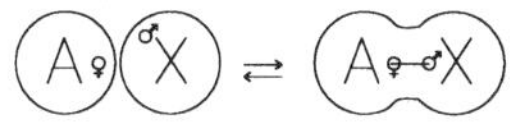

In this section we discuss the resulting thermodynamics, including the associated solvent reorganization. We also show that the kind of solvation that includes molecular-complex formation is outside the realm of enthalpy–entropy compensation.

Composition Tree (T-11.2)

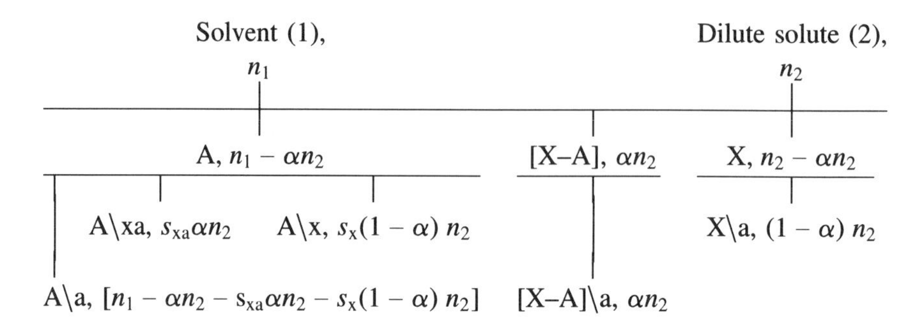

The relevant composition tree is (T-11.2). It has two components, just like the earlier (T-11.1), and three molecular species. Yet the visual impact of (T-11.2) is one of great complexity—markedly greater than that of (T-11.1)—largely because of the substantial number of solvent subspecies. The qualitative lesson is that even in simple cases, the inclusion of environmental subspecies and solvent reorganization fills up a composition model with so much detail that one needs courage to tackle the thermodynamics. The environmental subspecies can be omitted if one will limit the treatment to partial free energies and other first derivatives of G, because the contributions from changes in molecular environments then are dormant.

Actually, the analysis of (T-11.2) is quite tractable because the solution is dilute. G, H, S, ... are functions of $\{n_1, n_2, T, P\}$ and of three progress variables: α for A + X $\rightarrow$ [X–A], s_x for the solvation of X, and s_{xa} for the solvation of [X–A]. Because m_2 is dilute and A is the solvent, $\{\alpha, s_x, s_{xa}\}$ are independent of m_2; therefore derivatives such as $(\partial s_x/\partial n_2)_{n_1, \{y\}=0}$ vanish. [See Eq. (11.9).] Among the corollaries, those based on Eqs. (7.10) and (7.13) are stated below.

At constant n_1, T, P and equilibrium

$$H_2 = (\partial H/\partial n_2)_{\{y\}=0} = (\partial H/\partial n_2)_{\alpha, s_x, s_{xa}}$$
$$S_2 = (\partial S/\partial n_2)_{\{y\}=0} = (\partial S/\partial n_2)_{\alpha, s_x, s_{xa}}$$

Thus, to obtain an expression for H_2, we begin with H and differentiate at constant (α, s_x, s_{xa}):

$$
\begin{aligned}
H_{\{y\}=0} &= [n_1 - \alpha n_2 - s_{xa}\alpha n_2 - s_x(1-\alpha)n_2]H_{A\backslash a} + s_{xa}\alpha n_2 H_{A\backslash xa} \\
&\quad + s_x(1-\alpha)n_2 H_{A\backslash x} + \alpha n_2 H_{[X-A]\backslash a} + (1-\alpha)n_2 H_{X\backslash a} \\
H_2 &= (1-\alpha)[H_{X\backslash a} + s_x(H_{A\backslash x} - H_{A\backslash a})] \\
&\quad + \alpha[H_{[X-A]\backslash a} + s_{xa}(H_{A\backslash xa} - H_{A\backslash a}) - H_{A\backslash a}] \\
&= [H_{X\backslash a} + s_x(H_{A\backslash x} - H_{A\backslash a})] + \alpha[H_{[X-A]\backslash a} - H_{A\backslash a} - H_{X\backslash a}] \\
&\quad + \alpha[s_{xa}(H_{A\backslash xa} - H_{A\backslash a}) - s_x(H_{A\backslash x} - H_{A\backslash a})] \qquad (11.30)
\end{aligned}
$$

The three bracketed terms on the right-hand side of Eq. (11.30) may be identified, respectively, as $(H_2)_{\alpha=0}$, the partial enthalpy of X in the absence of complexing; $\Delta_c H_n$, the enthalpy change for nominal complexing according to A(solvent) + X $\rightarrow$ [X–A]; and $\Delta_c H_e$, the enthalpy change for the related environmental equation. Equation (11.30) thus states that the partial enthalpy of a dilute solute is a sum (11.31a) whose leading term $(H_2)_{\alpha=0}$ expresses the partial enthalpy as it would be in the absence of solvation-complex formation and whose further terms (when α departs from zero) express $\alpha(\Delta_c H_n + \Delta_c H_e)$ for complex formation. An analogous treatment applied to S and S_2 yields Eq. (11.31b).

$$H_2 = (H_2)_{\alpha=0} + \alpha(\Delta_c H_n + \Delta_c H_e) \qquad (11.31a)$$

$$S_2 = (S_2)_{\alpha=0} + \alpha(\Delta_c S_n + \Delta_c S_e) \qquad (11.31b)$$

The concentration-dependence of S_2 in Eq. (11.31b) transfers into $(S_2)_{\alpha=0}$. $\Delta_c S_n$ is independent of concentration since A is the solvent. Enthalpy–entropy compensation applies to the solvent-reorganization part of $(H_2)_{\alpha=0}$ and $(S_2)_{\alpha=0}$ and to the environmental part of complex formation: $\Delta_c H_e = \Delta_c S_e$.

Equation (11.30) and its analog for S_2 can be simplified without major error by introducing the following approximations:

(i) $s_{xa} = s_x + s_a - 2$, where s_a is the neighbor number to a solvent molecule.

(ii) $s_{xa} H_{A\backslash xa} = (s_a - 1)H_{A\backslash a} + (s_x - 1)H_{A\backslash x}$.

In (i), formation of the A–X bond displaces one solvent molecule each from A and X. In (ii), the partial enthalpies of the solvent molecules depend on interactions with direct neighbors only. On introducing these approximations and some steps of algebra we obtain

$$H_2 = (H_2)_{\alpha=0} + \alpha[H_{[A-X]\backslash a} - (H_{X\backslash a} + H_{A\backslash x})] \qquad (11.32a)$$

$$S_2 = (S_2)_{\alpha=0} + \alpha[S_{[A-X]\backslash a} - (S_{X\backslash a} + S_{A\backslash x})] \qquad (11.32b)$$

The bracketed terms in Eqs. (11.32a) and (11.32b) represent, respectively, the nominal ΔH and ΔS for the formation of the solvated complex [A–X] from the encounter pair [A, X].

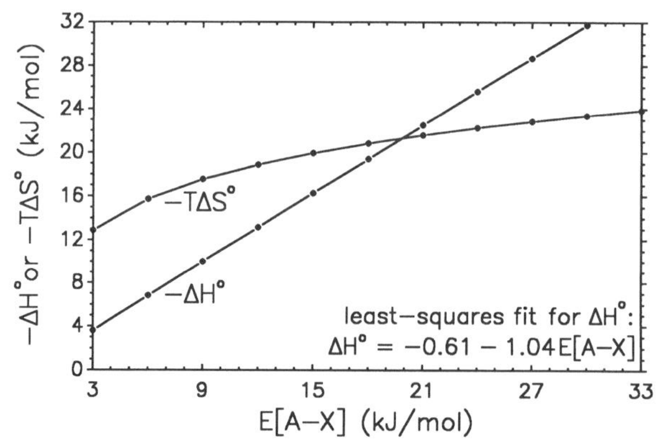

Figure 11.6. Propensity plots of ΔH^0 and $T\Delta S^0$ for the process, A\a + X\a $\rightleftarrows$ [A–X], versus E[A–X]. Based on Eqs. (10.13) and (10.14).

To show that ΔH_n and ΔS_n for the nominal process, [A, X]\a $\rightleftarrows$ [A–X]\a, are *not* constrained by enthalpy–entropy compensation, we will apply the propensity model developed in Chapter 10, using especially Eqs. (10.13) and (10.14). This model, it will be recalled, is designed for caged molecules in liquids and thus simulates the \a environments. The results of the calculations are plotted in Fig. 11.6. The plots show the following: (i) As the magnitude of the bond-dissociation energy E[A–X] increases, the changes in ΔH^0 are greater than those in $T\Delta S^0$. The average ratio, $\delta(T\Delta S^0)/\delta(\Delta H^0)$, is 0.35 in Fig. 11.6. (ii) ΔH^0 is a nearly linear function of E[A–X], with a slope near unity. (iii) Since $T\Delta S^0$ changes less than ΔH^0, there is only partial compensation, and ΔG^0 tends to track ΔH^0, which in turn tracks E[A–X].

SOLVATION MECHANISMS

Knowing a solvation mechanism is equivalent to knowing the solvation shell, just as knowing a reaction mechanism is equivalent to knowing the transition state. In the cage model, given the packing fraction f of the liquid, Eq. (9.8) specifies that the time-average cage is characterized by two variables: the mean neighbor number s (often used with a subscript, such as s_X) and the cage radius b. These variables relax to equilibrium on different time scales. The neighbor number relaxes on the time scale on which a neighbor molecule enters or leaves its own cage, which is about 10^{-10}–10^{-11} s. The cage radius, on the other hand, relaxes on the 10^{-12}–10^{-13} s time scale of the Brownian correlation time. Therefore, as s_X varies, b tracks along so that its value is the equilibrium value consistent with the instantaneous s_X, and $y_b = \partial G/\partial b = 0$. It is this property that allows us to omit b as a separate variable in the thermodynamic formulation. On the other hand, solvation mechanisms are more demanding, and there will be occasions when b and s_X are considered separately.

Solvation shells of course obey the laws and theorems of thermodynamics. Of special relevance is Eq. (11.15)—a manifestation of the Correspondence Theorems—which is repeated below.

At constant T and P and for $y_x = y_b = 0$,

$$G_1 = G_{A\backslash a} = G_{A\backslash x} \tag{11.15}$$

According to Eq. (11.15), the A\x subspecies of the solvent relaxes to a state in which $G_{A\backslash x}$ equals $G_{A\backslash a}$, whose variability is minor: At equilibrium, $G_{A\backslash a}$ equals G_1 of the formal solvent component; and in dilute solutions, $G_1 = G_1^0 - RTm_2/m_1$ [Eq. (3.12)], with $m_2 \ll m_1$. In short, $G_{A\backslash x}$ relaxes essentially to G_1^0, the fixed standard free energy of the pure solvent.

This constraint is remarkable. To see why, consider a solute–solute cluster consisting of one solute molecule X and a few solvent molecules A in the gas phase. Such a cluster may reasonably be described as a "solvation shell" transposed to the gas phase and, when it can be characterized experimentally, as a model for a solvation shell of the same composition in a liquid. But that reasoning is wrong! Even though the few molecules of A adjacent to X in a gas-phase cluster are analogous to A\x in a liquid, their partial free energy $G_{A\backslash x}$ (cluster) relaxes *without constraint*, to whatever value minimizes the free energy of the cluster. While in a liquid, $G_{A\backslash x}$ must relax to $G_{A\backslash a}$, whose value is predetermined and essentially fixed, being equal to $G_1 \approx G_1^0$ of the solvent.

Figure 11.7 tries to explore the ramifications of this constraint in a liquid.

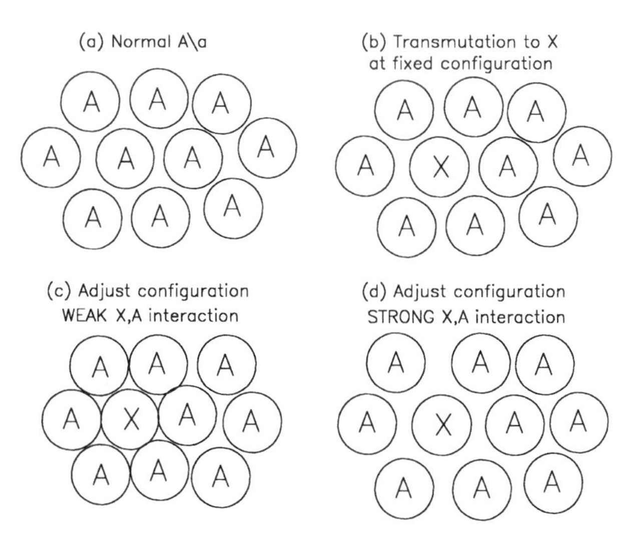

Figure 11.7. Implications of the constraint, $G_{A\backslash x} = G_{A\backslash a}$: Conceptual steps in the formation of A\x and X\a cages. (See text.) Note that the weak interaction produces the dense solvation shell.

The figure shows two hypothetical processes, (a) $\to$ (b) $\to$ (c) and (a) $\to$ (b) $\to$ (d). Part (a) shows a representative configuration for A\a. To proceed to (b), we commission a wizard to transmute one of the A molecules to an X molecule of similar size, *without otherwise changing the configuration*! Since the arrival of X in place of A changes the van der Waals interaction, the configuration attained in (b) is not an equilibrium and normal thermal relaxation ensues, leading either to (c) or (d). Given the propensities in Fig. 11.6 we will assume that the direction of the change in free energy follows that of the change in potential energy. Thus, if the A, X interaction in (b) is weaker than the corresponding A, A interaction in (a), $G_{A\backslash x}$ right after transmutation lies above $G_{A\backslash a}$, and the configuration must change so that the potential energy decreases. This is accomplished in part (c) by letting $r_{A,X}$, the distance between the centers of A and X, decrease. Conversely, if the A, X interaction in (b) is stronger than the corresponding A, A interaction in (a), $G_{A\backslash x}$ right after transmutation lies below $G_{A\backslash a}$ and $r_{A,X}$ increases, as indicated in part (d).

These predictions reflect displacements on the pair-potential in Fig. 11.8. The initial $r_{A,A}$ distance is placed near the inflection point. If, after the fast transmutation from (a) to (b), the unrelaxed $G_{A,X}$ is too high—that is, if the active van der Waals center in X is weaker than that in A—$G_{A,X}$ decreases to $G_{A,A}$ by a reduction in pairwise distance. If the unrelaxed $G_{A,X}$ is too low, $G_{A,X}$ rises to $G_{A,A}$ by an increase in pairwise distance. These schemes are practical when the disparity between the unrelaxed value of $G_{A,X}$ and the equilibrium value of $G_{A,A}$ is moderate or small. When the disparity is great, alternative mechanisms may be required. For example, in the case of a strong linkage interaction between X and A, the interacting centers might form a molecular complex. That would satisfy the affinity for enhanced interaction and allow the remaining van der Waals interactions to relax to equilibrium by small adjustments of the type (b) $\to$ (c)

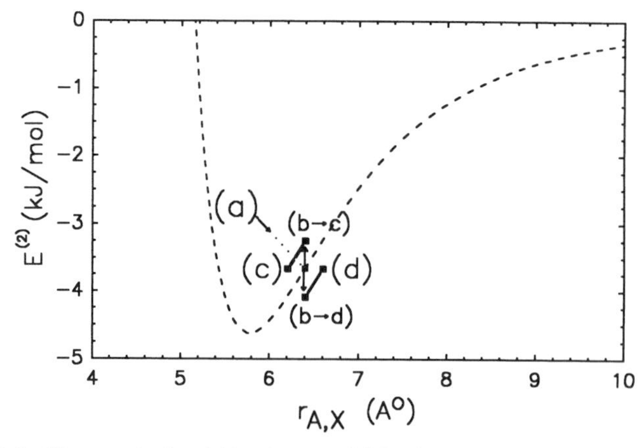

Figure 11.8. Changes in the A,X pair potential for the conceptual steps in Fig. 11.7.

or (b) → (d). On the whole, Fig. 11.7 opposes the axiom that weak interactions yield long equilibrium distances and that strong interactions yield short equilibrium distances. Here they yield just the opposite!

Partial Volumes

Since the solvation mechanisms inferred from the Correspondence Theorem (11.15) stress contractions and expansions of solvation shells, evidence from partial volumes of dilute solutes might be informative. Table 11.2 therefore lists the partial volumes V_2 for several dilute solutes in a number of solvents. According to ideal-solution theory, V_2 should be independent of the solvent, being always equal to V_2^0, the molar volume of the pure liquid solute at the given T and P—but that theory clearly does not fit the data. For a solvation model based on composition tree (T-11.1), a derivation analogous to that of Eq. (11.11) yields Eq. (11.33):

$$V_2 = V_{X\backslash a} + s_X(V_{A\backslash x} - V_{A\backslash a}) \qquad (11.33)$$

Table 11.2. Solvent effects on V_2 for selected solutes, 298 K

Solute: Cryptand-2,2,2.[h] *Solvent*, V_2 (ml/mol)[a]: Water, 316.5[a], 315.1[b]; methanol, 322.5; ethanol, 323.0; 1-propanol, 322.1; 2-propanol, 325.4; 1-butanol, 329.1; $CHCl_3$, 333.8; CH_3CN, 329.9; $(CH_3)_2SO$, 340.8; butylmethylketone, 345.1; benzene, 343.9; cyclohexane, 346.3; cyclohexylbenzene, 350.7; tetrahydronaphthalene, 341.5; hexane, 326.9; heptane, 336.5; octane, 341.1; decane, 352.9.

Solute: Argon. *Solvent*, V_2 (ml/mol)[c,d]: Water, 32; benzene, 43; CCl_4, 44; CS_2, 45; isooctane, 50; perfluoroheptane, 54.

Solute: Methane. *Solvent*, V_2 (ml/mol)[c–f]: Water, 37; methanol, 52; acetone, 55; methyl acetate, 53; benzene, 52; chlorobenzene, 49; CCl_4, 52; CS_2, 56; *n*-heptane, 55; isooctane, 57; perfluoroheptane, 68.

Solute: Ethane. *Solvent*, V_2 (ml/mol): Water, 41[g]; benzene, 73[g], 67[f]; CCl_4, 61[g], 67[f]

[a] Letcher and Mercer-Chalmers [1993].
[b] Zielenkiewicz et al. [1993].
[c] Cited by Hildebrand and Scott [1950].
[d] Smith and Walkley [1962].
[e] Kritchevsky and Ilinskaya [1945].
[f] Horiuti [1931].
[g] Cited by Ben-Naim [1974].
[h] Cryptand-2,2,2 = $N\begin{matrix} \diagup (C_2H_4O)_2C_2H_4 \diagdown \\ -(C_2H_4O)_2C_2H_4- \\ \diagdown (C_2H_4O)_2C_2H_4 \diagup \end{matrix}N$

Accordingly, solvation affects V_2 through $V_{X\backslash a}$ and $V_{A\backslash x}$, both of which decrease when the A,X distances in the solvation shell contract. A substantial reduction in V_2 therefore signals a tight solvation shell. In terms of Figs. 11.7 and 11.8, V_2 should contract if, after the rapid transmutation a $\rightarrow$ b, the solvent–solute interaction is weaker than the original solvent–solvent interaction. In agreement with this prediction, the experimental results for V_2 listed in Table 11.2 for cryptand-2,2,2, argon, methane, and ethane consistently are small in water and the alcohols—solvents in which the solvent–solvent interaction is relatively strong due to hydrogen-bonding.

In water, for example, V_2 is smaller by 15–30 ml/mol than in solvents that form regular solutions. When the molar-shift model is rendered as a solvation model (Fig. 11.4 $\rightarrow$ Fig. 11.5), the water molecules adjacent to a nonpolar solute molecule such as CH_4 are represented as W\4w + x\, and this is the water species whose apparent volume is curtailed. Note that the CH_4 molecule is a weak center of van der Waals attraction.

When solutions of methane or benzene in water are simulated in mechanical models consisting of one nonpolar molecule and 100–500 water molecules, and thermodynamic properties are calculated by statistical–mechanical averaging over statistically probable configurations, the regions around the methane or benzene molecules have a strikingly high density of water molecules [Alagona and Tani, 1980; Swaminathan et al., 1978; Ravishanker et al., 1984]. Although the statistical–mechanical approach does not classify the solvent molecules into molecular species and thus does not introduce Eq. (11.15) explicitly, the essence of Eq. (11.15) is clearly implied.

12

THERMODYNAMICS OF IONIC SPECIES

To a chemist, an ion is an atom or molecule bearing an electric charge. The bonds in molecular ions follow the same rules as those in molecules, and ions and molecules enjoy equal logical status in structure–reactivity correlations. Those properties that distinguish ions from uncharged molecules—the ability to transport electric charge and the relatively long range of charge–charge interactions—are basically physical.

While ionic species fit comfortably into chemistry, they do not sit easily in chemical thermodynamics. One might say that the partnership between chemistry and thermodynamics is here under strain. The reason is that normal macroscopic electrification (by addition or removal of ions or electrons) takes place on a scale that is well below the detection limit of analytical chemistry—the violations of conservation of charge go unnoticed. Electrifications that might be stoichiometrically significant generate unsustainably high voltages—voltages that are well above the breakdown thresholds for earthly matter.

This electrification paradox introduces two constraints that chemists tend to find awkward. First, regardless of the operational time scale, formal components must be electrically neutral; ionic species cannot serve as components. Second, transfer processes involving ionic species must be concerted transfers of two or more species so that stoichiometric electroneutrality is preserved. These constraints make it impossible to *measure* the thermodynamic properties of individual ionic species, and thus they deny us the kind of information for ionic species that is in principle available for electrically neutral species.

In this chapter we will examine the constraints and explore models through whose use their impact can be cushioned. We shall also examine the manifestations of interionic attractions.

THERMODYNAMIC COUPLING

We begin by considering the magnitude of normal macroscopic electrification. In electrochemistry and physiology, a change in potential of 100 mV is highly significant. To translate that change into numbers of ions and electrons, let us charge a capacitor consisting of two concentric spheres: (1) an inner sphere with a radius of 0.01 m and (2) an outer sphere with a radius of 0.1 m. The air capacitance of this device is 1.23 pF. To produce a potential difference of 100 mV then requires an electric charge of 1.23×10^{-13} coulomb ($0.1 \times 1.23 \times 10^{-12}$), which is equivalent to 770,000 univalent ions ($1.23 \times 10^{-13} \times 6.02 \times 10^{23}/96485$). By comparison, detecting the endpoint in an acid–base titration with a sensitive color indicator requires the transfer of at least 10^{17} protons, or more than 10^{-2} coulombs. Thus, in terms of ions and molecules, electrification is significant at a level that is at least 11 orders below the level of chemical detection.

Conversely, if the charge on the capacitor were large enough to be chemically detectable, the proportional voltage would be 10^{10} V. This is enormous, much greater than any known dielectric-breakdown threshold. Indeed, an ion accelerator with a capability of 10^{10} eV is an atom smasher. Even if the ionic component were added only in infinitesimal amount, the "infinitesimal" would be in the macroscopic sense, and the electrification would still be intolerable. It is clear, therefore, that formal components must be electrically neutral. When ions exist in a chemical system, they are not added directly and individually, but are generated in situ by ionic dissociation or other ionogenic reactions of the components.

Modes of Coupling

To derive the modes of thermodynamic coupling for electroneutrality, we begin with a simple example: a solution of potassium chloride in water. The composition tree is (T-12.1).

Composition Tree (T-12.1)

Components:	Water (1, n_1)	Potassium chloride (2, n_2)	
Molecular species:	H_2O, n_1	K^+, $n_+ = n_2$	Cl^-, $n_- = n_2$

To represent the free energy in (T-12.1) on the level of components we write $G = n_1 G_1 + n_2 G_2$.

To represent it on the level of molecular species we write

$$G = n_1 G_{H_2O} + n_2 G_{K^+} + n_2 G_{Cl^-}$$

Since n_1 and n_2 are independent variables in the two equations for G, their coefficients must be equal—hence Eqs. (12.1):

$$G_1 = G_{H_2O} \tag{12.1a}$$

$$G_2 = G_{K^+} + G_{Cl^-} \tag{12.1b}$$

Equation (12.1b) illustrates the coupling of ionic species for electroneutrality. Even though K^+ and Cl^- react and transport electricity as separate entities, their partial free energies are coupled and add up to the observed partial free energy of the potassium chloride component.

Coupling for electroneutrality pervades the thermodynamics of ionic species. For example, the partial enthalpy and entropy of the KCl component in (T-12.1) are expressed by Eqs. (12.2), which are mathematically analogous to (12.1b).

$$H_2 = H_{K^+} + H_{Cl^-} \tag{12.2a}$$

$$S_2 = S_{K^+} + S_{Cl^-} \tag{12.2b}$$

The need for coupling eliminates some otherwise admissible ionic processes. For example, the ionic phase transfer

$$Na^+(g) \rightleftarrows Na^+(\text{methanol})$$

disrupts electroneutrality in both phases and is inadmissible. On the other hand, either of the coupled processes

$$\text{(i)}\quad Na^+(g) + CH_3S^-(g) \rightleftarrows Na^+(\text{methanol}) + CH_3S^-(\text{methanol})$$

or

$$\text{(ii)}\quad Na^+(g) + Ag^+(\text{methanol}) \rightleftarrows Na^+(\text{methanol}) + Ag^+(g)$$

preserves electroneutrality in both phases and can be realized, at least in principle. The respective constraints are

$$\text{(i)}\quad [G_{Na^+} + G_{CH_3S^-}](g) = [G_{Na^+} + G_{CH_3S^-}](\text{methanol})$$

$$\text{(ii)}\quad [G_{Na^+} - G_{Ag^+}](g) = [G_{Na^+} - G_{Ag^+}](\text{methanol})$$

In summary, the parent components for ionic species must be electrically uncharged salts or ionogens. Thermodynamic processes involving ions must preserve electroneutrality in each phase of the system. While thermodynamic coupling for electroneutrality couples the partial properties of ionic species, it does not couple the ions themselves: The ions continue to act as separate, chemically distinct entities.

ELECTROCHEMICAL POTENTIAL

The electrochemical potential (ECP) of an ionic species in a homogeneous phase at constant T and P is the ionic analog of the partial free energy of a nonionic species—that is, of the chemical potential of a nonelectrolyte. It expresses the work per mole of adding an infinitesimal quantity of the ionic species to the homogeneous phase at constant T and P. (To avoid adding a macroscopically infinitesimal quantity to a single phase, a microscopically infinitesimal quantity might be added to each phase in a canonical ensemble.)

The electrochemical potential $(\mathrm{ECP})_i$ of the ith ionic species is a sum consisting of electric work ($W_{i,\mathrm{el}}$) and chemical work ($W_{i,\mathrm{chem}}$), Eq. (12.3):

$$(\mathrm{ECP})_i = L[W_{i,\mathrm{el}} + W_{i,\mathrm{chem}}], \quad \text{per mole} \tag{12.3}$$

$$(\mathrm{ECP})_i = z_i F\phi + G_i \tag{12.4}$$

The electric work is a function of the electric potential ϕ and can be defined only subject to special conditions, which will be described. The chemical work, as we shall show, is identical to the partial free energy of the ion when the ion is coupled into a component or in a process to preserve electroneutrality. For example, for potassium ion, $W_{\mathrm{K^+,chem}} = G_{\mathrm{K^+}}$, the partial free energy attributed to potassium ion in Eq. (12.1b). The result for $(\mathrm{ECP})_i$ is stated in Eq. (12.4).

Electric Potential

The basic problem with electric potentials is the ever-presence of adventitious electrification. Charge is associated with natural events such as lightning, cosmic-ray showers, or terrestrial radioactivity. And charge results from human activity: For example, your personal electrification by random friction on a dry day can be strong enough for its discharge to be painful. The adventitious charge eventually disperses into the local ground, but the ground is only a sink. Its absolute potential is not known. The earth's surface is not a good enough conductor for random differences in local potentials to be quickly relaxed. Thus there is no globally valid fixed reference point for defining an absolute electric potential.

Differences in electric potential between points can of course be measured, provided that (1) the points are embedded in an electrically contiguous system and (2) adventitious fluctuations in the potential differences relax on a time scale that is short compared to the response time of the measuring instrument. When these conditions are met, any point within the system can serve as a reference point whose potential ϕ may be assigned a convenient fixed value, which may be zero. And ϕ expresses the work done when a unit positive charge is transferred reversibly from a point where $\phi = 0$, to the given point. The systems considered in the rest of this chapter will satisfy these conditions.

Returning to Eq. (12.3), the electric work $W_{i,\mathrm{el}}$ then equals $z_i e\phi$ per ion, and $LW_{i,\mathrm{el}}$ equals $z_i F\phi$ per mole, where z_i is the (algebraic) charge number of the ion, e is the elementary (proton) charge, and F is the Faraday constant, 96485 coulombs per mole of charge.

The Term for Chemical Work

We now wish to show that (1) the chemical work per mole, $LW_{i,\mathrm{chem}}$, is identical to G_i, the partial free energy of the ionic species defined in the preceding section, where ionic species are coupled to preserve electroneutrality, and (2) G_i is independent of the electric potential ϕ. Because of (2), G_i is fully analogous to the partial free energy of a nonelectrolyte.

PROOF THAT $LW_{i,\mathrm{chem}}$ IS G_i. Let component 2 be any univalent strong electrolyte C^+, X^-; and let the component exist in a solution where C^+, X^- is dissociated into free ions. Then $G_2 = G_{C^+} + G_{X^-}$; and because the ions are dissociated, $G_{C^+} = (\mathrm{ECP})_{C^+} = +F\phi + LW_{C^+,\mathrm{chem}}$ and $G_{X^-} = (\mathrm{ECP})_{X^-} = -F\phi + LW_{X^-,\mathrm{chem}}$. Hence

$$G_2 = G_{C^+} + G_{X^-} = L[W_{C^+,\mathrm{chem}} + W_{X^-,\mathrm{chem}}]$$

This proves that the sum of the ionic partial free energies is equal to the sum of the chemical-work terms. To show that this is true for each ionic species separately, we recall that the symbol C^+ may stand for any univalent cation, and X^- may stand for any univalent anion. Thus C^+ and X^- are independent variables. The cation terms and anion terms therefore must be independently equal, Eq. (12.5):

$$G_{C^+} = L(W_{C^+,\mathrm{chem}}); \quad G_{X^-} = L(W_{X^-,\mathrm{chem}}) \tag{12.5}$$

Q.E.D.

PROOF THAT THE G_i's OF IONIC SPECIES ARE INDEPENDENT OF ϕ. At equilibrium, the electric potential is uniform throughout the homogeneous phase (except for short-lived insubstantial gradients when adventitious charge arrives). The ions therefore experience no net electric force, *and do not sense the existence of electrification*. This is true regardless of the magnitude of the potential. Q.E.D.

ACTIVITY COEFFICIENTS

One of the distinctive features of ionic species is that charge–charge interactions among the ions are of relatively long range. As a result, deviations from dilute-solution behavior are significant at lower concentrations than for electrically uncharged solutes. The expression for the partial free energy G_i of an ionic solute therefore routinely includes a term involving the *activity coefficient*, γ_i, defined in Eq. (3.8a), which is repeated below as (12.6):

$$G_i = G_i^0 + RT \ln c_i + RT \ln \gamma_i \qquad (3.8a) = (12.6)$$

As indicated earlier in Fig. 3.4, G_i^0 is defined so that $\gamma_i \to 1$ as $c_i \to 0$. The term $RT \ln \gamma_i$ thus becomes negligible at high dilutions. The standard partial free energy G_i^0 for an ionic species therefore has the same significance as for an uncharged species: It expresses the limit of $(G_i - RT \ln m_i)$ as $m_i \to 0$ and thereby provides a normalized index of thermodynamic stability at constant T and P.

When Eq. (12.6) is applied to the ionic species K^+ and Cl^-, we obtain Eqs. (12.7): (12.7a) and (12.7b) for the ionic species and (12.7c) and (12.7d) for the partial free energy G_2 of the KCl component. [Compare Eq. (12.1a).] To obtain Eq. (12.7d) we note that $m_{K^+} = m_{Cl^-} = m_2$ and identify $(G_{K^+}^0 + G_{Cl^-}^0)$ with the standard partial free energy, G_2^0, of the KCl component.

$$G_{K^+} = G_{K^+}^0 + RT \ln m_{K^+} + RT \ln \gamma_{K^+} \qquad (12.7a)$$

$$G_{Cl^-} = G_{Cl^-}^0 + RT \ln m_{Cl^-} + RT \ln \gamma_{Cl^-} \qquad (12.7b)$$

$$G_2 = G_{K^+}^0 + G_{Cl^-}^0 + RT \ln(m_{K^+} m_{Cl^-}) + RT \ln(\gamma_{K^+} \gamma_{Cl^-}) \qquad (12.7c)$$

$$G_2 = G_2^0 + 2RT \ln m_2 + RT \ln(\gamma_{K^+} \gamma_{Cl^-}) \qquad (12.7d)$$

Although the molalities of K^+ and Cl^- in aqueous KCl are equal, the partial free energies are *not*: $G_{K^+} \neq G_{Cl^-}$. This inequality follows because the only active constraint (besides coupling for electroneutrality) is the Correspondence Theorem, which does not require that G_{K^+} and G_{Cl^-} be equal.

PROOF. Since the formal component 2 is KCl, the Correspondence Theorem states that $G_2 = G_{KCl}$. This is paradoxical because aqueous KCl is a strong electrolyte, and the stoichiometric presence of KCl molecules and K^+, Cl^- ion pairs is neglected in (T-12.1). The thermodynamic presence is real, however, and we may invoke the equilibrium: $K^+ + Cl^- \rightleftarrows KCl$. Therefore at chemical equilibrium, $G_{K^+} + G_{Cl^-} = G_{KCl} = G_2$. The Correspondence Theorem thus duplicates Eq. (12.1) without imposing new conditions. It constrains the *sum* of the partial free energies, but not the individual values. Q.E.D.

Given that $G_{K^+} \neq G_{Cl^-}$, it follows from Eqs. (12.7a) and (12.7b) that $[G^0_{K^+} + RT \ln \gamma_{K^+}] \neq [G^0_{Cl^-} + RT \ln \gamma_{Cl^-}]$. Furthermore, as $m_2 \to 0$, both $\ln \gamma_{K^+}$ and $\ln \gamma_{Cl^-}$ approach zero. The inequality $G_{K^+} \neq G_{Cl^-}$ therefore implies a similar inequality for standard partial free energies: $G^0_{K^+} \neq G^0_{Cl^-}$. For any univalent electrolyte C^+, X^-, the preceding results derived for KCl are restated in more general form in Eqs. (12.8):

$$G_{C^+} \neq G_{X^-} \tag{12.8a}$$

$$G^0_{C^+} \neq G^0_{X^-} \tag{12.8b}$$

Equations (12.8) state that the partial free energies of the ions in an electrolyte are independent and specific. On taking first or higher derivatives with respect to T and/or P we find that this is true also for the partial entropy, volume, heat capacity, and other partial properties of ionic species. Evidently, the coupling imposed by electroneutrality on the *processes* of ionic species does not extend to the *chemical work* terms in the ionic electrochemical potentials, nor does it extend to their partial derivatives. When such properties are available, they have the attributes of ordinary chemical properties, and we may use and interpret them just like chemical properties of nonelectrolytes.

Catchall Property

As stated previously, the partial free energies of ionic species are represented by Eq. (12.6) even in dilute solutions, because charge–charge interactions are of relatively long range and cause departures from the dilute-solution model even at low ionic concentrations. However, while the term $RT \ln \gamma_i$ in Eq. (12.6) indeed absorbs the effect of the charge–charge interactions, it also absorbs everything else that might cause a deviation from $G_i = G^0_i + RT \ln m_i$. For example, if the ions associate, dissociate, or otherwise depart from the molecular formula assigned to them, the ensuing free-energy error (with algebraic sign) simply adds to the effect of the charge–charge interactions and becomes part of $RT \ln \gamma_i$.

When gathered up in $RT \ln \gamma_i$, the various terms associated with departures from the dilute-solution or composition model have lost their identity and can be resolved only by theoretical prediction. Fortunately, the effects of charge–charge interactions can be predicted fairly accurately, especially in the domain where the Debye–Hückel theory applies. Large deviations of the experimental activity coefficients from prediction then indicate that other departure mechanisms are also present. In such cases, the *sign* of the unpredicted residue of $RT \ln \gamma_i$ may be revealing. If it is negative—if γ_i is too small—the ionic species is more stable than the predictive model assumes. If it is positive, the model omits a destabilizing factor.

Activities and Mean-Ion Properties of Electrolytes

The following definitions are largely due to G. N. Lewis and his colleagues [Lewis and Randall, 1921, 1923]. The *activity* a_i is defined by $a_i = m_i\gamma_i$, where m_i is the molal concentration of the ith species. In view of Eq. (12.6), G_i is therefore related to the activity a_i according to Eq. (12.9). The solvent may be a single component or a homogeneous mixture of fixed composition.

$$G_i = G_i^0 + RT \ln a_i \qquad (12.9)$$

Because $\gamma_i \rightarrow 1$ as the concentration of all solutes approaches zero, a_i approaches m_i at very high dilution. On comparing Eq. (12.9) with the dilute-solution model, $G_i = G_i^0 + RT \ln m_i$, we find that there is mathematical analogy between a_i (for a real solution) and m_i (in the corresponding dilute-solution model). This analogy allows us to translate the free-energy equations obtained in Chapter 3 for the dilute-solution model into the language of activities and activity coefficients, simply by writing a_i in place of m_i. The substitution of a_i for m_i streamlines the mathematics without, however, simplifying the real problem, because we must still know the values of, and the underlying mechanisms for, γ_i.

Because of thermodynamic coupling for electroneutrality, one does not measure the individual activity coefficients of ionic species, but only the products or quotients of activity coefficients for electrically neutral combinations. For example, Eq. (12.7d) relates the partial free energy of the KCl component to the product, $\gamma_{K^+}\gamma_{Cl^-}$. To describe the activity coefficients of the ions produced in solution from their salts it is convenient to introduce the *mean-ion activity coefficient* $\gamma_\pm$ and the *mean-ion molality* $m_\pm$. For ions from 1 : 1 salts (such as K^+ and Cl^- from KCl, or Mg^{2+} and SO_4^{2-} from $MgSO_4$), the mean-ion molality is defined by $m_\pm = (m_+m_-)^{1/2}$, and the mean-ion activity coefficient is defined by $\gamma_\pm = (\gamma_+\gamma_-)^{1/2}$. The "+" subscript denotes the cation, and "−" denotes the anion. In keeping with this notation, the cation activity is defined by $a_+ = m_+\gamma_+$, and the anion activity is defined by $a_- = m_-\gamma_-$. The mean-ion activity is defined by $a_\pm = (a_+a_-)^{1/2}$.

For unsymmetrical valence types the definitions are logical extensions. Let C_xA_y be the chemical formula of the salt, and let C_xA_y dissociate to $xC^{+y} + yA^{-x}$. The mean-ion molality then is defined by

$$m_\pm = [m_+^x \cdot m_-^y]^{1/(x+y)}$$

The mean-ion activity coefficient is defined by

$$\gamma_{\pm} = [\gamma_{+}^{x} \cdot \gamma_{-}^{y}]^{1/(x+y)}$$

And the mean-ion activity is defined by

$$a_{\pm} = [a_{+}^{x} \cdot a_{-}^{y}]^{1/(x+y)}, \quad \text{where } a_{+} = m_{+}\gamma_{+} \text{ and } a_{-} = m_{-}\gamma_{-}$$

By virtue of these definitions, the partial free energy of the component, C_xA_y, is given by Eqs. (12.10):

$$G_{C_xA_y} = xG_{C^{+y}} + yG_{A^{-x}}$$

$$= G^{0}_{C_xA_y} + (x+y)RT \ln m_{\pm} + (x+y)RT \ln \gamma_{\pm} \qquad (12.10a)$$

$$= G^{0}_{C_xA_y} + (x+y)RT \ln a_{\pm} \qquad (12.10b)$$

Because there is no restriction on the nature of C and A, nor on the values of x and y, Eqs. (12.10) are general. In the absence of common-ion salts, $m_{+} = xm_2$ and $m_{-} = ym_2$, where m_2 is the formality of the salt, in formula-weights per kilogram of solvent.

Some Results for Aqueous Solutions

Measurements of ionic activity coefficients are notably indirect. Physically independent methods are based on the melting point of the solvent in contact with the solution (freezing-point lowering), the vapor pressure of the solvent over the solution, the vapor pressure of a volatile ionogenic solute (such as HCl) over the solution, and the resting voltage of an electrochemical cell. These measurements are technically difficult, heightened by the fact that the definition ($\gamma_{\pm} \to 1$ as $m_{\pm} \to 0$) requires at least some very low concentrations. Many of the measurements on aqueous solutions date back to the pre-electronic age of tap switches and galvanometers, and attest to the problem-solving power of imaginative craftsmanship.

In addition to reporting measurements of their own, G. N. Lewis and his colleagues surveyed the results obtained by others and showed that rigorous thermodynamic calculations including activity coefficients unify the various results [Lewis and Linhart, 1919; Lewis and Randall, 1921]. Examples for NaCl in water are shown in Fig. 12.1. The activity coefficients based on freezing-point lowering are the composite of three laboratories. The lowering measures the partial free energy G_1 of the solvent component at the melting temperature, which then is interpreted by the Gibbs–Duhem equation ($n_1\, dG_1 + n_2\, dG_2 = 0$). Let θ denote the observed lowering, let λ be the colligative molar lowering (1.858°C · kg/mol for water), and let x, y be stoichiometric coefficients as defined in Eq. (12.10). Then ln $\gamma_{\pm}$ is obtained from the data by integrating the

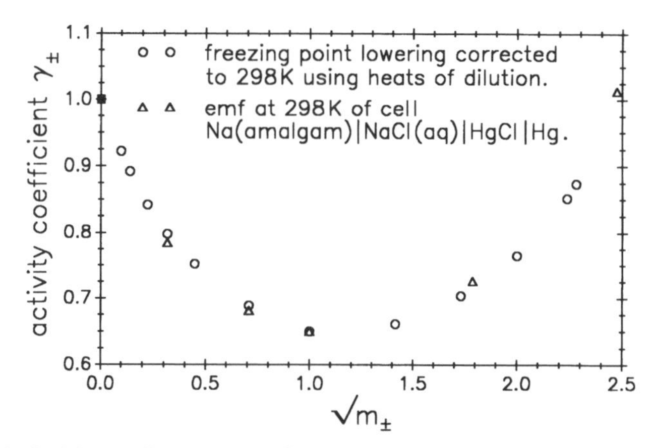

Figure 12.1. Activity coefficients of NaCl in water at 25°C: consistency of two methods [Lewis and Randall, 1921].

equation

$$d \ln \gamma_\pm = -d \ln m_2 + \frac{d\theta}{(x + y)\lambda m_2}$$

This procedure gives activity coefficients at the melting point, which are converted to the desired temperature by thermodynamic calculations employing heats of dilution [Lewis and Randall, 1921].

The freezing-point lowering is typically small, only 0.0343°C for the 0.01 formal NaCl solution in Fig. 12.1. To achieve an accuracy of 0.0001°C with a 25-junction copper–constantan thermopile, the salt solution in contact with crushed ice in a Dewar flask is placed side by side to pure water with crushed ice in a similar Dewar flask, and each natant liquid is pumped slowly over the melting ice. When temperature equilibrium is reached, a sample of the salt solution is withdrawn to measure the corresponding salt concentration [Adams, 1915].

The activity coefficients derived from emf data, in Fig. 12.1, are based on the cell Na(amalgam)|NaCl(aq)|HgCl|Hg. The normally explosive reaction of sodium with water is here suppressed by amalgamation. By convention the cell is written so that the anode, where oxidation occurs, appears on the left-hand side.

Electrochemical cells have the problem that the anode and its solution, and the cathode and its solution, are not at chemical equilibrium: They must be kept apart to eliminate spontaneous reaction. However, if the electrodes and their solutions are kept apart in separate containers, connected electrically by a salt bridge, precise interpretation is jeopardized by the vagaries of liquid junction potentials. This problem is bypassed by using cells without liquid junctions, such as the Na(amalgam) cell used here for NaCl. In cells without liquid junctions the anode and cathode are mounted in the same container and are in contact with the same solution. Mixing of anode and cathode materials by diffusion is slow (and can be minimized by keeping the apparatus as still as possible), but other problems arise. In the NaCl cell, for example, the HgCl material of the cathode is somewhat soluble, ~10^{-5} formal in pure water at 25°C. If dissolved material reaches the anode, the cell reaction HgCl(aq) + Na(amalgam) $\rightleftarrows$ NaCl(aq) + Hg occurs spontaneously and causes the emf to decrease. Promptness in making the measurements is important.

When the NaCl cell functions accurately, the cell emf is related to concentration by the Nernst equation, which is now written in the form

$$E = E^0 - (2RT/F)(\ln m_\pm + \ln \gamma_\pm)$$

where E^0 is the *standard* emf. To eliminate E^0 as an unknown, $\gamma_\pm$ was anchored to the freezing-point series at 1 formal NaCl. The other emf values in the series then gave independent results for $\gamma_\pm$. The values obtained for $\gamma_\pm$ in the two series differ by less than 0.2%.

The *shape* of the relationship between $\gamma_\pm$ and $m_\pm$ for NaCl in Fig. 12.1 is typical of many strong electrolytes. All electrolytes show an initial negative slope, which (we will find) is predicted accurately by the interionic attraction theory of Debye and Hückel. Like NaCl, many electrolytes show an increase in $\gamma_\pm$ at high concentrations. The increase in $\gamma_\pm$ signals a destabilization of the ions, partly because the ion–solvent interactions lower the dielectric constant, and partly because the supply of solvent molecules for ionic solvation shells becomes limited [Giese et al., 1970; Haggis et al., 1952; Stokes and Robinson 1948]. Conversely, when $\gamma_\pm$ continues to decrease at high formal concentrations, we may assume that ionic association to ion pairs or higher aggregates is significant, because the Stability Theorem (3.21) proves that the presence of additional species at equilibrium stabilizes a component.

By definition, as $m_\pm$ approaches zero, $\gamma_\pm$ approaches unity. Obtaining absolute values for mean-ion activity coefficients therefore requires an extrapolation to infinite dilution. In Fig. 12.1 this extrapolation is empirical and introduces a systematic error into $\gamma_\pm$ of about 0.5%. That error is largely eliminated when the extrapolation is guided by the Debye–Hückel theory within its tested range. Some mean-ion activity coefficients obtained in this way are compiled in Table 12.1 [Harned and Owen, 1958]. Among the univalent electrolytes, the results for KNO_3 decrease continuously and suggest significant ionic association.

Empirical Propensities

Among the propensities of mean-ion activity coefficients discovered before the advent of the Debye–Hückel theory, two are particularly important because they agree with key features of that theory [Lewis and Randall, 1921]. One is the

Table 12.1. Some mean-ion activity coefficients based on Debye–Hückel extrapolations, for various charge types in water at 25°C

m_2 [a]	LiCl	NaBr	KI	KNO_3	Na_2HPO_4	$BaCl_2$	Na_3PO_4
0.01	0.905	0.903	0.903	0.900	—	0.723	—
0.05	0.831	0.824	0.821	0.831	—	0.559	—
0.1	0.792	0.781	0.776	0.733	0.480	0.492	0.293
0.2	0.761	0.739	0.731	0.659	0.392	0.436	0.216
0.5	0.742	0.695	0.675	0.542	0.277	0.390	0.134
1	0.781	0.687	0.646	0.441	0.200	0.389	—
2	0.931	0.732	0.641	0.327	—	—	—
4	—	0.938	0.678	—	—	—	—

[a] Formality, in formula weight of electrolyte per kilogram water.
Source: Harned and Owen [1958].

ionic-strength principle; the other is the *square-root relationship* at low ionic strengths.

Ionic strength I is defined by Eq. (12.11), where m_i and z_i are the molality and algebraic charge number of any ith ionic species, and the summation extends over all ionic species.

$$I = \frac{1}{2} \sum m_i z_i^2, \quad \text{summed over all ionic species} \tag{12.11}$$

For example, assuming complete dissociation into ions, 0.01 f NaCl consists of 0.01 m $Na^+(z_+ = 1)$ and 0.01 m $Cl^-(z_- = -1)$; hence $I = \frac{1}{2}(0.01 + 0.01)) = 0.01$. Similarly, 0.001 f $BaCl_2$ consists of 0.001 m $Ba^{++}(z_+ = 2)$ and 0.002 m $Cl^-(z_- = -1)$; hence $I = \frac{1}{2}(0.004 + 0.002) = 0.003$.

Given Eq. (12.11) for the ionic strength, the *ionic-strength principle* states that at low ionic strengths ($I \lesssim .05$ m in water), the mean-ion activity coefficient $\gamma_\pm$ is nearly a colligative function of the ionic strength; that is, it is nearly independent of the nature of the ions that make up the ionic strength. For fully dissociated univalent salts, the ionic strength is equal to the formality. Thus at a formality of $\lesssim 0.05$, $\gamma_\pm$ for univalent strong electrolytes is expected to be a function of the formality and virtually independent of the nature of the electrolyte. The data in Table 12.1 support this.

For higher-valent salts where the ionic strength is a multiple of the formality, the values of $\gamma_\pm$ in Table 12.1 correspond to those of univalent salts at a multiple of the higher-valent formality. A closer comparison shows, however, that deviations from the ionic-strength principle are now more significant. This illustrates that the ionic-strength principle is only a propensity, not a law. While it paints the basic picture, it tends to miss the subtle shadings.

The square-root dependence is a rule for extrapolation to infinite dilution. It states that at low ionic strengths, $\gamma_\pm$ (and especially $\log \gamma_\pm$) are linear functions of the square-root of the ionic strength. An initial use of the rule is the square-root plot for NaCl in Fig. 12.1, which becomes linear at low formality. A more detailed test, involving the mean-ion solubility of thallous chloride, TlCl, in aqueous salt solutions is presented in Fig. 12.2 [Lewis and Randall, 1921].

By definition, the mean-ion solubility $s_\pm$ equals $(m_{Tl^+} \cdot m_{Cl^-})^{1/2}$ in the saturated solution, *even when a common-ion salt is present*, in which case $m_{Tl^+} \neq m_{Cl^-}$. The following proof shows that $1/s_\pm$ = constant $\cdot$ $\gamma_\pm$ for TlCl in the saturated solution.

PROOF. At equilibrium with solid TlCl at constant T and P,

$$\begin{aligned} G^0(\text{TlCl, solid}) &= G_+(\text{Tl}^+, \text{aq}) + G_-(\text{Cl}^-, \text{aq}) \\ &= G_+^0(\text{Tl}^+, \text{aq}) + G_-^0(\text{Cl}^-, \text{aq}) + 2RT\ \ln(s_\pm\gamma_\pm). \end{aligned}$$

The G^0's and RT are constant. Therefore, $s_\pm\gamma_\pm$ = constant. Q.E.D.

The plot of $-\log s_\pm$ versus $\sqrt{I}$ in Fig. 12.2 is therefore tantamount to a plot of $\log \gamma_\pm$ versus $\sqrt{I}$. It supports the principle of the ionic strength if the data points for all salts define the same relationship, and it supports the square-root dependence if that relationship is linear. In fact, the plot of Fig. 12.2 is both nearly monotonic and nearly linear, even though the nature of the added salt varies widely. Some are univalent; others are uni-bivalent, either with a bivalent cation or a bivalent anion. Some are common-ion salts that add either Tl^+ or Cl^-. Equally important, the plot is nearly linear and monotonic—but only if the abscissa is the square root of the ionic strength. When the same values of log

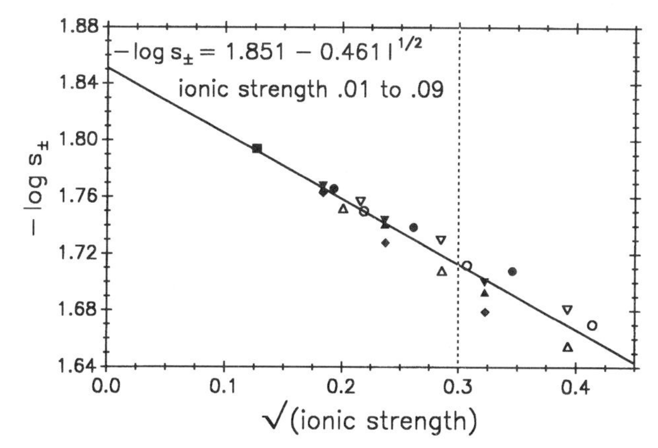

Figure 12.2. Mean-ion solubility, at 25°C, of thallous chloride in aqueous salt solutions. Symbol, added salt: Solid square, no added salt; solid circles, KNO_3; open circles, K_2SO_4; other solid symbols, uni-univalent common-ion salts (KCl, HCl, $TlNO_3$); other open symbols, uni-bivalent common-ion salts ($BaCl_2$, Tl_2SO_4) [Lewis and Randall, 1921].

$s_{\pm}$ are plotted against the molality or normality of the added salt, the deviations from a monotonic relationship are decisively greater.

DEBYE–HÜCKEL THEORY

Theoretical models in which the solvent is a continuous "medium" whose macroscopic properties, such as temperature, density, viscosity and electric polarization, can be defined are essentially macroscopic models. Infinitesimal volume elements in such models are infinitesimal on a macroscopic scale, because they include enough molecules so that statistically meaningful local averages exist [Mason and Weaver, 1929; Debye, 1945]. Accordingly, the work done on, or by, specific ions and molecules in such models is macroscopic work. As a result, electrostatic work is *free-energy work* rather than potential-energy work: $W_{\text{electrostatic}} = -\delta G$. Electrostatic work is the work of introducing electric charges or poles into a system, of moving and relocating electric charges within a system, and of creating, increasing or decreasing the intensities of electric fields [Duckworth 1960; Böttcher 1952; Frank 1955].

In this section we will consider the interionic-attraction theory as formulated by P. Debye and E. Hückel [1923; see also Robinson and Stokes, 1955]. The ions exist at macroscopic concentrations in a continuous medium with a definable temperature, density and dielectric constant; electrostatic work is identified with $-\delta G$. In later sections we will consider electrostatic models in which an ion transfers from one medium to another, and models in which a pair of ions forms an electrostatic ion pair.

Digression on Electrical Units

The three principal systems of electrical units are: practical, electrostatic, and electromagnetic. A brief overview is given in Table 12.2. Electrostatic units are anchored on Coulomb's law and employ the cgs system. Unit charge is defined so that the force f between point charges q and q' at a distance r in empty space is given by $f = qq'/r^2$. When f is in dynes and r is in centimeters, q and q' are in franklins, also called statcoulombs.

Practical (SI) units are defined for practical convenience and are in the mks system. The practical unit of charge, the coulomb, is relatively huge, amounting to three billion franklins, while the practical volt is only 1/300 statvolt. Coulomb's law in practical units is f (in empty space) $= (1/4\pi\eta_0) \cdot qq'/r^2$, where f is in newtons (1 newton $= 10^5$ dynes), r in meters, and q, q' in coulombs. The conversion factor is written in the form $1/4\pi\eta_0$, where η_0 is the *permittivity of free space*. It can be shown that $\eta_0 = 10^7/4\pi c_0^2 = 8.854 \times 10^{-12}$ farad m^{-1}; c_0, the speed of light in vacuum, is 2.9979×10^8 m s^{-1}.

When the interaction takes place in a dielectric with dielectric constant ε, the force is reduced to $f = qq'/\varepsilon r^2$ in esu, or to $f = (1/4\pi\eta_0) \cdot qq'/\varepsilon r^2$ in practical units. The traditional dielectric constant ε of chemistry, as used in this

Table 12.2. Electrical units and their conversion

Property	Practical (SI) Units (mks)	Electrostatic Units (cgs)	Electromagnetic Units (cgs)
Current (i)	1 ampere	$= 2.998 \times 10^9$ statamp	$= 10^{-1}$ emu-amp or biot
Charge (q)	1 coulomb	$= 2.998 \times 10^9$ franklin	$= 10^{-1}$ emu-coulomb
Potential (ϕ)	1 volt	$= 1/299.8$ statvolt	$= 10^8$ emu-volt
Capacitance ($C = q/\delta\phi$)	1 farad	$= 8.987 \times 10^{11}$ statfarad	$= 10^{-9}$ emu-farad
Work (W)	1 joule	$= 10^7$ erg	$= 10^7$ erg

Relationships for a Spherical Charge Distribution

Property	Electrostatic Units (cgs)	Practical Units (mks)
Electric displacement	$\mathcal{D} = q/r^2$	$\mathcal{D} = q/r^2$
Electric field	$\mathcal{E} = q/\varepsilon r^2$	$\mathcal{E} = (1/4\pi\eta_0) \cdot q/\varepsilon r^2$
Electric displacement at any point in a liquid	$\mathcal{D} = \varepsilon\mathcal{E}$	$\mathcal{D} = (4\pi\eta_0)\varepsilon\mathcal{E}$
Energy density at that point	$dW/dV = \varepsilon\mathcal{E}^2/8\pi$	$dW/dV = \eta_0\varepsilon\mathcal{E}^2/2$

book, is dimensionless and identical to the relative permittivity ε_r of physics. The relative permittivity of a medium M is the ratio of the permittivity of M to that of free space.

A few illustrations of the conversion from esu to practical units are given in Table 12.2. The electric displacement $\mathcal{D}$ is independent of the dielectric constant; for a point charge q, or outside the locale of a spherical charge distribution with total charge q, $\mathcal{D} = q/r^2$. The electric field $\mathcal{E}$, on the other hand, depends on the difference between the total charge q and the opposite polarization charge or screening charge induced by q in a narrow layer of the surrounding dielectric. For a point charge, $\mathcal{E} = q/\varepsilon r^2$ in esu, or $(1/4\pi\eta_0)q/\varepsilon r^2$ in practical units. The region occupied by the screening charge is called the *screening layer.*

Electric Potential Due to an Ion Atmosphere

In the Debye–Hückel model, ionic charge distributions have spherical symmetry. One ion, the subject ion with charge $z_j e$, defines the center of the coordinate system. The other ions, with a net charge $-z_j e$ (for electroneutrality), are distributed around the subject ion and define an ion atmosphere whose time average also has spherical symmetry. The model treats the electrostatic interaction between the subject ion and the ion atmosphere.

The electric potential at a point in a charge-bearing region is related to the local charge density ρ_q by the Poisson equation (12.12a):

$$\nabla^2\phi = -4\pi\rho_q/\varepsilon \quad (\text{esu, cgs}) = -(1/\eta_0)\rho_q/\varepsilon \quad (\text{SI, mks}) \qquad (12.12a)$$

The original paper by Debye and Hückel [1923] uses esu, and we shall do so here. Because of the spherical symmetry, $\nabla^2\phi$ in polar coordinates is simply a function of r, the distance from the center of the subject ion, and Eq. (12.12a) becomes (12.12b).

$$\frac{1}{r^2}\left(\frac{d}{dr}r^2\frac{d\phi}{dr}\right) = -\frac{4\pi\rho_q}{\varepsilon} \quad \text{(esu, cgs)} \tag{12.12b}$$

The local charge density at r is calculated by assuming a Boltzmann distribution for each species of ions. Thus, for the ith species, the molecular density $n_i(r) = n_i \exp(-z_i e\phi(r)/kT)$ ions/cm^3, where n_i is the average for the solution (as $r \to \infty$). The charge density $\rho_i(r) = z_i e n_i(r)$, and the overall $\rho_q(r)$ is the sum $\sum z_i e n_i(r)$ over all ionic species, including the jth species. Equation (12.12b) therefore takes the explicit form (12.13), which is known as the *Poisson–Boltzmann equation*:

$$\frac{1}{r^2}\frac{d}{dr}\left(r^2\frac{d\phi(r)}{dr}\right) = -(4\pi/\varepsilon)\sum [z_i e n_i \exp(-z_i e\phi(r)/kT)] \tag{12.13}$$

Equation (12.13) does not have a mathematical solution in closed form that applies at all values of r. There are two solutions that apply in different ranges of r. At sufficiently low ionic strengths these ranges overlap and the electric potential due to the ionic atmosphere can be evaluated. The result is called the *Debye–Hückel limiting law*.

To arrive at a solution that applies at small values of r, consider the idealized model shown in Fig. 12.3. The subject ion, of radius r_a, is here surrounded by a charge-free shell, of radius x, which in turn is surrounded by an electrified zone whose time average is an ion atmosphere of spherical symmetry. We shall show that inside the charge-free shell, the electric potential is given by Eq. (12.14), in which the term $z_j e/\varepsilon r$ is the potential due to the central ion, and $\phi_{\text{ion atm}}$ is a constant term due to the ion atmosphere.

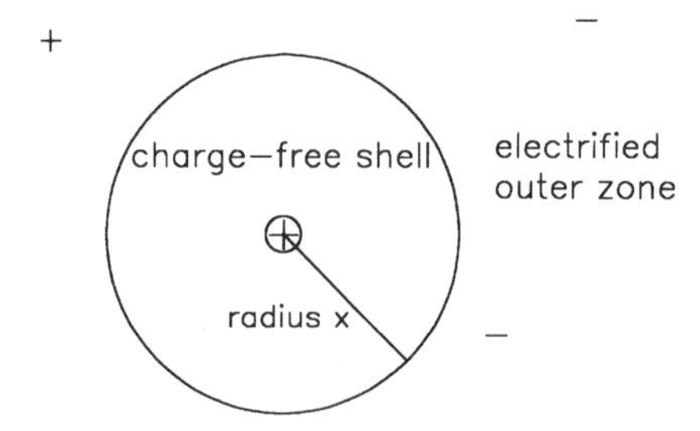

Figure 12.3. Electrical model for Eq. (12.14). The radius of the central cation is r_a.

For the model of Fig. 12.3, when $r_a \le r \le x$

$$\phi = z_j e/\varepsilon r + \phi_{\text{ion atm}}; \qquad \phi_{\text{ion atm}} = \text{constant} \qquad (12.14)$$

PROOF. In Fig. 12.3, for $r_a \le r \le x$, the Poisson equation (in esu) takes the form $(1/r^2)d/dr[r^2(d\phi/dr)] = 0$. On letting $u = d\phi/dr$, this becomes $du/dr + 2u/r = 0$. Hence $u = d\phi/dr = C_1/r^2$, and $\phi = -C_1/r + C_2$, where C_1 and C_2 are constants of integration. C_1 and C_2 are independent of r but may depend on z_j and the nature of the surrounding ion atmosphere.

C_1 is evaluated by applying Gauss' theorem, which in this case states that the electric field $(-d\phi/dr)$ in a charge-free shell around the spherical charge distribution inside r_a is given by Coulomb's law [Duckworth, 1960]. Thus in the range $r_a \le r \le x$, $4\pi\varepsilon C_1 = -4\pi r^2 \varepsilon\, d\phi/dr = 4\pi z_j e$ (esu). Hence $\phi = z_j e/\varepsilon r + C_2$, where $z_j e/\varepsilon r$ is the potential due to the central ion alone. C_2 may be identified with the additional potential, $\phi_{\text{ion atm}}$, due to the ion atmosphere surrounding the charge-free shell, which is therefore independent of r. Q.E.D.

The model of Fig. 12.3 is relevant because Eq. (12.14) applies in good approximation even if the charge inside the spherical shell is not truly zero, so long as the amount is physically negligible. For a given shell-size extending from r_a to x, this can always be arranged merely by making the ionic strength in the overall solution small enough. For an ion of charge $z_j e$, electroneutrality requires that the total charge in the ion atmosphere be $-z_j e$. This is true at all ionic strengths, whether high or low. Suppose we specify that a time-average charge of $0.01 z_j e$ in the shell $r_a \le r \le x$ is tolerable. The inner radius r_a is fixed, but x is variable. As the ionic strength decreases to zero, x increases—virtually without limit if the amount of solution is large.

PROOF. The electric charge in the shell is the product of the mean charge density times the shell volume. In our case, then, $0.01 z_j e$ = (mean charge density) $\cdot V_{\text{shell}} = [\sum z_i e n_i(r)] \cdot V_{\text{shell}}$. As the ionic strength decreases, the mean charge density decreases as well, because then $n_i(r)$ decreases nearly proportionally for all i. Hence, V_{shell} (and x) can be increased at will. Q.E.D.

The second closed-form solution of Eq. (12.13) requires that the potential $\phi(r)$ be small enough so that $\exp(-z_i e\phi(r)/kT) \simeq 1 - z_i e\phi(r)/kT$, in good approximation. This is always possible, because $\phi(r)$ approaches zero as r becomes large. When r is large enough (say, $r \ge y$), Eq. (12.13) takes the form (12.13L):

$$\frac{1}{r^2}\frac{d}{dr}\left(r^2\frac{d\phi(r)}{dr}\right) = -(4\pi/\varepsilon)\sum(z_i e n_i[1 - z_i e\phi(r)/kT]), \qquad r \geq y \tag{12.13L}$$

This equation can be simplified by noting that $\sum z_i e n_i = 0$, because of electroneutrality. (Strictly, $\sum z_i e n_i = -z_j e \simeq 0$.) The equation can then be changed into standard form by changing the dependent variable from ϕ to $z = r\phi$ and by defining a parameter κ^2 according to (12.15).

$$\kappa^2 = \frac{4\pi e^2 \sum n_i z_i^2}{\varepsilon kT} \tag{12.15}$$

The equation then reduces to the standard form, $d^2Z/dr^2 = \kappa^2 Z$, which has the general solution: $Z = A\exp(-\kappa r) + B\exp(\kappa r)$, in which A and B are constants of integration, and κ is the *positive root* of κ^2. Substituting $Z = \phi r$, we obtain Eq. (12.16) for the range $r \geq y$:

$$\phi = A\frac{\exp(-\kappa r)}{r} + B\frac{\exp(\kappa r)}{r}, \qquad r \geq y \tag{12.16}$$

To evaluate B, we let $r \to \infty$, so that $\phi \to 0$; $\therefore B = 0$. To evaluate A we make the ionic strength low enough so that the outer shell radius x in Eq. (12.14) is equal to the minimum radius, y, in Eq. (12.16). The two equations then apply simultaneously. Thus, when $r = x$ we obtain

$$\phi(x) = z_j e/\varepsilon x + \phi_{\text{ion atm}} = A(\exp[-\kappa x])/x \tag{12.17a}$$

$$\exp[-\kappa x] \simeq (1 - \kappa x) \quad \text{at low ionic strengths} \tag{12.17b}$$

$$\therefore \quad z_j e/\varepsilon x + \phi_{\text{ion atm}} \simeq A/x - A\kappa \tag{12.17c}$$

The near-equality (12.17b) applies at low ionic strengths because, when Eq. (12.14) is valid, κx varies nearly as $\kappa^{1/3}$, and hence as $I^{1/6}$. [As defined in Eq. (12.15), κ^2 is proportional to $\sum n_i z_i^2$ and hence to the ionic strength.]

Since x is an independent variable, we may equate the terms in $1/x$ in (12.17c). This yields $A = z_j e/\varepsilon$. On equating the remaining terms we arrive at Eq. (12.18) for $\phi_{\text{ion atm}}$, the electric potential experienced by the subject ion due to its surrounding ion atmosphere. Equation (12.18) is the key to the Debye–Hückel limiting law.

$$\phi_{\text{ion atm}} = -(z_j e/\varepsilon)\cdot\kappa \quad \text{when } I \text{ is small} \tag{12.18}$$

The Activity Coefficient Due to an Ion Atmosphere

$\phi_{\text{ion atm}}$ differs in kind from the electric potential ϕ that appears in the electrical-work term for the electrochemical potential [Eq. (12.4)]. ϕ is a variable whose absolute value is unknown; at equilibrium it is uniform throughout the solution. $\phi_{\text{ion atm}}$, on the other hand, is intrinsic and independent of the external electrification of the solution. It depends on the chemical makeup of the solution, and, in principle, its absolute value is known. The effect of $\phi_{\text{ion atm}}$ on the electrochemical potential $(\text{ECP})_j$ of the jth ionic species is therefore incorporated in the expression for G_j. This effect is $\frac{1}{2}Lez_j\phi_{\text{ion atm}}$. The statistical factor of $\frac{1}{2}$ enters because, if that factor were omitted, we would be counting each ion twice: once as the subject ion and once as part of the ion atmosphere of other ions.

The full expression for G_j, including the effect of the ion atmosphere, is therefore Eq. (12.19a). Comparison with Eq. (12.6) then yields the limiting law (12.19b) for $RT \ln \gamma_j$.

$$G_j = G_j^0 + RT \ln m_j + \tfrac{1}{2}Lez_j\phi_{\text{ion atm}} \qquad (12.19a)$$

$$RT \ln \gamma_j = \tfrac{1}{2}Lez_j \cdot \phi_{\text{ion atm}} \qquad (12.19b)$$

Equation (12.19b) is transformed to familiar variables by use of Eqs. (12.15) and (12.18) and by expressing $\sum n_i z_i^2$ in terms of the ionic strength $I = \frac{1}{2}\sum m_i z_i^2$.* The result is (12.20)—note the negative sign.

$$\ln \gamma_j = -z_j^2\sqrt{\pi e^3}\left(\sum n_i z_i^2\right)^{1/2}/(\varepsilon kT)^{3/2}$$

$$\ln \gamma_j = -z_j^2\sqrt{I}e^3(2\pi\rho L/1000)^{1/2}/(\varepsilon kT)^{3/2} \qquad (12.20a)$$

$$\log_{10}\gamma_j = -z_j^2\sqrt{I}(e^3/2.3026)(2\pi\rho L/1000)^{1/2}/(\varepsilon kT)^{3/2} \qquad (12.20b)$$

Equation (12.20b) may be rewritten in the form $\log_{10}\gamma_j = -z_j^2 S_\gamma\sqrt{I}$, where S_γ is the *Debye–Hückel limiting slope*. S_γ gathers up all that is independent of I and takes the explicit form (12.20c). In water at 298 K, $S_\gamma = 0.510 \pm 0.001$.

$$S_\gamma = (e^3/2.3026)(2\pi\rho L/1000)^{1/2}/(\varepsilon kT)^{3/2} = 1.825\times 10^6\sqrt{\rho}/(\varepsilon T)^{3/2} \qquad (12.20c)$$

A quantitative test of the Debye–Hückel limiting law is described below. At

*$\sum n_i z_i^2 = (\rho L/1000)\sum m_i z_i^2 = (2\rho L/1000)I$, where ρ is the density. In the conversion from n_i to m_i we make the dilute-solution approximation that n_i is the number of i ions per cubic centimeter of *solvent* rather than solution, and that ρ is the density of the solvent.

this point we note that Eqs. (12.20) reproduce two experimental propensities: (1) the ionic strength principle and (2) the square-root relationship at low ionic strengths.

In addition to the limiting law, Debye and Hückel [1923] also obtained a first approximation, Eq. (12.21):

$$\ln \gamma_j = -\frac{(z_j^2 e^2) \cdot \kappa}{(2\varepsilon kT)(1 + \kappa a_j)} \tag{12.21a}$$

$$\log_{10}\gamma_j = -\frac{z_j^2 S_\gamma \sqrt{I}}{1 + a_j A\sqrt{I}}, \qquad A = 50.29 \times 10^8 \left(\frac{\rho}{\varepsilon T}\right)^{1/2} \text{cm}^{-1} \tag{12.21b}$$

In theory, a_j is a distance of closest approach. In practice, it is treated as a fitting parameter and thus embraces all deviations from the limiting law. Usually the fitted values of a_j are in the range $(3–6) \times 10^{-8}$ cm, which is a plausible magnitude for ion–ion distances of closest approach. Examples are known, however, in which a_j is negative.

Note that γ_j is the activity coefficient for an ionic species, but the theory readily yields mean-ion activity coefficients for electrolytes by substitution in the defining equations preceding Eq. (12.10). For univalent strong electrolytes in solvents of high dielectric constant such as water, Eq. (12.21b) reproduces mean-ion activity coefficients within experimental error up to about 0.05 m ionic strength. In solvents of intermediate dielectric constant such as ethanol, one may expect good fit up to ~0.01 m, after allowance is made for ion-pair formation.

A Quantitative Test

Tests of the Debye–Hückel limiting law (12.20) or first approximation (12.21) consist in measuring $\log \gamma_\pm$ for a strong electrolyte and, without theoretical bias, testing the square-root relationship and confirming the limiting slope. Testing the square-root relationship is important because adherents of competing theory expect a cube-root relationship [Bahe, 1972; Bahe and Parker, 1975; Chen and Choppin, 1995]. Given a square-root relationship, the decisive issue becomes whether the experimental limiting slope agrees with S_γ according Eq. (12.20). Both tests require data of utmost precision at very low concentrations, because the departures of $\gamma_\pm$ from unity then are quite small. The individuals who succeeded under such trying conditions are among the heroes of electrolytic solution chemistry.

Figure 12.4 shows the emf results of N. J. Anderson published in 1934, for electrochemical cells containing $H^+ + Cl^-$ at concentrations ranging from 25×10^{-6} to 0.003 m in water at 25°C [Anderson, 1934]. The cells may be formulated as Pt/H_2 (g, 1 atm)$|H^+(aq, m_+), Cl^-(aq, m_-)|AgCl/Ag(s)$ and are without

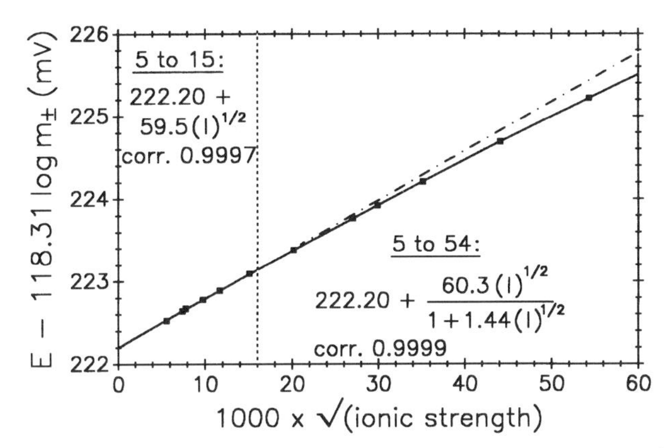

Figure 12.4. Electromotive force measurements for the cell Pt/H_2 (g, 1 atm)| H^+ (aq, m_+), Cl^-(aq, m_-)| AgCl/Ag(s), along with a test of the Debye–Hückel theory [Anderson, 1934].

liquid junctions. The mean-ion activity coefficient $\gamma_\pm$ of H^+ and Cl^- is obtained from the cell emf (E) by the equation

$$E\,(\text{in mV}) = E^0 - 118.31 \log m_\pm - 118.31 \log \gamma_\pm$$

E^0, the standard emf, is constant in these experiments. Thus Fig. 12.4 is essentially a plot of $-118.31 \log \gamma_\pm$ versus $\sqrt{I}$.

The data for ($E + 118.31 \log m_\pm$) cover a range of 2.8 mV and are precise to 5 μV. The feat was knowing the concurrent concentrations, because adsorption of ions on solid surfaces in the cell becomes significant at the lowest concentrations. So does the acid–base reaction of Pyrex with hydrogen ions, and Anderson substituted silica for Pyrex equipment. When cell equilibrium had been reached, he measured the H^+ + Cl^- concentration conductometrically, essentially in situ. The solubility of Ag^+ from the AgCl/Ag electrode was not quite negligible at the lowest concentrations, but according to control experiments this did not degrade the electromotive accuracy.

As shown in Fig. 12.4, the plot versus $\sqrt{I}$ is nicely linear, with $E^0 = 222.20$ mV and a slope of 59.5 mV per $m^{1/2}$. The plot up to the highest concentration fits the first approximation (12.21) almost perfectly, with the same $E^0 = 222.20$ mV and a limiting slope of 60.3 mV per $m^{1/2}$. At 298.15 K experimental values for the dielectric constant of water range from 78.30 to 78.54. Accordingly, the theoretical limiting slope S_γ is 0.5097 ± 0.0012, and 118.31 S_γ is 60.3 ± 0.1 mV per $m^{1/2}$, in very good agreement with experiment.

Some Corollaries

Beginning with Euler's equation: $G = \sum n_J G_J$ and introducing the Debye–Hückel limiting law, we obtain Eqs. (12.22a) and (12.22b). Thus, for any ionic species Z, G_Z is given by Eq. (12.22c), which is consistent with the prior Eqs. (12.19) and (12.20).

$$G = \sum n_J(G_J^0 + RT \ln m_J) - \sum n_J \cdot 2.3026RT\, z_J^2 S_\gamma \sqrt{I} \tag{12.22a}$$

$$\sum n_J d(G_J^0 + RT \ln m_J) - 2.3026 \sum n_J d(RT\, z_J^2 S_\gamma \sqrt{I}) = 0 \tag{12.22b}$$

$$G_Z = (\partial G/\partial n_Z)_{\{n \neq n_Z\}, T, P} = G_Z^0 + RT \ln m_Z - 2.3026RT\, z_Z^2 S_\gamma \sqrt{I} \tag{12.22c}$$

To obtain the partial entropy $S_Z = -(\partial G_Z/\partial T)_{P,\{n\}}$ and the partial volume $V_Z = (\partial G_Z/\partial P)_{T,\{n\}}$, we note that S_γ is a function of ε and T, which in turn is a function of T and P. Ordinarily the ionic strength $I = \frac{1}{2} \sum m_i z_i^2$ is independent of T and P, except when there are molar shifts in reactions that change the ionic strength. The following expressions for the partial entropy and the partial volume will therefore allow for molar-shift terms based on $\partial I/\partial T$ and $\partial I/\partial P$.

$$\begin{aligned} S_Z &= -(\partial G_Z/\partial T)_{\{n\},P} \\ &= S_Z^0 - R \ln m_Z + [2.3026RT\, z_Z^2 S_\gamma \sqrt{I}]\left(\frac{1}{T} + \frac{\partial \ln S_\gamma}{\partial T} + \frac{1}{2}\frac{\partial \ln I}{\partial T}\right)_{\{n\},P} \end{aligned} \tag{12.23a}$$

$$\begin{aligned} V_Z &= (\partial G_Z/\partial P)_{\{n\},T} \\ &= V_Z^0 - [2.3026RT\, z_Z^2 S_\gamma \sqrt{I}]\left(\frac{\partial \ln S_\gamma}{\partial P} + \frac{1}{2}\frac{\partial \ln I}{\partial P}\right)_{\{n\},T} \end{aligned} \tag{12.23b}$$

The ionic-strength-dependent terms in Eqs. (12.23a) and (12.23b) express the effect of interionic attraction in the approximation of the limiting law.

The Debye–Hückel theory treats only electrostatic charge–charge interactions, but its success does not mean that the more traditional van der Waals/London dispersion interactions among the ions are entirely negligible. The ion-atmosphere effects dominate at low concentrations because they vary as the square root of the ionic strength; while the traditional interactions, which vary as the first power of the concentration, become visible at higher concentrations. For example, when the London dispersion interactions between ions of like charge are strong enough, they outweigh the electrostatic repulsions, and anion–anion or cation–cation dimers or aggregates form. Classic examples are given by cyanocarbon anions [Boyd, 1961] and merocyanine cations [Jelley, 1936; Fidder et al., 1991; Higgins and Barbara, 1995]. Here the London dispersion interactions are above average in strength because the interacting electron distributions are delocalized. [London, 1942; Grunwald and Price, 1964].

ELECTROSTATIC IONIC SOLVATION

Solvation is defined as the transfer of a molecular species from the gas phase to a liquid solvent. For normal (nonpolymeric) nonelectrolytes, standard enthalpies of solvation are in the tens of kilojoules per mole. For many ionic species, however, they are in the hundreds or even thousands.

For example, ΔH^0 for the hydration of the ions of sodium chloride

$$Na^+(g) + Cl^-(g) \rightarrow Na^+(aq, dil) + Cl^-(aq, dil)$$

can be obtained with useful accuracy by combining calorimetric and spectroscopic measurements for vaporization, ionization, dissociation, electron capture, formation of the salt, and solution of the salt in water. The following data in kilojoules per mole at 298 K are from standard reference tables [JANAF, 1971, Wagman et al., 1982]:

$$\begin{array}{ll}
\text{(a)}\ Na(s) \rightarrow Na(g) \rightarrow Na^+(g) + e^-(g) & \Delta_f H^0 = +610.0 \\
\text{(b)}\ \tfrac{1}{2}Cl_2(g) \rightarrow Cl(g) \rightarrow Cl^-(g) - e^-(g) & \Delta_f H^0 = -234.2 \pm 2 \\
\text{(c)}\ Na(s) + \tfrac{1}{2}Cl_2(g) \rightarrow NaCl(\text{crystal}) & \\
\qquad \rightarrow Na^+(aq, dil) + Cl^-(aq, dil) & \Delta_f H^0 = -407.3 \\
\hline
\text{(c–a–b)}\ Na^+(g) + Cl^-(g) \rightarrow Na^+(aq, dil) + Cl^-(aq, dil) & \Delta_{hyd} H^0 = -783 \pm 2
\end{array}$$

The result, -783 ± 2 kJ/mol, is of the order of chemical bond energies, rather than of the van der Waals/hydrogen-bond energies that play a role in the hydration of nonelectrolytes. In view of the chemists' belief that ions are essentially atoms or molecules, and that the bonds formed by ions follow the same rules as those formed by molecules, this unequivocal discrepancy is a crisis that needs to be resolved.

Born's Explanation

It will be recalled that the energy of uncharged molecules is essentially mechanical energy, dominated by the potential energy of the interacting particles in the molecules. In first approximation this is *internal* energy, intrinsic to the uncharged molecules. Interactions of the molecules with their surrounding medium cause relatively modest perturbations.

Max Born [1920] showed that the energy of *ionic* molecules has an important additional component: The electrostatic energy of the ionic charge. This energy, he theorized, is mostly *external* to the ion, resident in the surrounding medium, and is dominated by the dielectric constant of the medium and the ionic charge and size. He was thus able to show that the enthalpy of hydration of the ions of sodium chloride is dominated by the electrostatic term, which successfully accounts for the observed order of magnitude.

Born's theory is founded on Michael Faraday's concept that the electrostatic energy (W) associated with a charge-distribution resides in the electric field created by that distribution, and which pervades all space. His model charge-distribution consists of an ion in an otherwise charge-free isotropic medium, so that W is the electrostatic energy per ion. Accordingly, the electrostatic term LW per mole of ions in macroscopic thermodynamics becomes part of the standard *free* energy of the ionic species.

In principle, the electric field is a vector quantity, but to apply Born's model we need only use its scalar magnitude $\mathcal{E}$. In a fluid medium, at a point where this magnitude is $\mathcal{E}$, the energy density $dW/dV = \varepsilon\mathcal{E}^2/8\pi$ (esu, cgs; see Table 12.2). Here ε is the scalar dielectric "constant," which in fact is a function of $\varepsilon(\mathcal{E})$ of the electric field (see below). Thus

$$W = \int \varepsilon\mathcal{E}^2 dV/8\pi \quad \text{(ergs/ion)} \tag{12.24a}$$

and, for solvation (g $\rightarrow$ M) of the ionic species in a liquid medium M,

$$L\Delta_{\mathrm{g}\rightarrow\mathrm{M}}W = L\int \Delta_{\mathrm{g}\rightarrow\mathrm{M}}(\varepsilon\mathcal{E}^2)dV/8\pi \tag{12.24b}$$

The integration in equations (12.24) is over all space, including the space occupied by the ion.

Born modeled an atomic ion as a conducting sphere with charge q and radius r_J embedded in a charge-free medium. We are interested also in molecular ions, so we shall replace the conducting sphere by a symmetric dielectric sphere with a central charge q and with radius r_J. The results of the two models turn out to be the same. Because of the spherical symmetry of both models and of their electric fields, the generalized function ε—i.e., $\varepsilon(\mathcal{E})$—in Eqs. 12.24 now becomes $\varepsilon(r)$, where r is the distance from the ionic center. For the ion modeled as a dielectric sphere we know little about $\varepsilon(r)$ inside the ion, but that is no problem because the integrand in Eq. (12.24b) is the *difference*, $\Delta_{\mathrm{g}\rightarrow\mathrm{M}}[\varepsilon(r)\mathcal{E}^2]$, and the awkward part of the integral over the ionic sphere itself cancels.

Outside the ionic sphere, by Gauss' theorem, $\mathcal{E}$ equals $q/\varepsilon(r)r^2$ [Duckworth, 1960; Böttcher, 1962]. On introducing $q = z_\mathrm{J}e$ and $dV = 4\pi r^2 dr$, Eq. (12.24b) reduces to Born's result, Eq. (12.24c), for the solvation of one mole of ions. The subscript "g" denotes the gas phase and "M" the solvating liquid medium. Eq. (12.24c) is written in esu-cgs units.

$$L\Delta W = -L\frac{z_\mathrm{J}^2 e^2}{2}\int_{r_\mathrm{J}}^{\infty}\left(\frac{1}{\varepsilon_\mathrm{M}(r)} - \frac{1}{\varepsilon_\mathrm{g}(r)}\right)d(1/r) \tag{12.24c}$$

In the gas phase, $\varepsilon_\mathrm{g}(r)$ is effectively unity. The variation of ε_M with r will be considered in the next section. But in first approximation, $\varepsilon_M(r)$ is simply ε_M, the normal low-field value of the dielectric constant. On introducing this value and $\varepsilon_g \simeq 1$ in Eq. (12.24c), we obtain the *Born Equation* (12.25a) for the electrostatic contribution to ΔG of solvation. When r_J is in angstroms and $L\Delta W$ is in kilojoules per mole, Eq. (12.25a) takes the explicit form (12.25b).

$$L\Delta W = \frac{z_J^2 e^2}{2r_J}\left(\frac{1}{\varepsilon_M} - 1\right) \tag{12.25a}$$

$$L\Delta W = \frac{694.5 z_J^2}{r_J/\text{Å}}\left(\frac{1}{\varepsilon_M} - 1\right) \quad (\text{kJ/mol}) \tag{12.25b}$$

The Born equation reproduces a number of facts. First, for plausible ionic radii in the 1–5 Å range, the solvation energy is in the hundreds of kilojoules per mole for univalent ions and (being proportional to z_J^2) can rise into the thousands for higher-valent ions. Using ionic crystal radii and $\varepsilon_M = 78.3$ for water at 298 K, $L\,\Delta W$ for the hydration of Na^+ (1.02 Å) is -672 kJ/mol, and that for Cl^- (1.81 Å) is -379 kJ/mol. The sum, -1051 kJ/mol, deviates by 30% from the experimental ΔH^0 and fits the observed magnitude and sign.

Second, because z_J^2 is always positive and $\varepsilon_M > 1$, $L\Delta W$ for solvation is always negative. Third, because the electrostatic energy is essentially external-residing in the surrounding medium, ionic species dissolved in a fixed medium remain free to share the intrinsic rules for bond-formation and structure-reactivity correlation with nonelectrolytes, in agreement with experience. On the other hand, when the medium is variable, the electrostatic contribution ($L\Delta W$) to ΔG^0 may well dwarf the orthodox mechanical–potential energy contribution, as it does in the hydration of sodium chloride.

The Born theory applies to isolated ions. The mutual interaction of the ionic charges is treated by the Debye–Hückel theory.

The Born Radius

Ionic crystal radii (as used above in the approximate calculation of $L\Delta W$ for the hydration of sodium chloride) are smaller than the appropriate Born radii for Eq. (12.25). Let $(\text{rad})_J$ denote the ionic crystal radius and r_J the Born radius of ion J. Then

$$r_J = (\text{rad})_J + \delta_J \tag{12.26}$$

where the incremental thickness δ_J is typically a few tenths of an angstrom. By regarding the ionic crystal radius as a measure of the hard-sphere radius of the ion, we predict relationships of the form (12.26) by basically two approaches. (1) When the polar liquid solvent (M) is modeled as a dielectric continuum, a thickness δ_J^* of this magnitude results from dielectric saturation. (2) When the solvent (M) is modeled as an ensemble of polar molecules, a similar thickness δ_J' represents the thickness of the screening layer around the ion. We will now analyze these phenomena.

Dielectric saturation arises because the orientation polarization of polar liquids "saturates" in high electric fields, reaching a maximum as the solvent dipoles become fully aligned with the field. The local dielectric constant therefore drops off at short distances from an ion. For example, Booth [1951] extended Oster and Kirkwood's [1943] model of the dielectric constant of water to apply at high fields; Fig. 12.5a shows his predictions for $1/\varepsilon(r)$ as a function of r for a univalent ion in water. Thus he predicts that $\varepsilon(r)$ decreases to half the low-field value ($\varepsilon_M = 78.3$) at about 2.5 Å from the center of a univalent ion, where $\mathcal{E} \approx 5 \times 10^6$ V/cm, and that the decrease in ε becomes quite marked

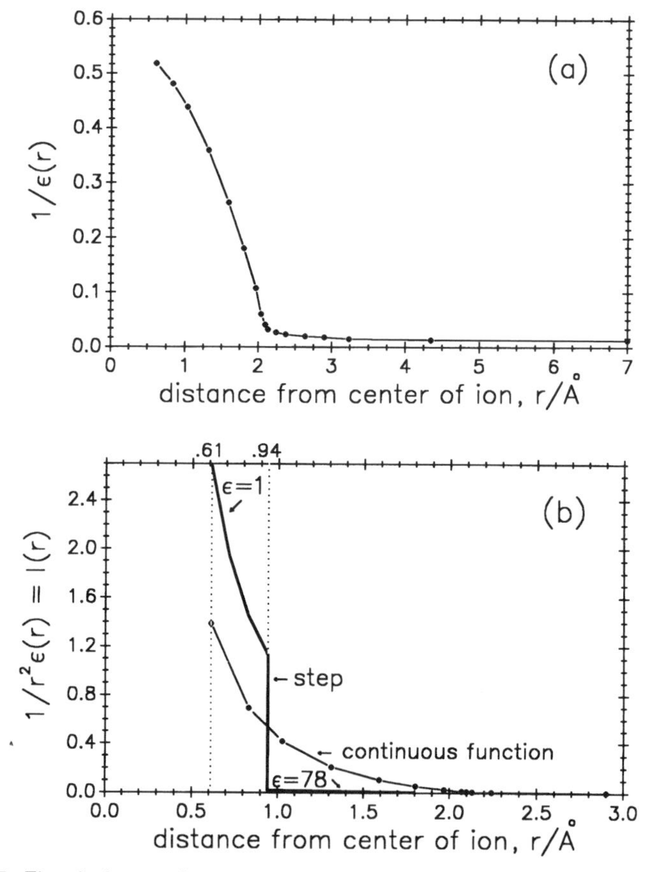

Figure 12.5. Electrical saturation near a univalent ion in water. (a) $1/\varepsilon(r)$ versus r, based on Booth's function [1951] for ε of water as a function of the electric field. (b) Graphical solution of Eq. (12.27).

when $r < 2$ Å. However, in the calculation of $L\Delta W_J$ it is convenient to represent the decrease in $\varepsilon(r)$ by an equivalent step function, indicated in Fig. 12.5b:

Equivalent step function for ε

$$\text{For} \quad (\text{rad})_J \le r \le (\text{rad})_J + \delta^*: \quad \varepsilon = 1$$

$$\text{For} \quad r > (\text{rad})_J + \delta^*: \quad \varepsilon = \varepsilon_M$$

The stepping distance δ^* is chosen so that the work $L\Delta W_J$ is the same for the step function as for the continuous function $\varepsilon(r)$. Let $W_J(\text{M})$ denote the electrical work based on the continuous $\varepsilon(r)$, then δ^* is the parameter in Eq. (12.27) that produces equality.

$$\frac{2W_J(\text{M})}{z_J^2 e^2} = \int_{(\text{rad})_J}^{\infty} \frac{dr}{r^2 \varepsilon(r)}$$

$$= \int_{(\text{rad})_J}^{(\text{rad})_J + \delta^*} \frac{dr}{r^2} + \int_{(\text{rad})_J + \delta^*}^{\infty} \frac{dr}{r^2 \varepsilon_M} \tag{12.27}$$

In the sample plot shown in Fig. 12.5b, $(\text{rad})_J$ is 0.611 Å, M is water, and δ^* is 0.33 Å. The advantage of the step function is that $[1/\varepsilon_M(r) - 1/\varepsilon_g(r)]$ now vanishes in the region $(\text{rad})_J \le r \le (\text{rad})_J + \delta^*$, and that ε assumes the constant value ε_M when $r > (\text{rad})_J + \delta^*$.

To a chemist, the molecular approach that involves a screening layer of polar solvent molecules around the ion is probably more congenial. A diagram defining the screening thickness δ' is given in Fig. 12.6a. Here the positive charge of the ion attracts the negative poles of the dipoles in adjacent solvent molecules. The complementary positive poles in the solvent molecules thus face into the bulk dielectric where they attract the negative poles of further solvent dipoles, and so on—to the effect that volume elements outside the effective screening thickness δ' are electrically neutral. That is, the dielectric continuum begins not at $(\text{rad})_J$ but at $(\text{rad})_J + \delta'$. The dielectric constant is again treated as a step function. Up to $(\text{rad})_J + \delta'$, $\varepsilon = 1$; and beyond $(\text{rad})_J + \delta'$, $\varepsilon = \varepsilon_M$. On the scale of ionic radii, δ' is significant. For example, Fig. 12.6b shows the electric charge distribution in Stillinger and Rahman's [1974] ST2 model of the water molecule, which is often used in computer simulations of liquid water. The model shows two positive poles of $0.236e$ at tetrahedral sites near the hydrogen atoms, as well as two negative poles of $0.236e$ at tetrahedral sites in the regions of the unshared electron pairs. A cation tends to interact with the centroid of the negative poles, while an anion interacts with a positive pole to form a hydrogen bond [Bockris and Reddy, 1973]. In either case, δ' is a specific parameter on the order of tenths of angstroms (Fig. 12.6c,d).

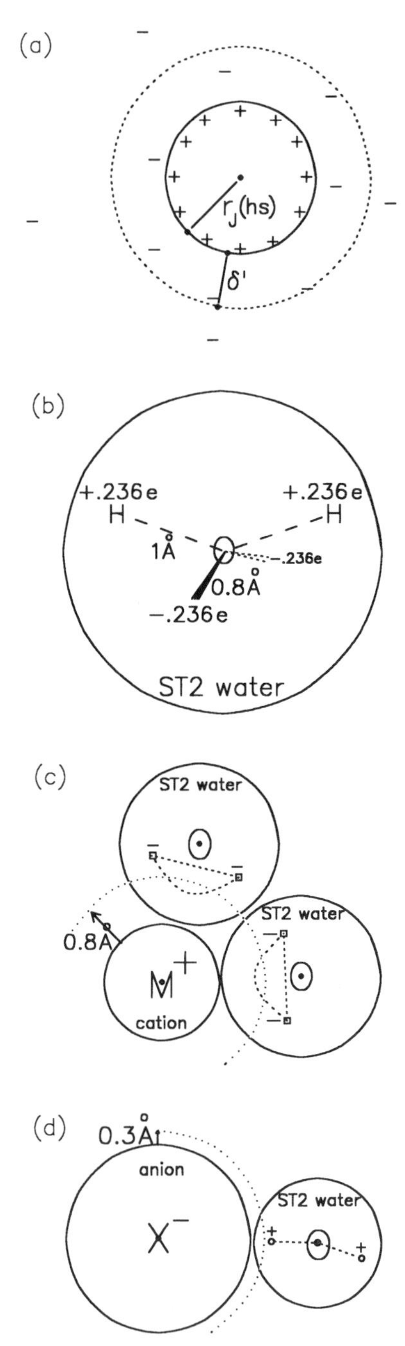

Figure 12.6. Screening of ionic charge in a polar medium. (a) Schematic. (b) ST2 model of the poles in a water molecule [Stillinger and Rahman, 1974]. (c) Author's conception of an alkali ion interacting with ST2 water; $\delta_+ \approx 0.8$ Å. (d) Author's conception of a halide ion interacting with ST2 water; $\delta_- \approx 0.3$ Å.

When δ^* and δ' are gathered up into a single parameter δ, Eq. (12.25b) takes the general form (12.28a), and the specific form (12.28b) for hydration.

$$L\Delta W(\varepsilon_M) = \frac{694.5 z_J^2}{[(\text{rad})_J + \delta_J]/\text{Å}} \left(\frac{1}{\varepsilon_M} - 1 \right) \quad (\text{kJ/mol}) \tag{12.28a}$$

$$L\Delta W(\text{hydration}) = -685.7/[(\text{rad})_J + \delta_J]/\text{Å} \quad (\text{kJ/mol}) \tag{12.28b}$$

Specific Dissection

Equations (12.28) and related theoretical equations will now be used to dissect experimental data for strong electrolytes so as to obtain specific values for the individual ionic species. In principle, the δ parameter is specific for an ionic species, but there is a school of thought that δ values should be similar within a chemical family. Thus Latimer, Pitzer, and Slansky in their historic [1939] paper showed that the experimental values of $\Delta_{\text{hyd}}G^0$ for electrically neutral combinations of alkali ions (M^+) and halide ions (X^-) are indeed sums, $\Delta_{\text{hyd}}G^0(M^+) + \Delta_{\text{hyd}}G^0(X^-)$, of ionic terms. And they dissected these sums on the basis of Eq. (12.28b) with one parameter, δ_+, for all alkali ions, and another parameter, δ_-, for all halide ions.

The actual calculation begins with a temporary dissection—*any* temporary dissection—of $\Delta_{\text{hyd}}G^0$ into $\Delta G^0_{M^+}$(temporary) + $\Delta G^0_{X^-}$(temporary). Let $(\text{rad})_+$, $(\text{rad})_-$ denote the respective ionic crystal radii, and let Y denote the error in the temporary dissection. Then

$$\Delta_{\text{hyd}}G^0(M^+) = \Delta G^0_{M^+}(\text{temporary}) + Y = \frac{-685.7}{(\text{rad})_+ + \delta_+} \tag{12.29a}$$

$$\Delta_{\text{hyd}}G^0(X^-) = \Delta G^0_{X^-}(\text{temporary}) - Y = \frac{-685.7}{(\text{rad})_- + \delta_-} \tag{12.29b}$$

The cation values are adjusted by $+Y$ and the anion values are adjusted by $-Y$, since the sum $(\Delta_{\text{hyd}}G^0_{M^+} + \Delta_{\text{hyd}}G^0_{X^-})$ is independent of the error in the temporary dissection. The data fitting therefore involves the adjustment of Y, δ_+, and δ_-. Figure 12.7 shows such a fit both for the original (1939) data and for more recent (1994) data [Latimer et al., 1939; Marcus, 1994]. In both cases the error of fit is near 10 kJ/mol, and the absolute values inferred for the ionic $\Delta_{\text{hyd}}G^0$'s agree quite well. The values obtained for δ_+(0.8 Å) and δ_-(0.2–0.3 Å) are plausible in view of the models shown in Fig. 12.6. Note, however, that the 10-kJ/mol error of fit exceeds both the random experimental errors and the 2.5-kJ/mol magnitude of RT by significant factors.

The addition of δ_+ and δ_- (δ for short) to the ionic hard-sphere radii is not trivial. While the success of Eqs. (12.29) shows that cations and anions define a single relationship of $\Delta_{\text{hyd}}G^0$ versus $[(\text{rad}) + \delta]$, the plot of $\Delta_{\text{hyd}}G^0$ versus the

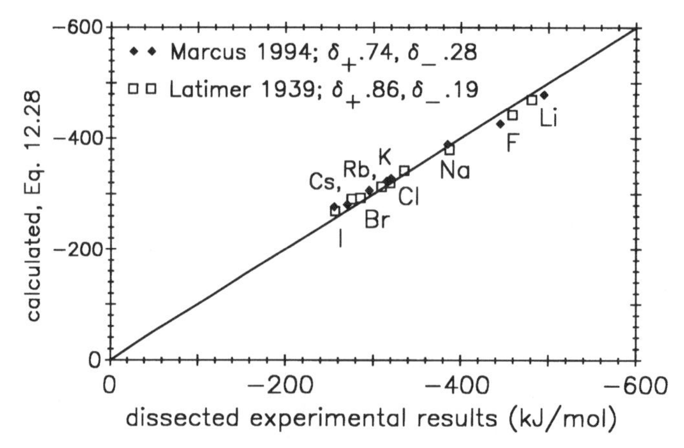

Figure 12.7. Ionic standard free energies of hydration. Dissected experimental results compared with values calculated by Eqs. (12.28).

original ionic crystal radius (rad) is dispersed into one relationship for cations and another, separated by 100 kJ/mol, for anions (Fig. 12.8). The dispersion can be minimized by using a different value for Y [see Eqs. (2.29)], but that hurts the error of fit and damages consistency with other theoretical approaches.

To prove consistency among different methods of dissection, we need to give results for only one ionic species, because values for other ionic species then are fixed by the experimental results for neutral combinations. Thus Table 12.3 lists results for the hydration of $H^+(g)$ based on various theoretical approaches. These approaches all involve the Born equation but differ in particular assumptions, which may entail elements of electrostatics, electrochemistry, physical

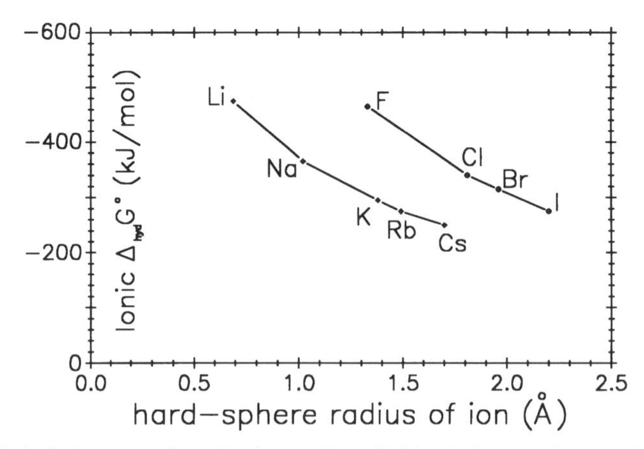

Figure 12.8. Ionic free energies of hydration (from Table 12.4) plotted versus the ionic crystal radius, showing dispersion into separate lines.

Table 12.3. $\Delta_{hyd}G^0$ and $\Delta_{hyd}H^0$ for H^+ (g; atm units) → H^+ (aq; molal units)

Source	$\Delta_{hyd}G^0$	$\Delta_{hyd}H^0$
Born [1920]		ΔU −1100
Latimer et al. [1939]; Eqs. 12.29	−1050 ± 10	−1100 ± 10
Noyes [1962]; Born equation with dielectric saturation	−1085	−1100
Halliwell and Nyburg [1963]; dominant charge-quadrupole interaction		−1090 ± 10
Conway [1978]; review of various methods	−1070 ± 20	−1100 ± 20
Reiss and Heller [1985]; electrochemical (absolute potential of H_2/H^+ electrode)	−1080 ± 10	
Marcus [1987]; physical–organic ($Ph_4As^+ = Ph_4B^-$)		−1100 ± 10

chemistry, and physical organic chemistry. As Table 12.3 shows, the results are in notable agreement, and the ionic $\Delta_{hyd}G^0$'s based on them are probably robust to 20 kJ/mol.

Results for the hydration of representative cations and anions, as well as for some model nonelectrolytes, are listed in Table 12.4. Most of these values are taken from a comprehensive review [Marcus, 1994]. Concentrations are specified in moles/liter, both in the gas phase and in aqueous solution, so that the listed properties of hydration are independent of concentration units. On the whole, the magnitudes of the properties are substantially greater for ions than for the model nonelectrolytes, which explains the emphasis on electrostatics in theoretical treatments.

A *sole* emphasis on electrostatics is not warranted, however. For example, the results for tetraphenylarsonium ion (Ph_4As^+) and tetraphenylboride ion (Ph_4B^-) in Table 12.4 (whose accuracy is comparable to that for the alkali and halide ions) differ substantially from prediction for a purely electrostatic medium effect. Indeed, $\Delta_{hyd}G^0$ for these ions is not even negative, and nonelectrostatic interactions appear to be important.

To interpret these results, we note that the tetraphenyl ions have closely similar radii, estimated at 4.2–4.3 Å, which are large enough to obviate dielectric saturation of the solvent. The Born charging energies should therefore be practically equal. Moreover, the charge-bearing central atom is embedded within the tetrahedrally configured phenyl groups, and thus it is not in contact with water. The nonelectrostatic (van der Waals) solvation is limited to the phenyl groups and, in view of the relatively large ionic radius, would hardly be influenced by the sign of the electric charge. One expects, therefore, that the Ph_4As^+ and Ph_4B^- ions will have nearly identical properties of solvation [Eq. (12.30)]:

Table 12.4. Thermodynamics of ionic hydration at 298 K[a]

Ion	$(\text{rad})_{\pm}$ (Å)	ΔG^0 (kJ/mol)	ΔH^0 (kJ/mol)	ΔS^0 (J/mol K)	ΔC_P^0 (J/mol K)	V_i (aq) (cm^3/mol)
			Cations			
Li^+	0.69	−475	−530	−161	−23	−6.4
Na^+	1.02	−365	−415	−130	−42	−6.7
K^+	1.38	−295	−330	−93	−72	3.5
Rb^+	1.49	−275	−305	−84	−94	8.6
Cs^+	1.70	−250	−280	−78	−108	15.8
Tl^+	1.50	−300	−335	−91	−103	5.4
H^+ (proton)		−1060	−1095	−120		
NH_4^+	1.48	−285	−325	−131	−29	12.4
Me_4N^+	2.80	−160	−215	−163	74	84.1
Et_4N^+	3.37	0	−205	−241	259	143.6
Ph_4As^+	4.25	50	−45	−321	803	295.2
Mg^{2+}	0.72	−1830	−1945	−350	−172	−32.2
Ca^{2+}	1.00	−1505	−1600	−271	−183	−28.9
			Anions			
F^-	1.33	−465	−510	−156	−59	4.3
Cl^-	1.81	−340	−365	−94	−70	23.3
Br^-	1.96	−315	−335	−78	−74	30.2
I^-	2.20	−275	−290	−55	−64	41.7
OH^-	1.33	−430	−520	−180	−91	−0.2
NO_3^-	1.79	−300	−310			34.5
ClO_4^-	2.50	−205	−245			49.6
Ph_4B^-	4.21	50	−45	−327	781	283.1
CO_6^{2-}	1.78	−1315	−1395	−264	−196	6.7
SiF_6^{2-}	2.59	−930	−980	−162	−161	50.5
$PtCl_6^{2-}$	3.95	−685	−740	−182		161.0
			Nonelectrolytes[b]			
Me_4C		2.51	−6.1	−28.9		
Benzene		−0.77	−7.1	−21.2		

[a]Concentrations are expressed in moles/liter in both phases [Marcus, 1994].
[b]*Source*: Me_4C: Wilhelm et al. [1977]; benzene: Franks et al. [1963]; Gill et al. [1976].

Model of hydration of $Ph_4As^+ + Ph_4B^-$

$$\Delta_{\text{hyd}}G^0(Ph_4As^+) = \Delta_{\text{hyd}}G^0(Ph_4B^-) = (1/2)\Delta_{\text{hyd}}G^0(Ph_4As^+ + Ph_4B^-)$$
$$\Delta_{\text{hyd}}H^0(Ph_4As^+) = \Delta_{\text{hyd}}H^0(Ph_4B^-)$$
$$\Delta_{\text{hyd}}C_P^0(Ph_4As^+) = \Delta_{\text{hyd}}C_P^0(Ph_4B^-)$$
and similarly for other properties of solvation (12.30)

In the present case, $\Delta_{\text{hyd}}H^0$ for the combination ($Ph_4As^+ + Ph_4B^-$) can be measured, and Eq. (12.30) lets us dissect it into separate ionic terms. Table 12.3 shows that the results based on this approach are consistent with those by other methods [Marcus, 1987].

Change of Solvent

The thermodynamics of transfer of ionic species from one solvent to another is essential for predicting ionic reactivity in a wide range of applications, including extraction technology, phase-transfer technology, and chemical synthesis. As a result, extensive data have accumulated over many decades. Current reviews are organized to show the transfer of ions from water to organic solvents [Parker, 1969; Marcus, 1983, 1985, 1986; Inerowicz et al., 1994]. The differences of such values represent the transfer from one organic solvent to another.

The transfer free energies ($\Delta_t G^0$) and enthalpies ($\Delta_t H^0$) of the ionic species are deduced in these reviews by applying the tetraphenylarsonium tetraphenylboride assumption (12.30). Grunwald et al. [1960] had originally proposed this method because tetraphenyl ions such as Ph_4As^+, Ph_4P^+, and Ph_4B^- are analogous to size-matched oil droplets with centrally located electric charges. Moreover, the radius is large enough so that the electric field produced in the surrounding solvent is moderate and the electrostatic work of transfer, $L\Delta_t W$, can be estimated by the Born equation. The additional nonelectrostatic (van der Waals) component of the free energy of transfer can be estimated via the model nonelectrolyte, tetraphenylmethane (Ph_4C), as in Eq. (12.31):

$$\Delta_t G^0(\text{tetraphenyl ion}) = \Delta_t G^0(Ph_4C) + L\Delta_t W(\text{Born}) \tag{12.31}$$

Equation (12.31) was tested in dioxane–water mixtures [Grunwald et al., 1960; see also Table 3.4]. Letting Z_1 denote the formula-weight fraction of water, the facts for 50 wt% dioxane–water at 298 K are as follows:

$$\partial G^0/\partial Z_1(Ph_4C) + L\,\partial W(\text{Born})/\partial Z_1 = (85 \pm 2.5) - 20 = 65 \pm 2.5\text{kJ/mol}$$
$$\partial G^0/\partial Z_1(\text{tetraphenyl ion}) = 62 \pm 3\text{ kJ/mol}$$

Note that the nonelectrostatic term, $\partial G^0/\partial Z_1(Ph_4C)$, is greater than the Born term and of opposite sign. The 3-kJ/mol departure from equality measures both the random experimental error and the determinate error of this approach. A conservative estimate of the latter, when tetraphenyl ions and Eqs. (12.30) are used, is therefore 3 kJ/mol. Meticulous studies show that further reduction of this determinate error through the use of alternative model ions is unlikely [Parker, 1969; Pirklbauer and Gritzner, 1993].

Standard free energies and enthalpies of transfer from water to a series of organic solvents for K^+, Cl^-, Li^+, ClO_4^-, and Ph_4As^+/Ph_4B^- are listed in

Table 12.5. Standard ionic free energies and enthalpies of transfer from water to a series of organic solvents, in kilojoules per mole[a]

Organic Solvent (ε_M)	$\Delta_t G^0$			$\Delta_t H^0$		
	K^+	Cl^-	Ph_4As^+	K^+	Cl^-	Ph_4As^+
Methanol, 32.5	+10	+13	−24	−19	+9	−2
PC[b], 65.0	+5	+40	−36	−22	+26	−13
DMSO[b], 46.7	−12	+39	−37	−36	+18	−11
Acetonitrile, 37.0	+9	+44	−33	−24	+20	−10
Pyridine, 12.4	+6	+37	−38	−23	+28	−23
DMTF[b]	+22	+22	−39	−24	0	0
	Li^+	ClO_4^-		Li^+	ClO_4^-	
PC[b], 65.0	+24	−3		+3	−16	
Acetonitrile, 37.0	+24	+3		(1)	−17	

[a]From reviews by Marcus [1983, 1985, 1986] and Inerowicz et al. [1994].
[b]PC = propylene carbonate; DMSO = dimethylsulfoxide; DMTF = *N*-dimethylthioformamide.

Table 12.5. The combined systematic and random errors of these values are about 3 kJ/mol. The values indicate that solvent effects are complicated.

For Ph_4As^+/Ph_4B^- the van der Waals solvation of the phenyl groups is more controlling than the ionic charge, because the transfer from water to any organic solvent—in spite of the decrease in dielectric constant—lowers the standard free energy. The entropy of transfer, as measured by $T\Delta_t S^0 = \Delta_t H^0 - \Delta_t G^0$, is positive.

For K^+ and Li^+, $\Delta_t G^0$ is positive, as expected for the transfer of an ionic charge to a less polar medium; but the magnitude varies widely, and one value is negative. The entropy of transfer, $T\Delta_t S^0 = \Delta_t H^0 - \Delta_t G^0$, is negative.

For Cl^-, $\Delta_t G^0$ is strikingly positive; for ClO_4^-, $\Delta_t G^0$ is close to zero. The entropy of transfer for both ions is negative.

For comparison with the free energies of transfer, Table 12.6 shows equilibrium constants for the association of the ions with water in dilute solutions in two different organic solvents. In these measurements, the dielectric constant of the medium is essentially constant. On the whole, the pattern displayed by the association constants in Table 12.6 correlates with that displayed by $\Delta_t G^0$ in Table 12.5.

In discussions of electrostatic medium effects, the emphasis is usually on the ionic crystal radius $(\text{rad})_J$ of the ion and the dielectric constant ε_M of the solvent, while the δ parameter of Eq. (12.28a), which expresses the effect of electrical saturation and screening, is treated with benign neglect. That is a hazardous omission, because the electrostatic work $L\Delta W$ is quite sensitive to changes in δ.

Table 12.6. Association of ions with water in organic solvents at 298 K[a]

Solvent	Ion	K_{ass}
Propylene carbonate (ε 65.0)	Li^+	6.5 ± 0.3
	K^+	0.4 ± 0.2
	Cl^-	6.5 ± 0.4
	ClO_4^-	0.3 ± 0.1
Acetonitrile (ε_M 37.0)	Li^+	4 ± 1
	K^+	1 ± 0.5
	Cl^-	10 ± 2
	ClO_4^-	1 ± 0.5

[a] $K_{ass} = [\text{Ion} \cdot H_2O]/[\text{Ion}][H_2O]\ \text{m}^{-1}$.
Source: Cogley et al. [1971] and Chantooni and Kolthoff [1967].

PROOF. Consider two solvents, M_1 and M_2, with dielectric constants ε_1 and ε_2. Let M_1 be a fixed reference solvent (such as water in Table 12.5) and let M_2 be variable. Consider the transfer of the ion J [ionic crystal radius $(\text{rad})_J$] from M_1 to M_2. Eq. (12.28a) then takes the form (12.32):

$$L\Delta W_{1\to 2} = \frac{Lz_J^2e^2}{2}\left(\frac{1/\varepsilon_2 - 1}{(\text{rad})_J + \delta_2} - \frac{1/\varepsilon_1 - 1}{(\text{rad})_J + \delta_1}\right) \tag{12.32}$$

Partial differentiation of Eq. (12.32) at constant $\{(\text{rad})_J, \varepsilon_1, \delta_1\}$ yields

$$\partial\Delta W_{1\to 2}/\partial\varepsilon_2 = -(z_J^2e^2/2)/\varepsilon_2^2((\text{rad})_J + \delta_2) < 0$$
$$\partial\Delta W_{1\to 2}/\partial\delta_2 = -(z_J^2e^2/2)[(1/\varepsilon_2) - 1]/((\text{rad})_J + \delta_2)^2 > 0$$

For definiteness, let $z_J = 1$, $(\text{rad})_J = 2.0$ Å, $\delta_2 = 0.5$ Å, and $\varepsilon_2 = 30$. Then $\partial\Delta W_{1\to 2}/\partial\varepsilon_2 = -0.31$ kJ/mol per unit increase in dielectric constant, while $\partial\Delta W_{1\to 2}/\partial\delta_2 = +107$ kJ/mol(!) per Å increase in δ_2. That is, a change in ε_2 by 10 units causes an electrostatic stabilization of ≈ 3 kJ/mol, while an increase in δ_2 by only 0.1 Å causes a destabilization of ≈ 11 kJ/mol. Clearly δ_2 is a major variable. Q.E.D.

Electrical saturation and electrostatic screening are obverse sides of the same coin, since they involve alignment of the same solvent dipoles. In the following we will focus on electrostatic screening, because the microscopic dimensions are more predictable. For an ion interacting with a solvent dipole, the screening may be tight or loose, as shown for a cation in Fig. 12.9. In tight screening the negative pole of the solvent dipole sits near the surface, and δ is small. For the same cation in loose screening, the negative pole is embedded, and δ is

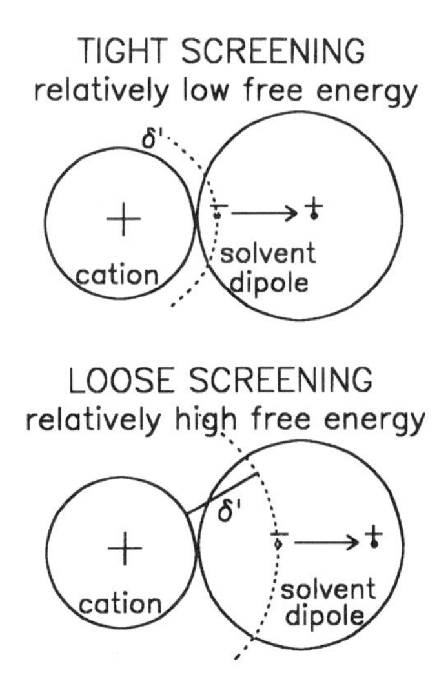

Figure 12.9. Contribution to the screening of a cationic charge by the negative end of a solvent dipole. The screening is tight when the negative end is exposed, but is loose when it is embedded within the steric contour of the solvent molecule.

larger. In both cases the complementary positive pole is effectively outside the screening layer.

For instance, in the dimethylsulfoxide molecule $(CH_3)_2(S^{+pole})(O^{-pole})$, the positive pole is sterically embedded within the methyl groups and is relatively distant from an anion even at close approach. The negative pole, on the other hand, is exposed and available for close interaction with a cation. Accordingly, dimethylsulfoxide is a good solvent for monatomic cations such as K^+ (note the negative value for $\Delta_t G^0$ in Table 12.5), while anions such as Cl^- are inefficiently solvated and relatively reactive [Parker, 1969]. When the anion is small (as in the following example), the resulting enhancement of reactivity can be astonishing.

In the transfer of 0.01 f $(CH_3)_4N^+OH^-$ from water to dimethylsulfoxide (0.05 f in H_2O), the proton-removing base strength of hydroxide ion increases by 14.2 orders of magnitude [Dolman and Stewart, 1967], so that $\Delta_t G^0$ is about +80 kJ/mol. Using a radius of 1.3 Å for hydroxide ion and a screening thickness δ_1 of 0.2 Å in water, $\Delta_t G^0$ will be +80 kJ/mol if the screening thickness δ_2 in dimethylsulfoxide (0.05 f in water) has the plausible value of 0.6 Å.

ION PAIRS

The ionic species considered so far are individual ions. In this section we will consider ionic aggregates, especially ion pairs.

An ion pair is a pair of oppositely charged ions that are bound by the attraction of their static charges, without forming a covalent bond. Examples are 1 : 1 pairs such as tetraisoamylammonium nitrate (i-Am_4^+, NO_3^-) or 2 : 2 pairs such as cupric sulfate (Cu^{2+}, SO_4^{2-}). Because charge–charge interactions are of relatively long range, there are two kinds of ion pairs: *loose ion pairs* in which the ions are separated by the solvent medium (which may or may not be modeled as a dielectric continuum) and *tight ion pairs* in which the ions are in direct contact. Ionic species consisting of individual ions are called *free* ions—"free" because such ions do not form kinetic units with other ions.

When an electrolyte forms both tight and loose ion pairs, the two are separated by an energy barrier and are distinguishable. The ions in tight ion pairs are bound not only by charge–charge interactions, but also by short-range van der Waals interactions, including the attraction resulting from their mutual polarization. In the mutual-polarization mechanism, each ion in the pair induces electric charges in the counterion so as to produce an additional electrostatic stabilization. This stabilization varies with the nature of the contacting ions and, in favorable cases, approaches the order required for the formation of molecular complexes [Grunwald et al., 1974].

While tight-ion pairs are distinguishable from loose ion pairs, the loose ion pairs may not be distinguishable from their free ions. In practice, such distinction rests on electrical conductivities, since the free ions conduct and the ion pairs, by definition, do not. On the other hand, loose ion pairs and free ions are not separated by a potential barrier in configuration space and thus lack ideal distinguishability.* In some views, the absence of ideal distinguishability is a crucial flaw, and free ions and loose ion pairs are treated as belonging to a single molecular species. But before considering this issue, let us examine some experimental facts.

Some Experimental Facts

The free energy associated with charge–charge interactions varies inversely as the dielectric constant. Thus, ion-pair formation tends to be negligible in solvents of high dielectric constant, and it becomes increasingly significant as ε_M decreases. In practice, this means that ionic compounds whose ions satisfy the octet rule behave like strong electrolytes in solvents such as water ($\varepsilon = 78$), but resemble weak electrolytes in solvents such as ethylene chloride ($\varepsilon = 10$). By "strong" we mean that in dilute solutions only long-range interionic effects proportional to $\sqrt{I}$ are present—the kind that can be predicted by the Debye–Hückel theory [1923] and by Onsager's [1927] related theory of electrolytic conductance.**

*In the absence of a potential barrier, the distinguishability index $\xi = 2\pi\tau|E_b^{(C)} - E_a^{(C)}|/Lh$ [Eq. (10.9)] can be made arbitrarily small simply by letting $E_b^{(C)}$ lie just below the defined boundary between the two molecular populations, and $E_a^{(C)}$ just above.

**For a univalent electrolyte, Onsager's limiting law takes the form $\Lambda = \Lambda^\circ - [0.8206 \times 10^6 \Lambda^\circ/(\varepsilon T)^{3/2} + 82.48/\eta(\varepsilon T)^{1/2}]\sqrt{c}$, where Λ = (1000 · conductivity/c), the equivalent conductance.

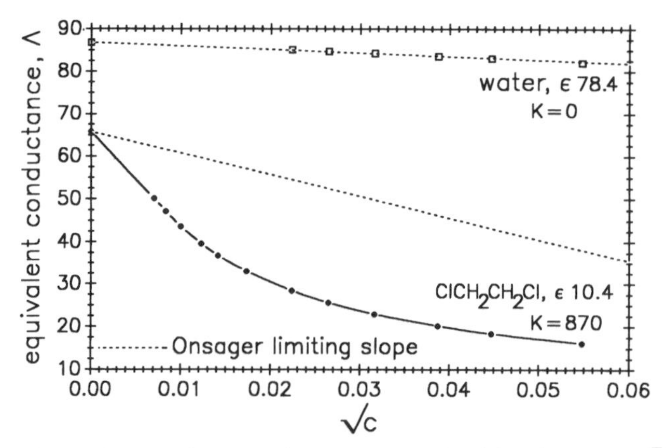

Figure 12.10. Equivalent conductance ($\Lambda = 1000 \cdot \text{conductivity}/c$) versus $\sqrt{c}$ for tetraisoamylammonium nitrate in water and ethylene chloride at 298 K [Kraus and Fuoss, 1933]. The dashed lines represent the Onsager [1927] limiting slopes. (See second footnote on p. 295.)

For example, Fig. 12.10 shows conductance data for tetraisoamylammonium nitrate, $i\text{-}Am_4N^+NO_3^-$, in water and ethylene chloride, at concentrations up to 0.003 formal [Kraus and Fuoss, 1933; Fuoss and Kraus, 1933]. The dashed lines represent Onsager's [1927] limiting law (see second footnote on p. 295) and are expected to fit the data when the solute is a strong electrolyte. On that basis, $i\text{-}Am_4N^+NO_3^-$ is a strong electrolyte in water but not in ethylene chloride. In the latter solvent, the conductance curve fits a model in which the electrolyte is an equilibrium mixture of conducting free ions and nonconducting ion pairs, Eq. (12.33):

$$M^+ + X^- \rightleftarrows M^+,X^-, \qquad K = [M^+,X^-]/[M^+][X^-] \cdot (\gamma_{+-}/\gamma_+\gamma_-) \qquad (12.33)$$

Here, K is the association constant for ion-pair formation, and the γ's are activity coefficients. According to the Debye–Hückel/Onsager limiting laws, the free ions (but not the electrically neutral ion pairs) create the ionic strength whose magnitude enters Eq. (12.20) and Onsager's expressions for the conductivity. On the same basis, since $z_{+-} = 0$, we have $\gamma_{+-} = 1$. In the concentration range of Fig. 12.10, the fit of this model with $K = 870$ is very good, as indicated by the solid curve.

Equation (12.33) normally fits only at low concentrations of dissolved electrolyte. As that concentration increases, in solvents of low dielectric constant, ionic association proceeds in further steps to yield triple ions ($M^+X^-M^+$ and $X^-M^+X^-$), quadruple ions or double ion pairs $(M^+X^-)_2$, and still higher ionic aggregates [Fuoss and Kraus, 1933, 1935; Batson and Kraus, 1934; Kraus and Vingee, 1934]. This demonstrates that ionic association differs from covalent bond formation, in that the valence (or binding capacity toward counterions) is flexible.

The fit of the model of Eq. (12.33) to conductance data indicates that there are free ions as well as nonconducting ion pairs, but it does not tell us whether the ion pairs are tight or loose. Our description now depends on whether we believe that loose ion pairs are distinguishable from free ions. If not, the loose ion pairs and free ions in principle are a single species, although we shall find that strong pairwise interactions among free ions can produce properties that mimic loose ion pairs.

To recognize the presence of tight ion pairs, a reliable technique is to look for perturbations of ionic properties when conductance data indicate the presence of ion pairs—perturbations such as changes in optical or nuclear magnetic resonance (NMR) absorption or changes in reactivity. When the properties of the ions in ion pairs are scarcely different from those of the free ions, it is reasonable to assume that most of the ion pairs are loose. For example, in acetic acid as solvent, electronic absorption spectra for a wide range of ions remain practically unchanged as free ions become ion pairs [Kolthoff and Bruckenstein, 1956]. Or in liquid sulfur dioxide as solvent, the reactivity of bromide ion decreases only slightly when bromide ion is paired with an alkali cation [Lichtin and Rao, 1961; Lichtin et al., 1967].

On the other hand, when the ion pairs are tight, each paired ion exists in the specific, strong, inhomogeneous electric field due to its counterion, and perturbations of the ionic spectra by what amounts to a Stark effect show up. This phenomenon depends specifically on the nature of the counterion. Convincing examples have been reviewed by Smid [1972] and Szwarc [1972].

Tight ion pairing also produces a significant effect on reactivity. For example, in the S_N1 mechanism of nucleophilic substitution described in Eq. (5.3b), the substrate first forms an ion pair, which then reacts with the nucleophile. The mechanism of the reaction of this short-lived ion pair with the nucleophile has been probed using optically active substrates, by parallel measurements of kinetic salt effects and racemization. The facts require not one but two ion-pair intermediates of greatly different reactivity. The substrate first forms a tight ion pair whose reaction rate with nucleophiles is negligible. The tight ion pair then loosens up to form a solvent-separated ion pair whose reactivity toward nucleophiles is high [Winstein et al., 1956].

Ionic Association in a Dielectric Continuum

In this section, ionic solutions are modeled as consisting of nonpolarizable charged spheres in a dielectric continuum. This model, which parallels the electrostatic models described earlier, is associated with key publications by Bjerrum [1926] and Fuoss [1935]. It reproduces the magnitude of ion-pair association constants and, through the Fuoss distribution function $J(r)$, describes loose ion pairs with unusual clarity. We will derive it for a univalent electrolyte.

We begin with Bjerrum's ion-pair concept. Let a denote the distance of closest approach of two hard spheres with charges z_+e and z_-e, and let the center-to-center distance r be greater than a. According to Coulomb's law, the electrostatic pair energy then is $-z_+z_-e^2/\varepsilon r$; and in an isotropic medium at thermal equilibrium, the probability density that the corresponding ions are separated by a distance r is given by $p(r) = 4\pi r^2 \exp(-z_+z_-e^2/\varepsilon rkT)$. Bjerrum pointed out that $p(r)$ goes through a minimum when $r = e^2|z_+z_-|/2\varepsilon kT$ and that, because population minima in configuration space define boundaries between molecular species, this minimum is a logical choice for the boundary between free ions and ion-pairs. The value of r at minimum $p(r)$ is called the *Bjerrum distance* r_B:

$$r_B = e^2|z_+z_-|/2\varepsilon kT \tag{12.34}$$

By this convention, the ions are free when $r > r_B$ and form an ion-pair when $r \leq r_B$. Note that when $r = r_B$, the Coulomb energy $Le^2|z_+z_-|/\varepsilon r = 2RT$ per mole, which is a significant binding energy.

Having defined the Bjerrum distance r_B, we shall now derive the Fuoss distribution function $J(r)$ for a univalent electrolyte. Let c be the formal concentration of the electrolyte (per liter) and let V be the solution volume (in cubic centimeters). Then, for a univalent electrolyte, $N = LcV/1000$ denotes the number of cations in the solution, which is equal to that of the anions. Now consider the pairing of cations and anions at statistical equilibrium. Let each positive ion be surrounded initially by a concentric sphere of radius $a/2$, a being the distance of closest approach, and let all the spheres expand at a uniform rate, while the ionic centers remain fixed in their statistically random sites. As soon as the sphere for a positive ion cuts the center of a negative ion, we will count the two ions as a pair, provided that the negative ion has not already been counted at a shorter distance as the partner of some other positive ion. In this way we assign one and only one partner to each ion.

On this basis the probability $J(r)\,dr$ that an ion will find its partner in a spherical shell between r and $r + dr$ is given by Eq. (12.35a), where $b = 2r_B = e^2/\varepsilon kT$ for a univalent electrolyte:

$$J(r)\,dr = \frac{4\pi N r^2\,dr}{V} e^{b/r} \left(1 - \frac{N-1}{N} \int_a^r J(x)\,dx\right) \tag{12.35a}$$

The factor $[1 - (N-1)/N \cdot \int J(x)\,dx]$ measures the probability that the nearest negative ion is not already the partner of another positive ion that happens to be closer. On expressing N/V in terms of c, letting $(N-1)/N = 1$, and eliminating dr, we obtain the working equation (12.35b), whose solution is (12.35c).

$$J(r) = \frac{4\pi L c r^2}{1000} e^{b/r} \left(1 - \int_a^r J(x)\,dx\right) \tag{12.35b}$$

$$J(r) = \frac{4\pi Lc}{1000} r^2 \exp\left(\frac{b}{r} - \frac{4\pi Lc}{1000} \int_a^r e^{b/x}\, dx \right) \tag{12.35c}$$

PROOF. Equation (12.35b) is solved by successive approximations.
First approximation:

$$J^{(1)}(r) = (4\pi Lc/1000) r^2 e^{b/r}$$

Second approximation:

$$J^{(2)}(r) = (4\pi Lc/1000) r^2 e^{b/r} \left(1 - \int_a^r J^{(1)}(x)\, dx \right)$$

Let $I = (4\pi Lc/1000) \int_a^r x^2 e^{b/x} dx$; then $dI = (4\pi Lc/1000) x^2 e^{b/x}\, dx$. Thus;

$$J^{(2)}(r) = (4\pi Lc/1000) r^2 e^{b/r} (1 - I).$$

Third approximation:

$$\begin{aligned} J^{(3)}(r) &= (4\pi Lc/1000) r^2 e^{b/r} \left(1 - \int_a^r J^{(2)}(x)\, dx \right) \\ &= (4\pi Lc/1000) r^2 e^{b/r} \left[1 - (4\pi Lc/1000) \int_a^r x^2 e^{b/x} (1 - I)\, dx \right] \\ &= (4\pi Lc/1000) r^2 e^{b/r} [1 - I + I^2/2] \end{aligned}$$

Fourth approximation:

$$J^{(4)}(r) = (4\pi Lc/1000) r^2 e^{b/r} [1 - I + I^2/2 - I^3/3!]$$

Convergence limit: As the order of approximation becomes infinite, $J(r) \to (4\pi Lc/1000) r^2 e^{b/r} e^{-I} = (4\pi Lc/1000) r^2 \exp[b/r - I]$ Q.E.D.

Plots of $J(r)$ versus r, with a = 5 Å, are given in Fig. 12.11 for $\varepsilon = 10$ (logarithmic scale) and in Fig. 12.12 for $\varepsilon = 30$ (linear scale). As expected, $J(r)$ shows a minimum at or near the Bjerrum distance r_B, but the effectiveness of that minimum at resolving the population into distinct subsets varies greatly. The resolution is good in Fig. 12.11, where $\varepsilon = 10$. Here $J(r)$ is small in a substantial region around r_B (27.6 Å), generating two populations that are separated by a wide, sparsely populated region in configuration space, and the concept of free ions as distinct from ion pairs is plausible.

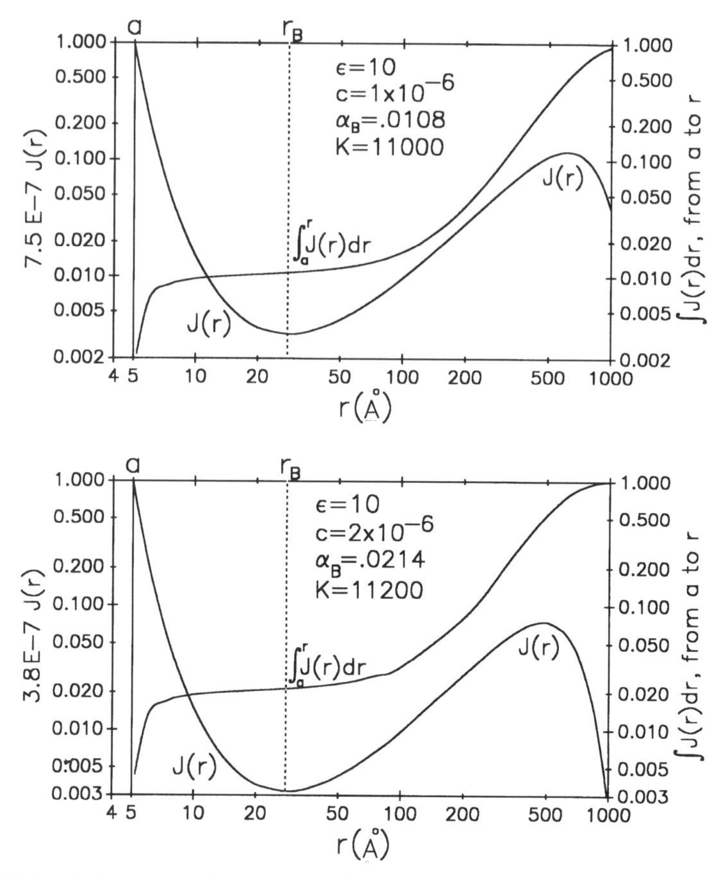

Figure 12.11. Note the logarithmic scales. The ion-pair distribution function $J(r)$ versus the cation-to-anion distance r at two concentrations at $\varepsilon = 10$; $a = 5$ Å. The peak above 500 Å occurs near the mean distance between free cations and anions.

In Fig. 12.12, at $\varepsilon = 30$, the distribution $J(r)$ is also bimodal, but the corresponding populations interpenetrate near the minimum at $r_B = 9.6$ Å, and there is no sparsely populated transition region to signal a changeover between two molecular species.

Properties of $J(r)$, and of the ion pairs based on $r \leq r_B$, are listed in Table 12.7. That the ion pairs defined in this way are "loose" is shown by their mean center-to-center distances $\langle r \rangle$, calculated by averaging $rJ(r)$ between a and r_B. The value of $\langle r \rangle$ exceeds that of a at all values of ε, but the relationship is not simple. As ε decreases and $(a + r_B)/2$ increases, $\langle r \rangle$ increases to a maximum at $\varepsilon \approx 15$ and then decreases. The decrease occurs as ε becomes small enough so that the exponential factor, $\exp(e^2/\varepsilon rkT)$, which favors short distances, becomes controlling over the geometric factor $4\pi r^2$ in the averaging region.

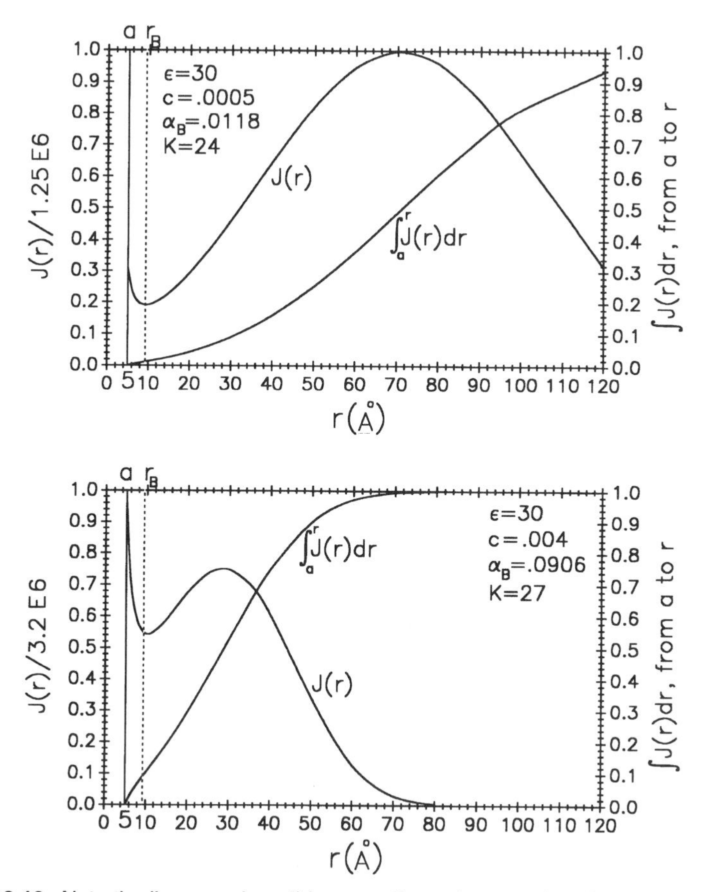

Figure 12.12. Note the linear scales. $J(r)$ versus the cation-to-anion distance r at two concentrations at $\varepsilon = 30$; $a = 5$ Å. The peak above 20 Å occurs near the mean distance between free cations and anions.

Table 12.7. Properties of ion pairs predicted by $J(r)$ at low values of c

ε	a/Å[a]	$\langle r \rangle$/Å[b]	r_B/Å	K (liters/mol)	ΔG^0 (kJ/mol)
30	5	6.99	9.28	24.2	−7.90
20	5	8.39	13.93	152	−12.45
15	5	8.76	18.57	640	−16.02
10	5	7.15	27.86	1.09×10^4	−23.05
7.5	5	5.83	37.15	2.64×10^5	−30.94

[a]This is a typical value.
[b]$\int r J(r)\,dr / \int J(r) dr$, integrated from a to r_B.

If we accept the Bjerrum distance r_B as the changeover mark from ion pairs to free ions, the distribution function $J(r)$ lets us predict ion-pair association constants. The relationship is $K = \alpha_B/[c(1 - \alpha_B)^2]$, where $\alpha_B = \int_a^{r_B} J(r)\,dr$. If K is indeed a genuine equilibrium constant, its values (for given a and ε) should be independent of c, and this is true when c is small. However, as c increases, K begins to increase, as illustrated for $\varepsilon = 30$ in Fig. 12.12. One reason for this is that the factor $[1 - \int J(x)\,dx]$ in Eq. (12.35) neglects all perturbations by ion pairs with center-to-center distances less than r that happen to encroach upon the sphere with radius r. This neglect is tantamount to neglecting double and higher-order ion pairs and is certain to introduce error at the higher concentrations [Stillinger and Lovett, 1968]. On the other hand, the constant values of K obtained at low concentrations, which are given in Table 12.7, should be all right.

The Bjerrum–Fuoss approach and the corresponding model of loose ion pairs account credibly for the dependence of log K on the dielectric constant. In particular, plots of log K versus $1/\varepsilon$ in a solvent series tend to be nearly linear. For example, Fig. 12.13 shows such a plot for cesium iodide in dioxane–water mixtures [Lind and Fuoss, 1961; Fuoss and Hsia, 1967], together with a similar plot based on the theoretical association constants listed in Table 12.7. The experimental data span 10 kJ/mol; the calculated association constants span 23 kJ/mol. Both plots are nearly linear and resemble each other with respect to magnitude of ΔG^0 and slope.

Since the electrostatic work of bringing opposite univalent charges from infinity to a distance l is given by $\Delta W = -Le^2/\varepsilon l$, Denison and Ramsey [1955] suggested that the slope of the plot of ΔG^0 versus $1/\varepsilon$ for ion-pair formation be

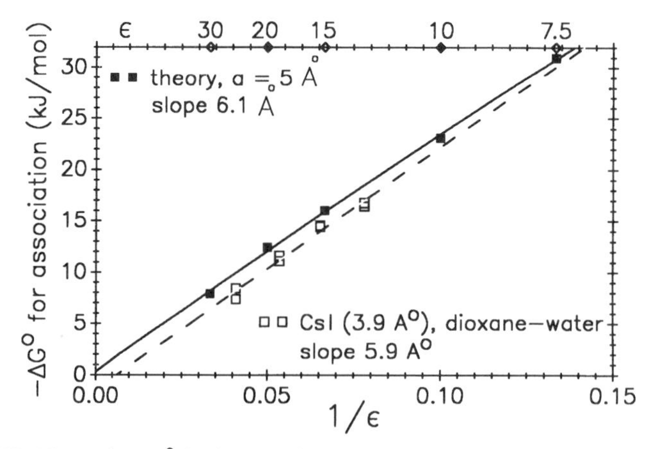

Figure 12.13. Plots of $-\Delta G^0$ for ion-pair formation versus $1/\varepsilon$. The solid line is based on integration from a to r_B of the distribution function $J(r)$, with $a = 5$ Å. The dashed line is obtained by least-squares fitting of highly accurate conductance data for CsI (radius sum 3.9 Å) in dioxane–water mixtures, using two different interionic attraction models (two sets of open squares) [Lind and Fuoss, 1961; Fuoss and Hsia, 1967].

identified with $-Le^2/l$, where l is the effective cation-to-anion distance in the ion pairs. Both plots in Fig. 12.13 give values for l that are distinctly greater than closest approach: For the theoretical plot, $l = 6.1$ Å and $a = 5$ Å. For the CsI plot $l = 5.9$ Å, while the ionic radius sum is 3.9 Å. Since the theoretical plot represents loose ion pairs, analogy suggests that the CsI ion pairs are loose. Note also that the data for CsI include relatively high dielectric constants, where free ions and ion pairs interpenetrate and are not separated by a sparsely populated transition region.

Ion Pairs Regarded as Molecular Species

In spite of the success of the Bjerrum–Fuoss ion-pair concept, the loose ion pairs that derive from it lack certain attributes that are integral to a molecular species: They lack a definite and distinctive equilibrium geometry, a distinctive set of molecular modes of motion, and their configuration space is not isolated from that of the free ions by a potential barrier. Fuoss and others recognized this problem and, beginning in the 1950s, tried to modify the ion-pair concept without compromising agreement with experiment.

Fuoss [1958, 1959] assumed that the ion pairs are tight, with the ions in contact at closest approach a. Thus $K = 4\pi La^3 \exp(e^2/\varepsilon akT)/3000$ [compare Eq. (9.3)], and γ_+ and γ_- are expressed by Eq. (12.21b) (with $a_j = a$). The interionic attraction theory for conductivity was revised accordingly [Fuoss and Onsager, 1958]. These changes caused modest changes in the values obtained for K while the agreement with experiment remained good. But because this approach expresses the interionic attractions to a higher-than-square-root order in the ionic strength, consistency required that other interactions that are first-order in ionic strength (or concentration) now be included [e.g., Fuoss and Hsia, 1967; Quint and Viallard, 1978]. As these additions were successively formulated and tested, it became clear that the fit of conductance data changed hardly at all, being excellent throughout, but that the values obtained for K were less robust and varied noticeably. For example, the two sets of points for CsI in Fig. 12.13 represent identical data interpreted by successive versions of theory.

There is another point: The equilibrium constant K is defined stoichiometrically by Eq. (12.33). The product species, denoted as M^+, X^-, therefore includes all entities formed from the free ions with 1 : 1 stoichiometry: tight ion pairs, loose ion pairs, and MX molecules. This is required by the Laws of Stoichiometry and the Correspondence Theorems.

Thus, even if we detect characteristic optical or NMR absorption due to tight ion pairs, we still don't know whether loose ion pairs are also present. To measure the fraction of tight ion pairs in the overall M^+, X^- species, we need to know absolute extinction coefficients. NMR band areas, which are basically nonspecific, are especially useful here.

When the electrolyte shows both absorption due to tight ion pairs and conductivity due to free ions, it is virtually certain that loose ion pairs, entities with $a < r \leq r_B$, are also present. This is a consequence of defining the boundary between free ions and ion pairs at r_B rather than at a. But even if we discard that definition, it is virtually certain that entities mimicking loose ion pairs are present.

The evidence again comes from statistical mechanics. In 1950, J. E. Mayer applied exact statistical mechanics to a model of charged hard spheres in a dielectric continuum. His procedure knew no Bjerrum distance and treated all charge–charge interactions by the same physical mechanism. To prove that mathematical solutions exist, the formulation became quite complicated, involving "cluster expansions" and "hypernetted chains." A "cluster" is a set of logically related terms of identical algebraic order, garnered in the power-series expansion of the original exponential equations. A "chain" is a set of such clusters, linked according to a recipe that cancels out divergent mathematical solutions. Mayer's formulation transcends the language of common mathematical operators, and he devised special operators expressly to solve this problem [Mayer, 1950].

Some years later, when adequate computers were available, H. L. Friedman [1962] and Rasaiah and Friedman [1968, 1969] mastered the special operators and solved Mayer's equations to a high approximation. Their thermodynamic results, which include activity coefficients, come close to representing exact statistical mechanics for charged hard spheres in dielectric continua.

A mathematically more tractable approximate version of this approach, called the *mean spherical model*, has been developed by Lebowitz and Percus [1966] and Blum [1975] and applied with growing confidence [Ebeling and Grigo, 1982; Cartailler et al., 1992; Kessling and De Maeyer, 1995]. The mean spherical model yields mathematical solutions in closed form. In common with Mayer's formulation, it reduces to the Debye–Hückel limiting law at high dilutions and does not single out the Bjerrum distance. It is more flexible, however, and allows inclusion of supplemental models. In particular, it can be solved to predict interionic effects in electrolytic conductance.

Figure 12.14 compares the nearly exact statistical mechanical results of Rasaiah and Friedman [1968, 1969] with results of the mean spherical model [Cartailler et al., 1992]. The osmotic coefficient ϕ of the solvent, plotted along the ordinate, is closely related to the activity coefficient $\gamma_{\pm}$, according to

$$\ln\ \gamma_{\pm}(m') = \phi - 1 + \int_0^{m'} (\phi - 1)\, d \ln\ m$$

Figure 12.14a compares results based on calculations that do not single out the Bjerrum distance. There is a substantial discrepancy. Figure 12.14b compares the same Rasaiah and Friedman results with a version of the mean spherical model that *does* single out the Bjerrum distance. Here the mean spherical model has been modified to include loose ion-pair formation, basically according to Eq. (12.33) but using an unorthodox K that combines elements of Bjerrum's theory with a mild concentration dependence. The discrepancy is now much reduced.

We interpret Fig. 12.14 as follows. The mean spherical model yields mathematically simple solutions in closed form because its assumptions downplay the

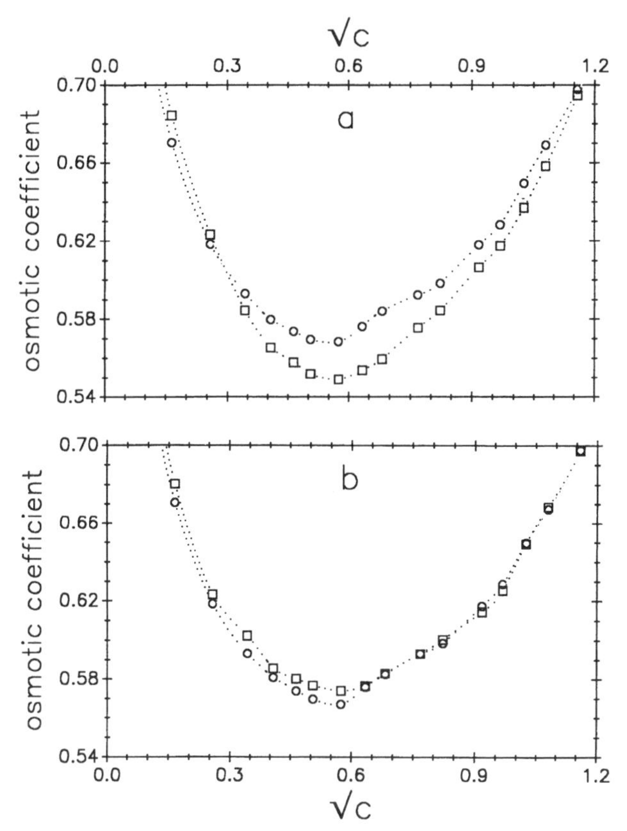

Figure 12.14. Osmotic coefficient of a 2 : 2 electrolyte composed of charged hard spheres of radius 4.2 Å in a continuum with the dielectric constant of water. Circles in both (a) and (b): nearly exact statistical mechanics. Squares, mean spherical model: (a) without and (b) with inclusion of loose ion pairs [Rasaiah and Friedman, 1968, 1969; Cartailler et al., 1992].

short-range interactions. That is why the results of this model in Fig. 12.14a differ substantially from those of the nearly exact calculation. In an attempt to compensate for the downplay, the mean spherical model in Fig. 12.14b is modified to include loose ion pairs. The improved agreement with the nearly exact calculation shows that when there are free ions, at concentrations at which ionic interactions are significant, the strong pairwise interactions that arise at the shorter distances mimic loose ion pairs. The Bjerrum–Fuoss model specifies the distances at which the mimicking is good enough. When it fits, we might as well use it.

Solvent-Separated Ion Pairs

In the world of discrete molecules, the interionic distance r at short ranges is not a continuous variable but, instead, increases in quanta that correspond to the progressive insertion of solvent molecules. In tight ion pairs the ions are nearest neighbors. When separated by a single solvent molecule the ions are next-nearest neighbors. When separated by n solvent molecules, with $n \geq 2$, the number of probable solvent configurations between the ions increases rapidly with n, so that the interionic distance r soon becomes a quasi-continuum.

The term *solvent-separated ion pair* usually denotes an ion pair in which the ions are separated by a single solvent molecule. When the contact distance a and the solvent diameter σ are both ~5 Å, the interionic distance in a solvent-separated ion pair is ~8 Å. Solvent-separated ion pairs are genuine molecular species with a definite stoichiometry and distinctive properties, and they populate an explicit potential minimum in configuration space.

The description of solvent-separated ion pairs as definite molecular species is illustrated in a classic study of Hogen-Esch and Smid [1966]. The aromatic hydrocarbon, fluorene, is acidic and forms alkali salts at the 9-position, Eq. (12.36):

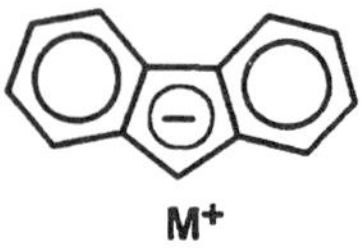

$$\text{Flu}^- \ \text{M}^+ \qquad (12.36)$$

In submillimolar solutions of a 9-fluorenylalkali in tetrahydrofuran ($\varepsilon = 7.39$ at 25°C), the optical absorption due to 9-Flu^- in the range 340–380 nm indicates an equilibrium between two molecular species: There are two isosbestic bands whose relative intensity varies with the temperature. The band that is dominant at 25°C appears to be due to a tight ion pair. The wavelength of its peak varies with the nature of the alkali ion, from 349 nm for Li^+ to 362 nm for Cs^+. The band that is dominant at −60°C might be due to solvent-separated ion-pairs, free ions, or both. Its peak is at 373 nm, independent of the nature of the alkali ion. Conductance data establish that free ions are present, but in the optical experiments their concentrations are relatively small, and the optical density at 373 nm accordingly is proportional to the first power, rather than the square root, of the fluorenide concentration. On the other hand, special experiments at very high dilutions show that free Flu^- ions also absorb with a peak at 373 nm.

Dimethylsulfoxide (DMSO) is a stronger solvating agent for alkali ions than is tetrahydrofuran. When DMSO is added to solutions of $\text{Flu}^-, \text{Li}^+$ in tetrahydrofuran at 25°C, the absorption band at 349 nm due to the tight ion pair decreases and that at 373 nm increases, showing that a DMSO-separated ion pair is formed. Studies of the effect of DMSO concentration show that this (DMSO) solvent-separated ion pair has the formula $\text{Flu}^-|\text{DMSO}|\text{Li}^+$. That is, the ions are separated by a single molecule.

The effect of temperature on the absorption spectra due to tight and solvent-separated fluorenide ion pairs shows that they are separated by a substantial enthalpy barrier. Similarly, the effect of temperature on conductance shows that there is a substantial enthalpy barrier between ion pairs and free ions [Hogen-Esch and Smid, 1966].

REFERENCES

J. L. M. Abboud, M. J. Kamlet, and R. W. Taft, **1981:** *Prog. Phys. Org. Chem.* **13,** 485.

M. H. Abraham, **1993:** *Chem. Soc. Rev.* **22,** 73.

L. E. Adams, **1915:** *J. Am. Chem. Soc.* **37,** 481.

G. Alagona and S. Tani, **1980:** *J. Chem. Phys.* **72,** 580.

R. A. Alberty, **1992:** *J. Phys. Chem.* **96,** 9614.

R. A. Alberty, **1994:** *Chem. Rev.* **94,** 1457.

B. J. Alder, **1955:** *J. Chem. Phys.* **23,** 263.

M. G. Alder and J. E. Leffler, **1954:** *J. Am. Chem. Soc.* **76,** 1425.

H. R. Allcock, **1978:** *Acc. Chem. Res.* **11,** 81.

N. J. Anderson, **1934:** "The Electromotive Force of the Cell Pt, H_2, HCl, AgCl, Ag with Extremely Dilute Electrolyte," Dissertation, University of Chicago.

B. M. Anderson and W. P. Jencks, **1960:** *J. Am. Chem. Soc.* **82,** 1773.

F. A. L. Anet and I. Yavari, **1978:** *J. Am. Chem. Soc.* **100,** 7814.

E. M. Arnett, **1952:** *J. Am. Chem. Soc.* **74,** 2027.

S. Arrhenius, **1889:** *Z. Phys. Chem. (Leipzig)* **4,** 226.

J. G. Aston, G. Szasz, H. W. Woolley, and F. G. Brickwedde, **1946:** *J. Chem. Phys.* **14,** 67.

V. Austel, **1983:** *Top. Curr. Chem.* **114,** 7–19.

L. W. Bahe, **1972:** *J. Phys. Chem.* **76,** 1062 and 1608.

L. W. Bahe and D. Parker, **1975:** *J. Am. Chem. Soc.* **97,** 5664–5670.

R. H. Baker and G. J. Boudreaux, **1967:** *Spectrochim. Acta* **23A,** 727.

F. M. Batson and C. A. Kraus, **1934:** *J. Am. Chem. Soc.* **56,** 2017.

G. Baughman and E. Grunwald, **1958:** *J. Am. Chem. Soc.* **80,** 3844.

R. P. Bell, **1980:** *The Tunnel Effect in Chemistry,* Chapman and Hall, London.

R. P. Bell and W. C. E. Higginson, **1949:** *Proc. R. Soc.* **A197,** 141.

D. Ben-Amotz and D. R. Herschbach, **1990:** *J. Phys. Chem.* **94,** 1038.

A. Ben-Naim, **1974:** *Water and Aqueous Solutions*, Plenum Press, New York, p. 320.

S. W. Benson, **1978:** *J. Am. Chem. Soc.* **100,** 5640.

S. W. Benson and E. D. Siebert, **1992:** *J. Am. Chem. Soc.* **114,** 4269.

T. H. Benzinger, **1971:** *Nature* **229,** 100.

T. H. Benzinger and C. Hammer, **1981:** *Curr. Top. Cell. Regul.* **18,** 475.

S. D. Bernal and R. H. Fowler, **1933:** *J. Chem. Phys.* **1,** 515.

C. F. Bernasconi, **1976:** *Relaxation Kinetics*, Academic Press, New York.

G. L. Bertrand and T. F. Fagley, **1976:** *J. Am. Chem. Soc.* **98,** 7944.

J. C. Bevington, **1955:** *Nature* **175,** 477.

N. Bjerrum, **1926:** *Kgl. Dansk. Vidensk. Selskab* **7**(9).

M. J. Blandamer, J. M. W. Scott, and R. E. Robertson, **1985:** *Prog. Phys. Org. Chem.* **15,** 149–196.

N. Bloembergen and M. D. Levenson, **1976:** *Top. Appl. Phys.* **13,** 315–369.

L. Blum, **1975:** *Mol. Phys.* **30,** 1529.

C. W. Bock, P. George, M. Trachtman, and M. Zanger, **1979:** *J. Chem. Soc. Perkins* (2), 26.

J. O'M. Bockris and A. K. N. Reddy, **1973:** *Modern Electrochemistry*, Vol. 1, Plenum Press, New York, Chapter 2.

M. Bodenstein and S. C. Lind, **1906:** *Z. Phys. Chem.* **57,** 168.

M. Bodenstein and F. Boes, **1922:** *Z. Phys. Chem.* **100,** 75.

C. J. F. Böttcher, **1952:** *Theory of Electric Polarization*, Elsevier, Amsterdam [especially Eqs. (2.43) and (5.13)].

P. D. Bolton, K. A. Fleming, and F. M. Hall, **1972:** *J. Am. Chem. Soc.* **94,** 1033.

F. Booth, **1951:** *J. Chem. Phys.* **19,** 391.

F. G. Bordwell and W. J. Boyle, **1972:** *J. Am. Chem. Soc.* **94,** 3907.

M. Born, **1920:** *Z. Phys.* **1,** 45.

R. H. Boyd, **1961:** *J. Phys. Chem.* **65,** 1834.

R. N. Bracewell, **1978:** *The Fourier Transform and Its Applications*, 2nd ed., McGraw-Hill, New York, Chapters 6 and 8 and pp. 85–87 and 168.

J. W. Breitenbach and A. Schindler, **1952:** *Monatsh. Chem.* **83,** 724.

J. N. Brønsted and K. J. Pedersen, **1924:** *Z. Phys. Chem.* **108,** 185.

P. Brüesch, **1982:** *Phonons, Theory and Experiments*, Springer-Verlag, Berlin.

A. D. Buckingham, A. C. Legon, and S. M. Roberts, eds., **1993:** *Principles of Molecular Recognition*, Blackie, London.

E. F. Caldin, **1964:** *Fast Reactions in Solution*, John Wiley & Sons, New York.

T. Cartailler, P. Turq, L. Blum, and N. Condamine, **1992:** *J. Phys. Chem.* **96,** 6766.

D. C. Champeney, **1973:** *Fourier Transforms and Their Physical Applications*, Academic Press, London.

M. K. Chantooni and I. M. Kolthoff, **1967:** *J. Am. Chem. Soc.* **89,** 1582.

J. Chen and G. R. Choppin, **1995:** *J. Solution Chem.* **24,** 465.

Y. Chiang, A. J. Kresge, and Y. S. Tang, **1984:** *J. Am. Chem. Soc.* **106,** 480.

Y. Chiang, M. Hojatti, J. R. Keeffe, A. J. Kresge, N. P. Schepp, and J. Wirz, **1987:** *J. Am. Chem. Soc.* **109,** 4000.

J. J. Christensen, J. Ruckman, D. J. Eatough, and R. M. Izatt, **1972:** *Thermochim. Acta* **3,** 203, 219, 233.

S. G. Christov, **1972:** *Ber. Bunsenges. Phys. Chem.* **76,** 507.

G. Claeson, G. Androes, and M. Calvin, **1961:** *J. Am. Chem. Soc.* **83,** 4357.

D. R. Cogley, J. N. Butler, and E. Grunwald, **1971:** *J. Phys. Chem.* **75,** 1477.

B. Cohen and S. Weiss, **1983:** *J. Phys. Chem.* **87,** 3606.

M. D. Cohen, G. M. J. Schmidt, and F. I. Sonntag, **1964:** *J. Chem. Soc.*, 2000.

J. P. Collman, L. S. Hegedus, J. R. Norton, and R. G. Finke, **1987:** *Principles and Applications of Organotransition Metal Chemistry*, University Science Books, Mill Valley, CA.

M. F. Colombo, D. C. Rau, and V. A. Parsegian, **1992:** *Science* **256,** 655.

B. E. Conway, **1978:** *J. Solution Chem.* **7,** 721.

E. G. Cox, **1958:** *Rev. Mod. Phys.* **30,** 159.

L. W. Daasch, C. Y. Liang, and J. R. Nielsen, **1954:** *J. Chem. Phys.* **22,** 1293.

A. F. Danil de Namor, M.-C. Ritt, M.-J. Schwing-Weill, F. Arnaud-Neu, and D. F. Lewis, **1991a:** *J. Chem. Soc. Faraday Trans.* **87,** 3231.

A. F. Danil de Namor, M.-C. Ritt, D. F. V. Lewis, M.-J. Schwing-Weill, and F. Arnaud-Neu, **1991b:** *Pure Appl. Chem.* **63,** 1435.

W. Dannhauser, **1968:** *J. Chem. Phys.* **48,** 1911.

M. Dantus, M. J. Rosker and A. H. Zewail, **1987:** *J. Chem. Phys.* **87,** 2395.

O. C. M. Davis, **1912:** *Z. Phys. Chem.* **78,** 353.

P. Debye, **1942:** *Trans. Electrochem. Soc.* **82,** 265.

P. Debye, **1945** (reprinted): *Polar Molecules*, Dover Publications, New York, Chapter 5.

P. Debye and E. Hückel, **1923:** *Physik. Z.* **24,** 305.

S. F. Dec and S. J. Gill, **1985:** *J. Solution Chem.* **14,** 417 and 827.

T. De Donder, **1922:** *Bull. Acad. R. Belg. Classe Sci.* **7**(5), 197 and 205. This reference, unfortunately, was not available to the author.

L. C. M. De Maeyer, **1971:** *Israel J. Chem.* **9,** 351.

J. T. Denison and J. B. Ramsey, **1955:** *J. Am. Chem. Soc.* **77,** 2615.

M. J. S. Dewar, R. Golden, and J. M. Harris, **1971:** *J. Am. Chem. Soc.* **93,** 4187.

B. DiBartolo and R. C. Powell, **1976:** *Phonons and Resonances in Solids*, Wiley-Interscience, New York.

D. Dolman and R. Stewart, **1967:** *Can. J. Chem.* **45,** 911.

T. Drakenberg and S. Forsen, **1971:** *J. Chem. Soc. D*, 1404.

H. E. Duckworth, **1960:** *Electricity and Magnetism,* Holt, Rinehart and Winston, New York.

D. J. Eatough, R. M. Izatt, and J. J. Christensen, **1972:** *Thermochim. Acta* **3,** 240.

W. Ebeling and M. Grigo, **1982:** *J. Solution Chem.* **11,** 151.

M. Eigen and L. De Maeyer, **1961:** Technique of Organic Chemistry (VIII), *Investigation of Rates and Mechanisms of Reactions*, Part II, A. Weissberger, ed., Interscience Publishers, New York, Chapter 18.

M. Eigen, **1964:** *Angew. Chemie Int. Ed.* **3,** 1.

A. Einstein, **1956** (reprinted): *Investigations on the Theory of the Brownian Movement*, A. D. Cowper, translator, Dover Publications.

E. L. Eliel, **1965:** *Conformational Analysis*, Interscience, New York.

C. A. Emeis and P. L. Fehder, **1970:** *J. Am. Chem. Soc.* **92,** 2246.

J. B. F. N. Engberts, **1979:** Mixed Aqueous Solvent Effects, in *Water, A Comprehensive Treatise*, Vol. 6, F. Franks, ed., Plenum Press, New York, Chapter 4.

J. F. J. Engbersen and J. B. F. N. Engberts, **1975:** *J. Am. Chem. Soc.* **97,** 1563.

V. Enkelmann, G. Wegner, K. Novak, and K. B. Wagener, **1993:** *J. Am. Chem. Soc.* **115,** 10390.

J. W. Erickson and P. J. Estrup, **1986:** *Surf. Sci.* **167,** 519.

K. M. Ervin, J. Ho, and W. C. Lineberger, **1989:** *J. Chem. Phys.* **91,** 5974.

A. Eucken and E. Lindenberg, **1942:** *Berichte* **175,** 1953.

D. F. Evans and H. Wennerström, **1994:** *The Colloidal Domain*, VCH, New York.

M. G. Evans and M. Polanyi, **1935:** *Trans. Faraday Soc.* **31,** 875.

A. H. Ewald and J. A. Scudder, **1970:** *Aust. J. Chem.* **23,** 1939.

O. Exner, **1964:** *Collection Czech. Chem. Commun.* **29,** 1094.

H. Eyring, **1935:** *J. Chem. Phys.* **3,** 107.

H. Eyring and M. S. Jhon, **1969:** *Significant Liquid Structures*, John Wiley & Sons, New York.

A. H. Fainberg and S. Winstein, **1957:** *J. Am. Chem. Soc.* **79,** 1597.

G. Feher and M. Weissman, **1973:** *Proc. Natl. Acad. Sci. USA* **70,** 870.

H. Fidder, J. Knoester and D. A. Wiersma, **1991:** *J. Chem. Phys.* **95,** 7880.

R. Fowler and E. A. Guggenheim, **1956:** *Statistical Thermodynamics*, Cambridge University Press, New York, Sections 802 and 808.

H. S. Frank, **1945:** *J. Chem. Phys.* **13,** 493.

H. S. Frank, **1955:** *J. Chem. Phys.* **23,** 2023.

H. S. Frank and M. W. Evans, **1945:** *J. Chem. Phys.* **13,** 507.

H. S. Frank and W.-Y. Wen, **1957:** *Discuss. Faraday Soc.* **24,** 113.

F. Franks and J. E. Desnoyers, **1985:** Alcohol–Water Mixtures Revisited, *Water Science Review*, Vol. 1, Cambridge University Press, New York.

F. Franks, M. Gent and H. H. Johnson, **1963:** *J. Chem. Soc.*, 2716.

F. Franks and D. J. G. Ives, **1966:** *Q. Rev. Chem. Soc.* **20,** 1–44.

H. L. Friedman, **1962:** *Ionic Solution Theory*, Interscience Publishers, New York.

T. Fujita and H. Iwamura, **1983:** *Top. Curr. Chem.* **114,** 119–57.

R. M. Fuoss, **1935:** *Chem. Rev.* **17,** 27.

R. M. Fuoss, **1958:** *J. Am. Chem. Soc.* **80,** 5059.

R. M. Fuoss, **1959:** *J. Am. Chem. Soc.* **81,** 2659, 6535.

R. M. Fuoss and K.-L. Hsia, **1967:** *Proc. Natl. Acad. Sci. USA*, **57,** 1550.

R. M. Fuoss and C. A. Kraus, **1933:** *J. Am. Chem. Soc.* **55,** 476 and 2387.

R. M. Fuoss and C. A. Kraus, **1935:** *J. Am. Chem. Soc.* **57,** 1.

R. M. Fuoss and L. Onsager, **1958:** *J. Phys. Chem.* **62,** 1339.

M. M. Gallo, T. P. Hamilton, and H. F. Schaefer III, **1990:** *J. Am. Chem. Soc.* **112,** 8714.

A. K. Galwey, **1977:** *Adv. Catal.* **26,** 247.

B. C. Garrett, D. G. Truhlar, A. J. C. Varandes, and N. C. Blais, **1986:** *Int. J. Chem. Kinet.* **18,** 1065.

E. Garrone, B. Fubini, E. Escalona Platero, and A. Zecchina, **1989:** *Langmuir* **5,** 240.

A. Geiger, A. Rahman, and F. H. Stillinger, **1979:** *J. Chem. Phys.* **70,** 263.

W. F. Giauque and J. D. Kemp, **1938:** *J. Chem. Phys.* **6,** 40.

J. W. Gibbs, **1948** (reprinted): *The Collected Works of J. Willard Gibbs*, Vol. 1, Yale University Press, New Haven, CT, pp. 70–89.

K. Giese, U. Kaatze and R. Pottel, **1970:** *J. Phys. Chem.* **74,** 3718.

R. G. Gilbert and S. C. Smith, **1990:** *Theory of Unimolecular and Recombination Reactions*, Blackwell: Oxford, p. 246.

S. J. Gill, N. F. Nichols, and I. Wadsö, **1975:** *J. Chem. Thermodyn.* **7,** 175.

S. J. Gill, N. F. Nichols, and I. Wadsö, **1976:** *J. Chem. Thermodyn.* **8,** 445.

G. Gilli and P. A. Borea, **1991:** *The Application of Charge Density Research to Drug Design*, G. A. Jeffrey and J. F. Piniella, eds., Plenum Press, New York, pp. 241–286.

P. Gilli, V. Ferretti, and G. Gilli, **1994:** *J. Phys. Chem.* **98,** 1515–1518.

S. Glasstone, K. J. Laidler, and H. Eyring, **1941:** *The Theory of Rate Processes*, McGraw-Hill, New York, Chapter 9.

E. Grunwald, **1986a:** *J. Am. Chem. Soc.* **108,** 5719.

E. Grunwald, **1986b:** *J. Am. Chem. Soc.* **108,** 5726.

E. Grunwald and A. L. Bacarella, **1958:** *J. Am. Chem. Soc.* **80,** 3840.

E. Grunwald and G. Baughman, **1960:** *J. Phys. Chem.* **64,** 933.

E. Grunwald and B. J. Berkowitz, **1951:** *J. Am. Chem. Soc.* **73,** 4939.

E. Grunwald and C. D. Brown, **1982:** *J. Phys. Chem.* **86,** 182.

E. Grunwald and A. F. Butler, **1960:** *J. Am. Chem. Soc.* **82,** 5647.

E. Grunwald and L. Comeford, **1988:** *Environmental Influences and Recognition in Enzyme Chemistry*, J. F. Liebman and A. Greenberg, eds., VCH Publishers, New York, pp. 81–107.

E. Grunwald and A. Effio, **1974:** *J. Am. Chem. Soc.* **96,** 423.

E. Grunwald and J. F. Haley, **1968:** *J. Phys. Chem.* **72,** 1944.

E. Grunwald and L. J. Kirschenbaum, **1972:** *Introduction to Quantitative Chemical Analysis*, Prentice-Hall, Englewood Cliffs, NJ, pp. 178–179.

E. Grunwald and K.-C. Pan, **1976:** *J. Phys. Chem.* **80,** 2929.

E. Grunwald and E. Price, **1964:** *J. Am. Chem. Soc.* **96,** 4517.

E. Grunwald and E. K. Ralph, **1971:** *Acc. Chem. Res.* **4,** 107.

E. Grunwald and C. Steel, **1993a:** *J. Phys. Chem.* **97,** 13326.

E. Grunwald and C. Steel, **1993b:** *Pure Appl. Chem.* **65,** 2543.

E. Grunwald and C. Steel, **1994:** *J. Phys. Org. Chem.* **7,** 734.

E. Grunwald and C. Steel, **1995:** *J. Am. Chem. Soc.* **117,** 5687.

E. Grunwald and S. Winstein, **1948:** *J. Am. Chem. Soc.* **70,** 846.

E. Grunwald, G. Baughman and G. Kohnstam, **1960:** *J. Am. Chem. Soc.* **82,** 5801.

E. Grunwald, J. Herzog and C. Steel, **1995:** *J. Chem. Educ.* **72,** 210.

E. Grunwald, S. Highsmith, and T.-P. I, **1974:** in *Ions and Ion Pairs in Organic Reactions*, M. Szwarc, ed.; Wiley-Interscience, vol. 2, Ch. 5.

E. Grunwald, S. P. Anderson, A. Effio, S. E. Gould, and K. C. Pan, **1976a:** *J. Phys. Chem.* **80,** 2935.

E. Grunwald, K.-C. Pan, and A. Effio, **1976b:** *J. Phys. Chem.* **80,** 2937.

S. R. Gunn, **1971:** *J. Chem. Thermodyn.* **3,** 19.

J. R. Haak, **1986:** *Water and Aqueous Nonelectrolyte Solutions*, Ph.D. Thesis, University of Groningen, p. 70.

F. Haber, **1974:** "Introduction to Information and Communication Theory," Addison-Wesley, Reading, MA, Chapter 2.

G. H. Haggis, J. B. Hasted, and T. J. Buchanan, **1952:** *J. Chem. Phys.* **20,** 1452.

D. N. Hague, **1971:** *Fast Reactions*, Wiley-Interscience, London.

H. F. Halliwell and S. C. Nyburg, **1963:** *Trans. Faraday Soc.* **59,** 1126.

L. P. Hammett, **1935:** *Chem. Rev.* **17,** 125.

L. P. Hammett, **1937:** *J. Am. Chem. Soc.* **59,** 96.

L. P. Hammett, **1940:** *Physical Organic Chemistry*, McGraw-Hill, New York, Chapter 2 and p. 84.

L. P. Hammett, **1970:** *Physical Organic Chemistry*, 2nd ed.; McGraw-Hill, New York.

G. S. Hammond, J. N. Sen, and C. E. Boozer, **1955:** *J. Am. Chem. Soc.* **77,** 3244.

J. P. Hansen and I. R. McDonald, **1990:** *Theory of Simple Liquids*, Academic Press, New York.

C. H. Hansch and A. Leo, **1979:** *Substituent Constants for Correlation Analysis in Chemistry and Biology*, Wiley-Interscience, New York, 1979.

H. S. Harned and B. B. Owen, 1958: *The Physical Chemistry of Electrolytic Solutions*, 3rd ed.; Reinhold, New York, p. 668.

N. C. Harris and E. M. Hemmerling, **1955:** *Introductory Applied Physics*, McGraw-Hill, New York, pp. 42–46.

J. B. Hendrickson, **1967:** *J. Am. Chem. Soc.* **89,** 7036 and 7047.

G. Herzberg, **1939:** *Molecular Spectra and Molecular Structure. I. Diatomic Molecules*, Prentice-Hall, New York, pp. 106, 109, 487.

D. A. Higgins and P. A. Barbara, **1995:** *J. Phys. Chem.* **99,** 3.

J. H. Hildebrand, **1971:** *Science* **174**, 490.

J. H. Hildebrand and R. H. Lamoreaux, **1972:** *Proc. Natl. Acad. Sci. USA* **69,** 3428.

J. H. Hildebrand and R. L. Scott, **1950:** *The Solubility of Nonelectrolytes*, Reinhold, New York, p. 247.

J. H. Hildebrand and R. L. Scott, **1962:** *Regular Solutions*, Prentice-Hall: Englewood Cliffs, NJ, Chapter 7, pp. 90–91.

J. H. Hildebrand, J. M. Prausnitz, and R. L. Scott, **1970:** *Regular and Related Solutions*, Van Nostrand Reinhold, New York.

J. F. Hinton, J. Q. Fernandez, C. S. Dikoma, W. L. Whaley, R. E. Koeppe, and F. S. Millett, **1988:** *Biophys. J.* **54,** 527.

J. O. Hirschfelder, C. F. Curtiss, and R. B. Bird, **1954:** *Molecular Theory of Gases and Liquids:* John Wiley & Sons, New York.

G. Hoffmann, D. Oh, Y. Chen, Y. M. Engel, and C. Wittig, **1989:** *Israel J. Chem.* **30,** 115.

T. E. Hogen-Esch and J. Smid, **1966:** *J. Am. Chem. Soc.* **88,** 307, 318.

H. A. J. Holterman and J. B. F. N. Engberts, **1979:** *J. Phys. Chem.* **83,** 443.

J. Horiuti, **1931:** *Sci. Papers Inst. Phys. Chem. Tokyo* **17,** 125.

J. K. Hurley, H. Linschitz, and A. Treinin, **1988:** *J. Phys. Chem.* **92,** 5151.

H. D. Inerowicz, W. Li, and I. Persson, **1994:** *J. Chem. Soc. Faraday Trans.* **90,** 2223.

C. K. Ingold, **1953:** *Structure and Mechanism in Organic Chemistry*, Cornell University Press, Ithaca, NY.

H. Inone and S. N. Timasheff, **1968:** *J. Am. Chem. Soc.* **90,** 1890.

International Union of Pure and Applied Chemistry, **1988:** *Quantities, Units and Symbols in Physical Chemistry*, Blackwell Scientific Publications, Oxford, England.

H. H. Jaffe, **1953:** *Chem. Rev.* **53,** 191.

JANAF, **1971:** *Thermochemical Tables*, 2nd ed., National Standards and Reference Data Series, National Bureau of Standards (U.S.), Vol. 37, U.S. Government Printing Office, Washington, D.C. 20402.

G. A. Jeffrey and R. K. McMullen, **1967:** *Prog. Inorg. Chem.* **8,** 43.

E. E. Jelley, **1936:** *Nature (London)* **38,** 1009.

P. C. Jordan, **1979:** *Chemical Kinetics and Transport*, Plenum Press, New York.

W. L. Jorgensen and D. L. Severance, **1990:** *J. Am. Chem. Soc.* **112,** 4768.

J. Jortner, **1992:** *Z. Physik [D]* **24,** 247.

L. S. Kassel, **1936:** *Chem. Rev.* **18,** 277–313.

J. R. Keeffe, A. J. Kresge, and N. P. Schepp, **1988:** *J. Am. Chem. Soc.* **110**, 1993.

G. Kessling and L. De Maeyer, **1995:** *J. Chem. Soc. Faraday Trans.* **91,** 303.

J. G. Kirkwood, **1939:** *J. Chem. Phys.* **7,** 911.

J. G. Kirkwood, **1946:** *Trans. Faraday Soc.* **42A,** 7.

F. F. Knudsen and K. R. Naqvi, **1995:** *J. Phys. Chem.* **99,** 6199.

D. A. V. Kliner and R. N. Zare, **1990:** *J. Chem. Phys.* **92,** 2107.

I. M. Kolthoff and S. Bruckenstein, **1956:** *J. Am. Chem. Soc.* **78,** 1 and 10.

H. G. Korth, P. Lommes, W. Sicking, and R. Sustmann, **1983:** *Int. J. Chem. Kinet.* **15,** 267.

G. Kortum and H. Wilski, 1954: *Z. Phys. Chem. [N.F.]* **2,** 264.

C. A. Kraus and R. M. Fuoss, **1933:** *J. Am. Chem. Soc.* **55,** 21.

C. A. Kraus and R. A. Vingee, **1934:** *J. Am. Chem. Soc.* **56,** 511.

I. Kritchevsky and A. Ilinskaya, **1945:** *Acta Physicochim. U.R.S.S.* **20,** 327.

M. M. Kreevoy and K.-C. Chang, **1976:** *J. Phys. Chem.* **80,** 259.

J. Laane, **1970:** *Appl. Spectrosc.* **24,** 73.

K. L. Laidler and K. J. Laidler, **1987:** *Chemical Kinetics*, 3rd ed., Harper-Collins, New York.

F. W. Lampe and R. M. Noyes, **1954:** *J. Am. Chem. Soc.* **76,** 2140.

Landolt–Börnstein, **NS IV/4:** *Tabellen, New Series* IV/4, G. Beggerow, compiler, p. 215.

A. Lapworth, **1904:** *J. Chem. Soc.* **85,** 30.

J. W. Larson and T. B. McMahon, **1983:** *J. Am. Chem. Soc.* **105,** 2944.

W. M. Latimer, K. S. Pitzer, and C. M. Slansky, **1939:** *J. Chem. Phys.* **7,** 108.

J. L. Lebowitz, and J. K. Percus, **1966:** *Phys. Rev.* **144,** 251.

J. E. Leffler, **1955:** *J. Org. Chem.* **20,** 1202.

J. E. Leffler and E. Grunwald, **1989** [reprint]: *Rates and Equilibria of Organic Reactions*, Dover Publications, New York.

A. C. Legon and D. J. Millen, **1986:** *Chem. Rev.* **86,** 635.

A. C. Legon, E. J. Campbell, and W. H. Flygare, **1982:** *J. Chem. Phys.* **76,** 2267.

T. M. Letcher and J. D. Mercer-Chalmers, **1993:** *J. Solution Chem.* **22,** 851.

C. S. Leung and E. Grunwald, **1970:** *J. Phys. Chem.* **74,** 687.

R. D. Levine and R. B. Bernstein, **1987:** *Molecular Reaction Dynamics and Chemical Reactivity*, Oxford University Press, New York.

G. N. Lewis and G. A. Linhart, **1919:** *J. Am. Chem. Soc.* **41,** 1951.

G. N. Lewis and M. Randall, **1921:** *J. Am. Chem. Soc.* **43,** 1112.

G. N. Lewis and M. Randall, **1923:** *Thermodynamics*; McGraw-Hill, New York.

G. N. Lewis, M. Randall, K. S. Pitzer, and L. Brewer, **1961:** *Thermodynamics*, 2nd ed.; McGraw-Hill, New York, Chapters 13 and 27.

F. M. Lewis and M. S. Matheson, **1949:** *J. Amer. Chem. Soc.* **71,** 747.

J. C. M. Li and K. S. Pitzer, **1956:** *J. Am. Chem. Soc.* **78,** 1077.

N. N. Lichtin and K. N. Rao, **1961:** *J. Am. Chem. Soc.* **83,** 2417.

N. N. Lichtin, M. S. Puar, and B. Wasserman, **1967:** *J. Am. Chem. Soc.* **89,** 6677.

J. E. Lind and R. M. Fuoss, **1961:** *J. Phys. Chem.* **65,** 1414.

R. Lindemann and G. Zundel, **1972:** *J. Chem. Soc. Faraday Trans II* **68,** 979.

R. Lindemann and G. Zundel, **1977:** *J. Chem. Soc. Faraday Trans II* **73,** 788.

R. L. Lipnick, **1989:** *Envir. Toxicology and Chem.* **8,** 1–12.

S. J. Lippard and J. M. Berg, **1994:** *Principles of Bioinorganic Chemistry*, University Science Books, Mill Valley, CA, pp. 285ff.

B. Liu, **1984:** *J. Chem. Phys.* **80,** 581.

I. Loeff, S. Goldstein, A. Treinin, and H. Linschitz, **1991:** *J. Phys. Chem.* **95,** 4423.

F. London, **1942:** *J. Phys. Chem.* **46,** 305.

K. Lonsdale, **1959:** *Z. Kristallographie* **112,** 188.

J. P. Lowe, **1968:** *Prog. Phys. Org. Chem.* **6,** 1.

W. A. P. Luck and M. Fritzsche, **1992:** *J. Molec. Liquids* **52,** 215.

R. Lumry, **1995:** *Energetics of Biological Macromolecules, Methods in Enzymology*, Vol. 259, M. Johnson and G. Ackers, eds., Academic Press, New York, Chapter 29.

R. Lumry and S. Rajender, **1970:** *Biopolymers* **9,** 1125–1227.

R. Lumry and R. B. Gregory, **1986:** in *The Fluctuating Enzyme*, G. R. Welch, ed., Wiley-Interscience, New York, 1986.

V. Majer and R. H. Wood, **1994:** *J. Chem. Thermodyn.* **26,** 1143.

R. A. Marcus and M. E. Coltrin, **1977:** *J. Chem. Phys.* **67,** 2609.

Y. Marcus, **1983:** *Pure Appl. Chem.* **55,** 977.

Y. Marcus, **1985:** *Pure Appl. Chem.* **57,** 1102.

Y. Marcus, **1986:** *Pure Appl. Chem.* **58,** 1721.

Y. Marcus, **1987:** *J. Chem. Soc. Faraday Trans. [I]* **83,** 339.

Y. Marcus, **1994:** *Biophys. Chem.* **51,** 111.

H. Margenau and G. M. Murphy, **1943:** *The Mathematics of Physics and Chemistry*, Van Nostrand, New York, p. 7.

M. Mason and W. Weaver, **1929:** *The Electromagnetic Field*, University of Chicago Press, Chicago, IL.

T. Matsui, H. C. Ko, and L. G. Hepler, **1974:** *Can. J. Chem.* **52,** 2906.

J. E. Mayer, **1950:** *J. Chem. Phys.* **18,** 1426.

J. M. McBride, B. E. Segmuller, M. D. Hollingsworth, D. E. Mills, and B. A. Weber, **1986:** *Science* **234,** 830.

H. M. McConnell, **1958:** *J. Chem. Phys.* **28,** 430.

M. A. McCool, A. F. Collings, and L. A. Woolf, **1972:** *J. Chem. Soc. Faraday Trans.* **168,** 1489.

D. H. McDaniel and H. C. Brown, **1958:** *J. Org. Chem.* **23,** 420.

W. F. McDevit and F. A. Long, **1952:** *J. Am. Chem. Soc.* **74,** 1773.

C. W. McGary, Y. Okamoto, and H. C. Brown, **1955:** *J. Am. Chem. Soc.* **77,** 3037.

S. Meiboom, **1960:** *Z. Elektrochem.* **64,** 50–53.

N. Menninga and J. B. F. N. Engberts, **1976:** *J. Org. Chem.* **41,** 3101.

J. V. Michael, **1990:** *J. Chem. Phys.* **92,** 3394.

J. V. Michael and J. R. Fisher, **1990:** *J. Phys. Chem.* **94,** 3318.

J. V. Michael, J. R. Fisher, J. M. Bowman, and Q. Sun, **1990:** *Science* **249,** 269.

F. J. Millero, C.-H. Wu, and L. G. Hepler, **1969:** *J. Phys. Chem.* **73,** 2453.

D. N. Mitchell and D. J. LeRoy, **1973:** *J. Chem. Phys.* **58,** 3449.

S. Mizushima, **1954:** *Structure of Molecules and Internal Rotation*, Academic Press, New York, Chapters 2 and 3.

S. Mizushima and Y. Morino, **1938:** *Proc. Ind. Acad. Sci.* **8,** 315.

S. Mizushima, Y. Morino, and K. Higasi, **1934:** *Sci. Pap. Inst. Phys. Chem. Res. Tokyo* **25,** 159.

S. Mizushima, Y. Morino, and S. Noziri, **1936:** *Sci. Pap. Inst. Phys. Chem. Res. Tokyo* **29,** 63.

S. Mizushima, T. Shimanouchi, I. Harada, Y. Abe, and H. Takeuchi, **1975:** *Can. J. Phys.* **53,** 2085.

J. Morgan and B. E. Warren, **1938:** *J. Chem. Phys.* **6,** 666.

P. W. Mui and E. Grunwald, **1982:** *J. Am. Chem. Soc.* **104,** 6562.

A. H. Narten and H. A. Levy, **1969:** *Science* **165,** 447.

A. H. Narten and H. A. Levy, **1971:** *J. Chem. Phys.* **55,** 2263.

S. F. Nelsen and G. R. Weisman, **1976:** *J. Am. Chem. Soc.* **98,** 3281.

R. M. Noyes, **1961:** *Prog. React. Kinet.* **1,** 129.

R. M. Noyes, **1962:** *J. Am. Chem. Soc.* **84,** 513.

R. Noyori and M. Kitamura, **1991:** *Ang. Chemie Int. Ed. (Engl.)* **30,** 49–69.

Y. Okamoto and H. C. Brown, **1957:** *J. Org. Chem.* **22,** 485.

Y. Okamoto and H. C. Brown, **1958:** *J. Am. Chem. Soc.* **80,** 4979.

L. Onsager, **1927:** *Phys. Z.* **28,** 277.

L. Onsager, **1936:** *J. Am. Chem. Soc.* **58,** 1486.

G. Oster and J. G. Kirkwood, **1943:** *J. Chem. Phys.* **11,** 175.

C. G. Overberger, M. T. O'Shaughnessy, and H. Shalit, **1949:** *J. Am. Chem. Soc.* **71,** 2661.

K.-C. Pan and E. Grunwald, **1976:** *J. Phys. Chem.* **80,** 2932.

A. J. Parker, **1969:** *Chem. Rev.* **69,** 1.

H. J. Parkhurst and J. Jonas, **1975a:** *J. Chem. Phys.* **63,** 2698.

H. J. Parkhurst and J. Jonas, **1975b:** *J. Chem. Phys.* **63,** 2705.

R. K. Pathria, **1972:** *Statistical Mechanics*, Pergamon Press, Toronto, section 13.5 and p. 462.

K. I. Petersen and W. Klemperer, **1984:** *J. Chem. Phys.* **81,** 3842.

W. D. Phillips, C. E. Looney, and C. P. Spaeth, **1957:** *J. Molec. Spectroscopy* **1,** 35.

J. R. Pierce, **1980** [reprint]: *Introduction to Information Theory: Symbols, Signals and Noise*, Dover Publications, New York.

G. C. Pimentel and A. L. McClellan, **1960:** *The Hydrogen Bond*, W. H. Freeman, San Francisco, pp. 236–238.

G. C. Pimentel and A. L. McClellan, **1971:** *Annu. Rev. Phys. Chem.* **22,** 347–385.

W. H. Pirkle and T. C. Pochapsky, **1989:** *Chem. Rev.* **89,** 347–362.

M. Pirklbauer and G. Gritzner, **1993:** *J. Solution Chem.* **22,** 585.

G. Porter, **1961:** *Technique of Organic Chemistry: VIII. Investigation of Rates and Mechanisms of Reactions*, Part II, A. Weissberger, ed., Interscience, New York, Chapter 19.

C. M. Preston and W. A. Adams, **1979:** *J. Phys. Chem.* **83,** 814.

I. Prigogine and R. Defay, **1954:** *Chemical Thermodynamics*, English translation by D. H. Everett, Longmans Green & Co., New York, Chapters 1 and 4.

J. Quint and A. Viallard, **1978:** *J. Solution Chem.* **7,** 533.

E. Rabinowitch and W. C. Wood, **1936:** *Trans. Faraday Soc.* **32,** 1381.

E. Rabinowitch and W. C. Wood, **1937:** *Trans. Faraday Soc.* **33,** 1225.

R. W. Ramirez, **1985:** *The FFT: Fundamentals and Concepts*, Prentice-Hall, Englewood Cliffs, NJ.

J. C. Rasaiah and H. L. Friedman, **1968:** *J. Chem. Phys.* **48,** 2742.

J. C. Rasaiah and H. L. Friedman, **1969:** *J. Chem. Phys.* **50,** 2965.

G. Ravishanker, P. K. Mehrotra, M. Mezei, and D. L. Beveridge, **1984:** *J. Am. Chem. Soc.* **106,** 4102.

H. Reisler and C. Wittig, **1986:** *Annu. Rev. Phys. Chem.* **37,** 307.

H. Reiss and A. Heller, **1985:** *J. Phys. Chem.* **89,** 4207.

P. M. Rentzepis, **1970:** *Science* **169,** 239.

W. Rhodes, **1991:** *J. Phys. Chem.* **95,** 10246.

F. G. Riddell, **1980:** *Conformational Analysis of Heterocyclic Compounds*, Academic Press, London.

J. A. Riddick and W. B. Bunger, **1970:** *Organic Solvents*, Wiley-Interscience, New York.

R. A. Robinson and R. A. Stokes, **1955:** *Electrolyte Solutions*, Butterworths Scientific Publications, London, Chapters 4 and 9.

P. J. Rossky and M. Karplus, **1979:** *J. Am. Chem. Soc.* **101,** 1913.

F. J. W. Roughton and B. Chance, **1961:** *Technique of Organic Chemistry: VIII. Investigation of Rates and Mechanisms of Reactions*, Part II, A. Weissberger, ed., Interscience Publishers, New York, Chapter 14.

E. G. Sander and W. P. Jencks, **1968:** *J. Am. Chem. Soc.* **90,** 6154.

P. W. Schmidt, **1987:** *McGraw-Hill Encyclopedia of Science and Technology*, sixth edition, **4,** 160.

F. Sciortino, A. Geiger, and H. E. Stanley, **1992:** *J. Chem. Phys.* **96,** 3857.

R. Shaw, **1969:** *J. Chem. Eng. Data* **14,** 461.

X. Shi and L. S. Bartell, **1988:** *J. Phys. Chem.* **92,** 5667.

J. Shorter, **1973:** *Correlation Analysis in Organic Chemistry*, Clarendon Press, London.

P. Siegbahn and B. Liu, **1978:** *J. Chem. Phys.* **68,** 2457.

R. T. Skodje, D. G. Truhlar, and B. C. Garrett, **1981:** *J. Phys. Chem.* **85,** 3019.

J. Smid, **1972:** in *Ions and Ion Pairs in Organic Reactions*, Vol. 1, M. Szwarc, ed., Wiley-Interscience, Ch. 3.

E. B. Smith and J. Walkley, **1962:** *J. Phys. Chem.* **66,** 597.

M. v. Smoluchowski, **1916:** *Phys. Z.* **17,** 557 and 585.

J. I. Steinfeld, J. S. Francisco, and W. L. Hase, **1989:** *Chemical Kinetics and Dynamics*, Prentice-Hall, Englewood Cliffs, NJ, Ch. 10, p. 337.

E. O. Stejskal and J. E. Tanner, **1965:** *J. Chem. Phys.* **42,** 288.

F. H. Stillinger and R. Lovett, **1968:** *J. Chem. Phys.* **49,** 1991.

F. H. Stillinger and A. Rahman, **1974:** *J. Chem. Phys.* **60,** 1545.

R. H. Stokes and R. A. Robinson, **1948:** *J. Am. Chem. Soc.* **70,** 1870.

A. Streitwieser, **1963:** *Molecular Orbital Theory for Organic Chemists*, John Wiley & Sons, New York.

C. G. Swain, S. H. Unger, N. R. Rosenquist, and M. S. Swain, **1983:** *J. Am. Chem. Soc.* **105,** 492.

S. Swaminathan, S. W. Harrison, and D. L. Beveridge, **1978:** *J. Am. Chem. Soc.* **100,** 5705.

M. Szwarc, **1972:** in *Ions and Ion Pairs in Organic Reactions*, Vol. 1, M. Szwarc, ed., Wiley-Interscience, New York, Chapter 1.

R. W. Taft, **1956:** in *Steric Effects in Organic Chemistry*, M. S. Newman, ed., John Wiley & Sons, New York, Chapter 13.

S. N. Timasheff, **1993:** *Annu. Rev. Biophys. Biomol. Struct.* **22,** 67–97.

R. C. Tolman, **1938:** *The Principles of Statistical Mechanics*, Oxford University Press, pp. 165 and 414.

J. K. Trautman, A. P. Shreve, C. A. Violette, H. A. Frank, T. G. Owens, and A. C. Albrecht, **1990:** *Proc. Natl. Acad. Sci. USA* **87,** 215.

D. G. Truhlar and C. J. Horowitz, **1978:** *J. Chem. Phys.* **68,** 2466.

Y. Tsuno, T. Ibata, and Y. Yukawa, **1959:** *Bull. Chem. Soc. Japan* **32,** 960.

J. J. Valentini and D. L. Phillips, **1989:** in *Biomolecular Collisions*, M. N. R. Ashfold and J. E. Bassott, eds., Royal Society of Chemistry, London, xvii, chap. 1.

D. D. Wagman, W. H. Evans, V. B. Parker, R. H. Schumm, I. Halow, S. M. Bailey, K. L. Churney, and R. L. Nuttall, **1982:** *J. Phys. Chem. Ref. Data* **11,** Suppl. 2.

C. Walling, **1954:** *J. Polymer Sci.* **14,** 214.

C. Walling, **1957:** *Free Radicals in Solution*, John Wiley & Sons, New York.

G. E. Walrafen, M. S. Hokmabadi and W.-H. Yang, **1986a:** *J. Chem. Phys.* **85,** 6964.

G. E. Walrafen, M. R. Fisher, M. S. Hokmabadi, and W.-H. Yang, **1986b:** *J. Chem. Phys.* **85,** 6970.

B. E. Warren, **1969:** *X-Ray Diffraction*, Addison-Wesley, Reading, MA.

J. A. Waters, G. A. Mortimer, and H. E. Clements, **1970:** *J. Chem. Eng. Data* **15,** 174.

S. Weiner and G. S. Hammond, **1968:** *J. Am. Chem. Soc.* **90,** 1659.

A. Weller, **1968:** *Pure Appl. Chem.* **16,** 115.

W. E. Wentworth, **1965:** *J. Chem. Educ.* **42,** 96.

A. A. Westenberg and N. de Haas, **1967:** *J. Chem. Phys.* **47,** 1393.

E. Wilhelm and R. Battino, **1973:** *Chem. Rev.* **73,** 1.

E. Wilhelm, R. Battino and R. J. Wilcock, **1977:** *Chem. Rev.* **77,** 219.

R. K. Williams, **1981:** *J. Phys. Chem.* **85,** 1795.

S. Winstein, E. Clippinger, A. H. Fainberg, R. Heck, and G. C. Robinson, **1956:** *J. Am. Chem. Soc.* **78,** 328.

S. Winstein, E. Grunwald and H. W. Jones, **1951:** *J. Am. Chem. Soc.* **73,** 2700.

R. Wolfenden, **1978:** *Biochemistry* **17,** 201.

R. Wolfenden, P. M. Cullis, and C. C. F. Southgate, **1979:** *Science* **206,** 575.

K. A. Wood and H. L. Strauss, **1990:** *J. Phys. Chem.* **94,** 5677.

R. H. Wood, **1989:** *Thermochimica Acta* **154,** 1.

J. D. Worley and I. M. Klotz, **1966:** *J. Chem. Phys.* **45,** 2868.

C. A. Wulff, **1963:** *J. Chem. Phys.* **39,** 1227.

J. Wyman, **1964:** *Adv. Protein Chem.* **19,** 223.

J. Wyman and S. J. Gill, **1990:** *Binding and Linkage*, University Science Books, Mill Valley, CA.

Y. B. Zhao and R. Gomer, **1990:** *Surf. Sci.* **239,** 198.

W. Zielenkiewicz, O. V. Kulikov, and I. Kulis-Cwikla, **1993:** *J. Solution Chem.* **22,** 963.

J. Zimmerberg, F. Bezanila, and V. A. Parsegian, **1990:** *Biophys. J.* **57,** 1049.

G. Zundel and M. Eckert, **1989:** *J. Mol. Struct.* **200,** 73–92.

INDEX